Islands and the Sea

Essays on Herpetological Exploration in the West Indies

Islands and the Sea

Essays on Herpetological Exploration in the West Indies

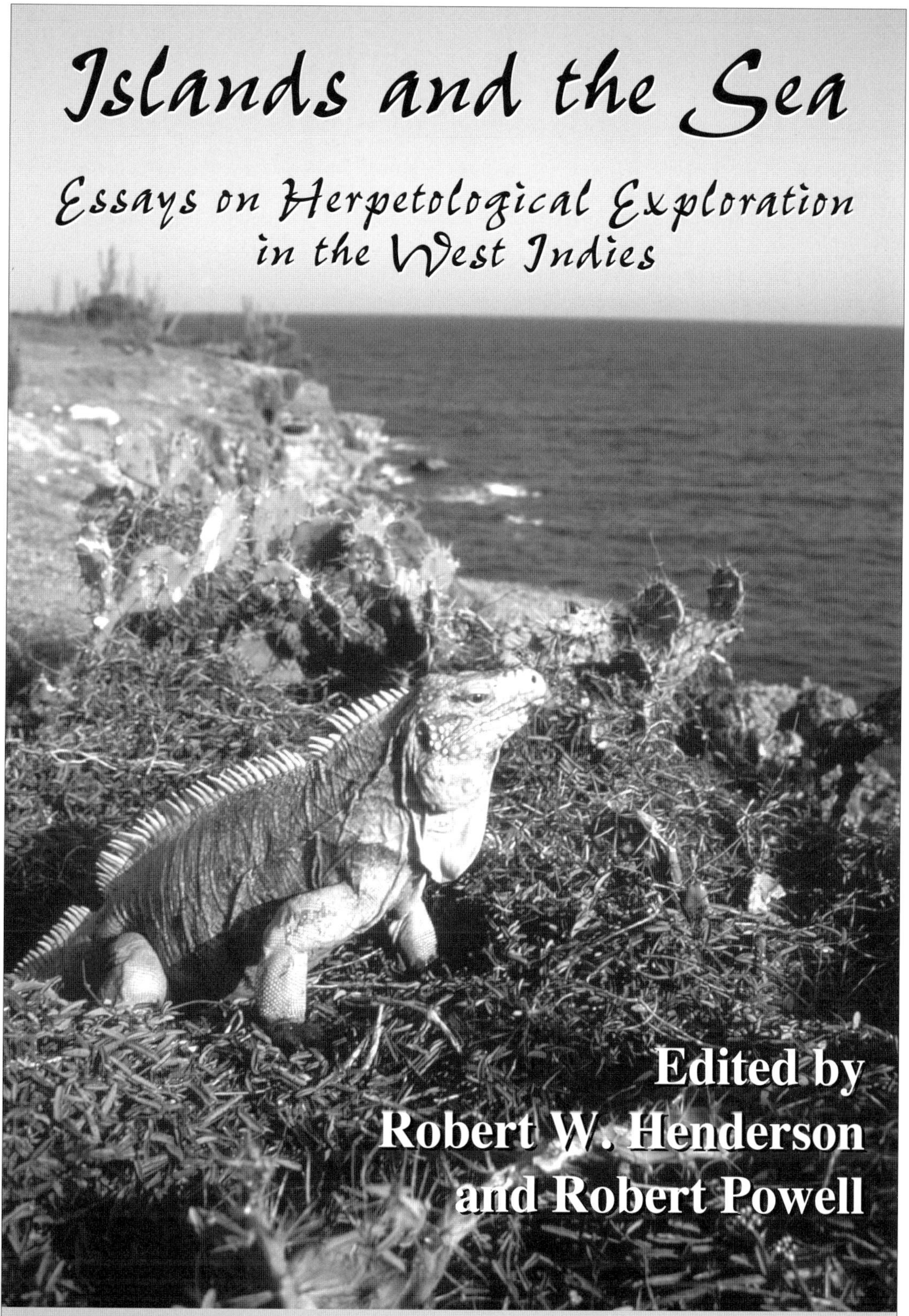

Edited by
Robert W. Henderson
and Robert Powell

Society for the Study of Amphibians and Reptiles

Series Editor's Note

The Society is pleased to issue this unusual compendium of personal experiences in the West Indies. We hope it will inspire new generations of students to be drawn to careers involving field studies of herpetofauna in the West Indies and other places. We thank the volume editors for their cooperation in producing this volume and the various authors for their support.

The drawing of an iguana, used as a headpiece throughout this book, is of *Cyclura cychlura inornata*, drawn by John Bendon. The frog, *Eleutherodactylus abbotti*, used as a tailpiece, was computer generated by Robert Powell from an image published in *Herpetological Review*, volume 32, number 3, page 180, 2001. The photograph on the title page, by John Phillips, depicts an adult male *Cyclura nubila* in its natural habitat at Guantanamo Bay, Cuba; see pages 67–68 in this book for additional information.

Publication costs for this book have been partially funded from The Joseph R. Bailey Endowment.

Contributions to Herpetology, Volume 20

Kraig Adler, *Editor* Timothy D. Perry, *Associate Editor*

Volumes in the *Contributions to Herpetology* series can be purchased from the Publications Secretary: Breck Bartholomew, P. O. Box 58517, Salt Lake City, Utah 84158–0517, USA (*telephone and fax*: area code 801, 453–0489; *e-mail*: ssar@herplit.com). A list of all Society publications, including those of The Ohio Herpetological Society and the *Catalogue of American Amphibians and Reptiles*, is printed at the end of this book. Volumes in the *Contributions* series are published irregularly and ordered by separate subscription, although Society members receive a pre-publication discount. Authors who wish to have manuscripts considered for publication in the *Contributions* series should contact the Editor: Kraig Adler, Cornell University, Department of Neurobiology and Behavior, Seeley G. Mudd Hall, Ithaca, New York 14853–2702, USA.

Members of the Society receive a quarterly technical journal (*Journal of Herpetology*), a quarterly news-journal (*Herpetological Review*), and, depending upon the category of membership, the *Catalogue of American Amphibians and Reptiles*. Society members receive discounts on *Herpetological Circulars* and on books in the *Facsimile Reprints*, *Contributions to Herpetology*, and *Herpetological Conservation* series. The *Catalogue* also is available by separate subscription. Apply to the Society's Treasurer: Theodora Pinou, Yale University, Peabody Museum of Natural History, P. O. Box 208118, New Haven, Connecticut 06520–8118, USA (*telephone*: area code 203, 432–5028; *fax*: 203, 432–5176; *e-mail*: theodora.pinou@yale.edu; *SSAR web address*: http://www.ssarherps.org/). Overseas customers can make payments in USA funds or by International Money Order. All persons may charge to MasterCard or VISA (include account number and expiration date). Details concerning SSAR membership may also be obtained from Dr. Pinou.

 ISBN: 0–916984–62–1. LC: 2003103563.
Production specifications, including place and date of publication, are given at the end of this book.

Acknowledgments

We have incurred many debts of gratitude while assembling this collection of essays, most notably for the donation of photographs and other illustrations. For illustrations that we solicited, we offer thanks to Kraig Adler, the American Museum of Natural History (via Mark Katzman), Hans Boos, Joe Burgess, Suzanne L. Collins, James Forester, Alain Fossé, Glenn Gerber, Richard Glor, Blair Hedges, Nathan Kraucunas, Anita Malhotra, John Murphy, David Nieves, Louis Porras, Richard Sajdak, Lourdes Rodríguez Schettino, the Albert Schwartz Estate, Florence Sergile, Donald Smith, Margaret Stewart, University Press of Florida (via Ken Scott), Philip Walwyn, and Alan Winstel. For assistance with processing of photographs and interlibrary loans, we thank, respectively, Joanne Peterson and Judith Turner, both of the Milwaukee Public Museum.

The volume editors gratefully acknowledge the support of the individuals and organizations listed below. Their help made possible the expedient publication of this book.

Patrons

Anonymous
Michel Breuil
Suzanne L. and Joseph T. Collins
S. Blair Hedges
Julian C. Lee
Mikael Lundberg
A. Stanley Rand
Margaret M. Stewart
Richard Wahlgren

Organizations

Avila University
Bibliomania!
The Center for North American Herpetology
Earlham College
Kansas City Herpetological Society
Milwaukee Public Museum
Museum of Comparative Zoology – Harvard University
Proyecto Coquí Inc.
Southeastern Louisiana University
United States Department of the Interior – United States Geological Survey
University of Florida Foundation
Zoological Society of San Diego

THIS SNAKE HUNTER WAS PHOTOGRAPHED ON ST. LUCIA IN 1888. THE SNAKE IS *BOTHROPS CARIBBAEUS* AND IT MAY HAVE BEEN killed for a bounty, or merely because it was a snake and potentially dangerous. Although attitudes of most West Indians toward snakes have not changed dramatically since this photograph was taken, herpetological research in the West Indies has experienced a revolution — progressing from incidental collecting and anecdotal natural history observations during the latter half of the 19th century to the identification of new species based on molecular methods, decades-long projects focused on the ecology of single species, and a new focus on conservation. However, 19th-century naturalists and 21st-century scientists who have worked in the region share a fascination with amphibians and reptiles, a sense of adventure, and the thrill of discovery (photograph courtesy of Philip Walwyn).

Table of Contents

Introduction and Historical Perspectives

The Bahamas

Cuba

Jamaica

Hispaniola

Puerto Rico

Introduction

The Lesser Antilles

Indices

Introduction and Historical Perspectives

Introduction | Bahamas | Cuba | Jamaica | Hispaniola | Puerto Rico | Lesser Antilles | Indices

Introduction

Robert W. Henderson[1] and Robert Powell[2]

Mention the Caribbean, and most folks think of palm trees, sun-drenched white sand beaches, calypso, rum, steel band music, and a great place to relax for a week or two (Figure 1). To scientists who study amphibians and reptiles, the West Indies are all of those and much more. What most casual visitors to the region don't realize is that the several hundred islands, ranging in area from something about the size of your backyard to over 100,000 km^2, harbor one of the richest assortments of amphibians and reptiles found anywhere in the world. Nearly 600 species have been described, and new species are discovered every year, mostly from Cuba and Hispaniola, but also from smaller islands in the Lesser Antilles and Virgin Islands. The islands are, in essence, a natural laboratory in which biologists study ecology, behavior, biogeography, evolution, and conservation biology; seeking answers to questions about when, how, and why; and using frogs, lizards, and snakes as the focal subjects of their research.

The Idea

In June 2000, we were on Anguilla introducing nine undergraduate students to the joys of working with frogs, lizards, and snakes in the West Indies. In the course of that adventure, we spent pleasant hours discussing the appeal of doing herpetological field work in the Caribbean. We found that, in some instances, the same things had attracted and continue to attract us to the region. However, our lists differed on other germane factors (most likely due to the fact that one of us prefers to study lizards, whereas the other focuses most of his efforts on the snakes that eat them). The attractions we noted were many and varied: for example, diversity of the herpetofauna; palm-fringed shorelines; phenomenally high population densities of some species; a usually friendly and helpful human population; amazing geographic variation exhibited by many species; beaches; provocative problems in biogeography; proximity of the islands to one another; a long history of human colonization of the islands and its impact on the herpetofauna; resplendent tropical vegetation; sunshine; habitat diversity; and the undeniable romance of a tropical island. We also agreed that temperate-dwelling colleagues likely would put a different spin on the West Indian experience, and almost certainly one that was different from those of our colleagues who were born, raised, and continue to live and work in the islands.

Figure 1. A "typical" West Indian beach on Isla Saona, Dominican Republic (photograph by Robert W. Henderson).

The Historical Record

Written references to reptiles in the West Indies date to over 500 years ago. Columbus's first landfall was in the Bahamas, and he made mention in his journal of sea turtles and iguanas. A bit later, in 1564, French explorer René Laudonniere (*in* Hakluyt 1904) recounted an episode that occurred on Dominica:

"... it happened on a day that certaine of our men desirous to see some new things in these strange countries, walked

[1] Section of Vertebrate Zoology, Milwaukee Public Museum, Milwaukee, Wisconsin 53233, USA.

[2] Department of Biology, Avila University, Kansas City, Missouri 64145, USA.

through the woods: and following still the litle rivers side, they spied two serpents [almost certainly *Boa constrictor*, Figure 2] of exceeding bignes, which went side by side overthwart the way. My souldiers went before them thinking to let them from going into the woods: but the serpents nothing at all astonied at these gestures glanced into the bushes with fearful hyssings: yet for all that, my men drew their swords and killed them, and found them afterward 9 great foote long, and as big as a mans leg."

FIGURE 2. *Boa constrictor* from Cabrits National Park, Dominica (photograph by Anita Malhotra).

In 1851, Philip Gosse published his delightful *A Naturalist's Sojourn in Jamaica*, in which he described some of Jamaica's lizards with a keen eye and an evocative writing style:

"One feature with which a stranger [to Jamaica] cannot fail to be struck on his arrival in the island, and which is essentially tropical, is the abundance of the Lizards that everywhere meet his eye. As soon as he ever sets foot on the beach . . . he sees on every hand the beautifully-coloured and meek-faced Ground Lizard (*Ameiva dorsalis*) scratching like a bird among the sand. . . .

"If he looks into the outbuildings of the estates, the mill-house, or the boiling house, or the cattle sheds, a singular croaking sound above his head causes him to look up. . . . It is the gecko, or Croaking Lizard [*Aristelliger praesignis*; Figure 3], a nocturnal animal in its chief activity, but always to be seen in these places, or in hollow trees, even by day.

"The stranger walks into the dwelling-house. Lizards, lizards, still meet his eye. The little Anoles (*A. iodorus* [*A. grahami*], *A. opalinus* &c) are chasing each other in and out between the jalousies, now stopping to protrude

FIGURE 3. *Aristelliger praesignis* from the Discovery Bay Marine Lab in St. Ann, Jamaica (photograph by Richard E. Glor).

from the throat a broad disk of brilliant colour, crimson or orange, like the petal of a flower, then withdrawing it, and again displaying it in coquettish play. . . . Another is running up and down on the plastered wall, catching the ants as they roam in black lines over its whited surface; and another leaps from the top of some piece of furniture upon the back of the visitor's chair, and scampers nimbly along the collar of his coat. . . ."

Similarly, Frederick Ober (1880), in a travelogue entitled *Camps in the Caribbees: the Adventures of a Naturalist in the Lesser Antilles*, wrote that:

"The islands, especially the shores, are teeming with lizards of every color, of every variety of marking, and of all sizes."

Although primarily interested in birds, Ober described iguanas (". . . in the West Indies there is [no lizard] of more hideous appearance. . . ."), a Boa Constrictor, and a Fer-de-lance, but, like Gosse, he devoted most of his attention to lizards:

"In a country where you must keep every sense on the alert, to guard against sudden surprise by serpents or poisonous insects, it is very annoying, often startling, to

FIGURE 4. Raymond L. Ditmars, long-time curator at the Bronx Zoo in New York City, with an *Iguana iguana* (photograph courtesy of Kraig Adler).

FIGURE 5. Fer-de-lance (*Bothrops lanceolatus*) from Martinique (photograph by Hans E. A. Boos).

be so frequently disturbed by these active creatures. . . . [Lizards] are not poisonous, though repulsive to many, and though some of them will bite severely, they do not inflict dangerous wounds."

Raymond Ditmars (Figure 4) took a different approach. In a short essay (1946), he wrote from, more or less, the perspective of a Fer-de-lance (*Bothrops lanceolatus*, Figure 5) on Martinique prior to the eruption of Mt. Pelee in 1902 (Figure 6):

"It is night over the tropical luxuriance of Martinique — the night born only in the Caribbean. . . . Slowly the undergrowth begins to stand in silhouette against the eastern sky and the tropic moon appears to call forth the creatures who love and await her light. The leaves of the stately palms soon glow blue and against the purple sky the bat takes wing. Crawling slowly from the tree the snake flattens in the moonlight. . . . There is a quivering flash of the black, forked tongue and, following a clearing through the grove, the snake starts off in gliding undulations. . . . Strangely the creature's green and yellow tints blend with the vegetation over which it glides. If it were motionless, some thoughtless creature might approach its doom with little warning."

Thomas Barbour (Figure 7), long-time director of the Museum of Comparative Zoology at Harvard

FIGURE 6. St. Pierre, Martinique in 1888, 14 years before the eruption of Mt. Pelee (photograph courtesy of Philip Walwyn).

University, was the most prolific and influential herpetologist working in the West Indies during the first half of the 20th century. Although his work has, at times, been criticized, he wrote affectionately and evocatively in *A Naturalist in Cuba* about people, places, and reptiles on the largest island in the West Indies. In our estimation, however, the best descriptions of the West Indian experience are in Archie Carr's (Figure 8) *The Windward Road* (which was dedicated to Thomas Barbour):

FIGURE 7. Thomas Barbour (photograph courtesy of Kraig Adler).

"To the Caribs it was Iere — the place where the hummingbirds were. It is also the land of the asphalt lake and where steel band music started. It is the last of the islands going south, and the happiest of all, and there is sweet country in it that there will not be time to tell about. But to me that misty dawn, heading eastward toward Toco on the windward road, Trinidad was most of all the home of the paradox frog.

I was glad of the way the world was, that morning, and it was not the tropical day beginning or the thought of the batalí or the long turtle beaches ahead. There was going to be sand to walk all right, but that was not what buoyed me. . . .

It was just a notion of course, a sort of half-professional caprice, and not the real reason for my being there at all. It was the sea turtles [Figure 9] that brought me to Trinidad, and the good beaches and the places the turtle fishermen lived set my course and determined my stops. But that morning . . . the air had a fine faraway smell, and the lazy waves of early morning climbed the gray rocks quietly, and I was feeling good; and a good part of it was

FIGURE 8. Archie Carr, a conservationist whose work with sea turtles is world famous (photograph courtesy of Kraig Adler).

the prospect of seeing and of hearing the song of the legendary Pseudis, the paradox frog [Figure 11].

I have always liked frogs. . . . I like the looks of frogs, and their outlook, and especially the way they get together in wet places on warm nights and sing about sex. The music frogs make at night is a pleasant thing, full of optimism and inner meaning."

A Missed Opportunity

Rather than wax poetic, the two most influential biologists working on Caribbean amphibians and reptiles during the latter half of the 20th century, Albert Schwartz (Figure 11) and Ernest Williams (Figure 12), promoted and encouraged herpetological research in the Antilles by way of their detailed and often thought-provoking scientific publications. Schwartz, although primarily a herpetologist, embraced the entire West Indian fauna, publishing on butterflies, birds, and mammals in addition to frogs and reptiles. Williams, on the other hand, focused his energy on the rich anoline lizard fauna of the West Indies and the Neotropical mainland, and influenced many students to do the same. Despite having cumulatively produced hundreds of technical publications, we have no written record of why the West Indies were so important to either of them. The herpetofauna was, almost surely, the primary attraction, but were other factors germane in promoting their continued fascination with these remarkable island faunas? In conversation, Schwartz would speak emotionally about certain islands (Cuba, Saba), and describe dramatically the discovery of particular new species (e.g., *Anolis fowleri*), but, like Williams, he left no written record of any emotional ties to the islands. That they existed, we have no doubt. It's a pity that, unlike predecessors such as Barbour, neither Schwartz nor Williams recorded on paper their impressions of the West Indian experience.

FIGURE 9. Leatherback Sea Turtle (*Dermochelys coriacea*) from Matura Beach, Trinidad (photograph by John C. Murphy).

Purpose

We sought to avoid similar omissions by contemporary herpetologists by providing a venue for those who have worked in the Antilles over the past half-century to document their West Indian experiences. We solicited from our colleagues essays in which they could take either a broad approach, recollections or musings based on many years of field work, or record a vignette (or series of vi-

FIGURE 10. The Paradox Frog (*Pseudis paradoxa*) from Trinidad (photograph by John C. Murphy).

FIGURE 11. Albert Schwartz, examining a series of giant anoles (photograph courtesy of Kraig Adler).

gnettes) based on memorable experiences. The latter might deal with a favorite species or place, a particularly vexing problem, an entire fauna, a single day or a memorable expedition, a conservation concern, a philosophical issue, etc. Each essay was to be personal, to have a herpetological element to it, and to convey some emotion elicited by being in the West Indies and studying its herpetofauna.

Our specific goals for compiling the following essays were multifaceted: We have long been fascinated with the histories recorded by our predecessors and by the comparisons we have been able to make between the islands they experienced and those with which we have become familiar. In addition, we have benefited substantially and repeatedly from the lessons they learned. Consequently, our primary goal was to document the constantly evolving history of research in the region. In addition, we sought to educate, and we know from our years 'in the business' that knowledge often is conveyed most effectively in the context of one's own experiences. Being able to record the experiences of many scientists should only enhance the learning opportunities, both for those already working in the region and for those who have yet to experience the region's many wonders. Furthermore, we are guilty of blatantly promoting the West Indies as an ideal venue for biological research. Knowing that the questions vastly outnumber the scientists who seek to address them and selfishly interested in learning the answers to far more inquiries than we could ever hope to tackle ourselves, we wish to encourage more research by more scientists, particularly those with a herpetological bent. A related concern lies in the sad reality that the islands are changing at ever-increasing rates, and most of those changes figure to adversely affect native faunas. In other words, if we can't encourage more research now, many questions will remain forever unanswered. Finally, field research is notoriously unpredictable, especially when conducted in developing nations and, even moreso, when practiced in what often are the most remote regions of those nations. So, we anticipated that the recollections of scientists engaged in the region would include experiences that were exciting, stupid, frustrating, dangerous, and humorous, as well as informative. Consequently, we hope the last (but not the least) of our goals, to entertain, is realized.

THE CONTRIBUTORS

The essays presented herein were written by biologists ranging in age from their early 20s to over 90. They come from graduate students relatively new

FIGURE 12. Ernest E. Williams, biologist at the Museum of Comparative Zoology, Harvard University (photograph courtesy of Kraig Adler).

to the region, U.S.- and European-based scientists who have concentrated their research in the Antilles for decades, and Caribbean nationals that have spent their entire professional lives in their homelands. The authors are associated with universities, museums, zoos, conservation organizations, and national trusts; most live in the U.S., but also represented are France, Cuba, Puerto Rico, Jamaica, the Dominican Republic, and Anguilla. We were gratified to receive essays from biologists (e.g., Estrada and Rodríguez Schettino) who made mention of meeting and/or being inspired by the work of, for example, Orlando Garrido and Stan Rand, and then receiving essays from those sources of inspiration as well.

Despite the disparity in ages and backgrounds, the variety of approaches taken by the authors in telling their stories, the many islands about which they chose to write, or whether they were Caribbean nationals or ephemeral visitors, two common threads are expressed eloquently in every contribution: a deeply felt interest in pure science (a quest for knowledge without concern for practical applications) and a fondness for the amphibians and reptiles we study.

The adventures recounted may entail the discovery of a species new to science, an understanding of the nesting ecology of a Bahamian Rock Iguana, amazement at the relative abundance of snakes on Grenada, or the sad awareness that some perturbation of the environment is threatening the future of some frogs in Puerto Rico. These biologists have experienced, among other events, harrowing boat trips through rough seas, a rock slide through a campsite, the kindness of strangers, and a reminder from a most unlikely source what their passion is all about.

The contributors have for decades discovered, described, hypothesized, and philosophized while searching for, collecting, conserving, and observing the rich amphibian and reptilian faunas of the West Indies. Cumulatively, they have published in excess of a thousand technical papers which have appeared in scientific journals throughout the world. For most of them, however, this will mark the first (and, in some cases, the last) attempt to write for an audience not comprised only of their colleagues. What follows may provide insights into why this remarkable chain of islands and its equally remarkable herpetofauna elicit such devotion from such a diverse cadre of biologists. Maybe Bill Watterson, in our favorite comic strip (Figure 13), best explained this fascination with animals in a wonderful place.

FIGURE 13. Bill Watterson's Calvin and Hobbes say it all (Calvin & Hobbes © Watterson. Reprinted with permission of Universal Press Syndicate. All rights reserved).

Suggested Reading

Barbour, T. 1945. *A Naturalist in Cuba*. Little, Brown and Co., Boston.

Carr, A. 1979 [original edition published in 1955]. *The Windward Road*. Univ. Presses of Florida, Gainesville.

Ditmars, R. L. 1946. *Reptiles of the World*. Macmillan Co., New York.

Laudonniere, R. 1904. The voiage of Captaine René Laudonniere to Florida in 1564. *In* R. Hakluyt (ed.), *The Principal Navigations Voyages Traffiques & Discoveries of the English Nation*. Vol. 9. Reprint ed., Maclehose, Glasgow.

Gosse, P. H. 1851. *A Naturalist's Sojourn in Jamaica*. Longman, Brown, Green, & Longmans, London.

Ober, F. A. 1880. *Camps in the Caribbees: the Adventures of a Naturalist in the Lesser Antilles*. Lee & Shepard, Boston.

Henderson, R. W. and R. Powell. 2003. Introduction, pp. 3–8. *In* R. W. Henderson and R. Powell (eds.), Islands and the Sea: Essays on Herpetological Exploration in the West Indies. Society for the Study of Amphibians and Reptiles, Ithaca (New York). Contributions to Herpetology, volume 20.

Some Historical Perspectives

Robert Powell[1] and Robert W. Henderson[2]

The allure of the West Indies has attracted the attention of naturalists and scientists for centuries. Although the first European visitors sought wealth, only a few years had passed after Columbus's arrival before nature's siren call began to make itself felt. The first writers to record their impressions of the natural wonders of the region had not come primarily for that purpose, but their accounts of vast forests and an incredible wildlife soon attracted those for whom the flora and fauna were lure enough.

Following is a collection of excerpts from the writings of naturalists and scientists who, in some cases, spent prolonged periods in the islands, sometimes of several years' duration. Their works were published between the mid-19th and the mid-20th centuries and provide a yardstick with which to compare the essays and experiences written by contemporary scientists.

Differences in the style of writing may be the most notable distinction between the historical prose and the contemporary counterparts later in this volume. This is most obvious in the selections by Gosse and Ober, but also is evident to lesser degrees in the writings of Noble and Barbour. The earlier works have a more formal flavor and the writing is very descriptive, often "flowery." Reading what was written more than a century ago, when the West Indies were still largely *terra incognita*, one senses the excitement of Gosse and Ober as they approached a new landfall, reached a destination after hours of hiking, or observed some unusual insect, frog, or reptile. Perhaps because of the descriptive approach to their writing, it is, in general, more evocative and romantic than that of contemporary writers. We often have wished that we could be magically transported to the West Indies of a hundred or more years ago, or even to see photographs of the islands as they appeared to Gosse and Ober, and of a West Indies that is no more (Figure 1).

The essays included in this volume tend to be more concise than those excerpted in this chapter, are usually devoid of flowery descriptions, and are notably more casual. The modern style, however, should not be taken to imply that biologists working in the region today are casual about their work. Whereas Gosse and Ober each lived in the West Indies for well over a year, biologists visiting the Antilles today are often on tight schedules and even tighter budgets. Their visits may last only weeks to, at most, a few months. They almost certainly arrived by air, in contrast to Gosse, Ober, Noble, and Barbour, all of whom leisurely reached their destinations by boat. Because of their abbreviated visits, contemporary field biologists put in long hours, often working day and night to frantically accumulate sufficient data to test hypotheses, more accurately assess an island's biodiversity, contribute more information to the knowledge base pertaining to West Indian species, or develop conservation action plans for the increasing number of threatened and endangered species (Figure 2).

Sadly, many of the species that exert such a strong attraction on biologists live in habitats that are being fragmented or disappearing, with much of the destruction or alteration attributable either to tourism-related development or subsistence agriculture. Additional populations are being decimated due to the introduction of alien competitors or predators (e.g., cattle and goats or cats, dogs, and mongooses; Figure 3). Perhaps the concise, less descriptive writing styles seen in the contemporary essays reflect the hurried race to learn as much as possible about as many species as possible in the relatively short span of time still available for such studies.

[1] Department of Biology, Avila University, Kansas City, Missouri 64145, USA.

[2] Section of Vertebrate Zoology, Milwaukee Public Museum, Milwaukee, Wisconsin 53233, USA.

Exempted from at least some of the time restrictions are those biologists who were born, raised, often educated, and who continue to work in their island nations. They have an intimacy with the culture and the fauna that a visitor can never achieve. However, they may face other obstacles to pursuing field studies, convincing governmental or private agencies that the work in which they are engaged can be justified in light of severely limited financial resources or coping with political ideologies that are not sympathetic to environmental issues and pure science.

FIGURE 1. The West Indies in 1888 (from top): a Jamaican sugar factory, an unidentified Jamaican waterfront, a silk cotton tree (*Ceiba pentandra*) on an unidentified Bahamian island, and a Trinidadian family (photographs courtesy of Philip Walwyn).

Despite different time frames, funding concerns, or political obstacles, both historical and contemporary naturalists have faced many of the same dangers. Hurricanes, rock slides, uncomfortable and risky boat rides on high seas, active volcanos, and insecure footing on precipitous, overgrown trails respect no one. Nevertheless, the centuries-long quest for some elusive frog, lizard, or snake continues.

Today, although collecting for scientific institutions is still done, and new species of West Indian amphibians and reptiles are described every year, many of the contemporary essays describe projects which focus on the ecology, behavior, and conservation of frogs and reptiles. Some 19th century chroniclers of West Indian natural history described frogs and reptiles as "hideous," "repulsive," and "vengeful"; in contrast, the essays produced by biologists currently involved in field work with Antillean frogs and reptiles convey a passion for the animals and a concern for their future. Today, we are armed with a knowledge that probably never occurred to scientists and naturalists working in the West Indies during earlier times; namely, that species, or even entire West Indian island faunas, are threatened with extirpation or extinction, and we are now in a race to learn as much as we can, as fast as we can, for as long as we can.

Philip H. Gosse

Philip H. Gosse (Figure 4) spent 18 months in Jamaica during 1844 and 1845. Unlike most visitors to Jamaica, he did not concentrate his time in Kingston, but made his base in the Bluefields/Content region of what is now Westmoreland Parish. Gosse devoted a considerable number of pages of *A Naturalist's Sojourn in Jamaica* (1851) to the physical description and natural history of frogs and reptiles, and one gets the impression that he, unlike many of his contemporaries, truly liked them. Gosse described 13 new species, several of which are still recognized as valid, and he is considered the father of Jamaican herpetology (Crombie 1999).

FIGURE 2. Today, the American Crocodile (*Crocodylus acutus*) is endangered, but in the early 19th Century, exploiting the species for hides and meat was widespread and "officially" sanctioned. Here, convicts skin crocodiles in the Dominican Republic in 1922 (photograph by G. K. Noble; courtesy of the Department of Library Services, American Museum of Natural History).

Blue Mountain Peak

"Dec. 4th. My first sight of Jamaica was one that I can never forget. The approach to the land of one's destination after a long voyage is always a season of pleasurable excitement; a peculiar flutter of the spirits, a disposition to hilarity, and to mutual gratulation pervades more or less every one on board, from cabin passenger to the black cook in the caboose.... I had risen early, and at dawn of day had seen Cape Tiburon, the west point of Hayti, dark, distinct, and palpable, though distant sixty or seventy miles. The brightness of the east was behind it, and threw it out into strong and bold relief; but the moment the sun was visible above its outline, it faded away and became entirely invisible....

"During the afternoon the mountains of Jamaica were seen, and gradually grew more distinct as we neared the island. Yet the cloudiness of the day had prevented my having had any satisfactory view of it until evening. About sunset, I was standing forward, when one by my side said, 'Look at the Peak!' I looked intently, directing my gaze to the neighbourhood of the horizon, where I supposed it was to be seen; but nothing but the dull white clouds met my eye. 'Up there!' said my informant; and his finger pointed up into the sky; and there indeed was its noble head (perhaps elevated by refraction), a conical mass, darkly blue, above the dense bed of clouds that hung around its sides, and enveloped all beneath its towering elevation. Yet it is situated far inland, and was then full forty miles distant from our ship.

FIGURE 3. A mongoose (*Herpestes javanicus*) from Savane-à-Mulets, Soufrière, Guadeloupe (photograph by Alain Fossé).

"But night soon fell, and, as we were somewhat anxiously watching for the light on Point Morant, I had the pleasure of first seeing it from the main rigging. We were soon abreast of it, and as we passed on before an increasing breeze, that tempered the tropical heat with its refreshing breath, we saw the coast dark and high only a few miles off. Many lights were seen in the scattered cottages, and here and there a fire blazed up from the beach, or a torch in the hand of some fisherman was carried from place to place. My mind was full of Columbus, and of his feelings on that eventful night, when the coast of Guanahani lay spread out before him with its moving lights, and proud anticipations. With curiosity and hope, somewhat analogous (*parva componere magnis*), did I contemplate the tropical island before me, its romance heightened by the indefiniteness and obscurity in which it lay. I was on deck several times during the night, and during intervals was still engaged, in dreams, in endeavouring to penetrate the darkness of the shore...."

FIGURE 4. Philip A. Gosse (image courtesy of Kraig Adler).

Gosse had a good eye and described Jamaican wildlife vividly and accurately. He devoted a number of pages to descriptions of *Anolis* lizards and their interesting and entertaining behavior.

The Venus Lizard

"The first time I met with that fine Iguaniform Lizard,

Dactyloa Edwardsii [= *Anolis garmani*; Figure 5], called Venus by the negroes, and sometimes Green Guana, was on this ridge [i.e., Bluefields Ridge]. And though I afterwards found it in lower situations, as on the moderate elevations of the St. Elizabeth's hills, and even in the pimento groves of Mount Edgecumbe that border the sea shore, it seems rather to affect the higher mountain-woods. The mode in which I formed an acquaintance with the species may be worthy of being related. One day in February, having ascended the ridge with a companion, my attention was arrested by a Lizard about a foot long, and of a lively green colour, on the trunk of a small tree, head downward, intently watching our motions as we stood near. My young friend suggested the possibility of capturing it by slipping a noose over its head, while its attention was engaged by whistling, I laughingly proceeded to try the spell; and having made a noose of small twine, which I tied to the end of a switch, I gently walked towards him, whistling a lively tune. To my astonishment he allowed me to slip the noose over his head, merely glancing his bright eye at the string as it passed. I jerked the switch; the music ceased; and the green-coated forester was sprawling in the air, dangling, greatly to his annoyance, at the end of my string. He was very savage, biting at every thing near; presently his colour began to change from green to blackish, till it was of an uniform bluish black with darker bands on the body, and a brownish black on the tail: the only trace of green was just around the eyes. I carefully secured, without injuring him, and brought him home in the collecting-basket, into which I had no sooner put him, than he fiercely seized a piece of linen in his teeth, and would not let it go for several hours. I transferred him to a wired cage, linen and all; and at length he suddenly let go his hold, and flew wildly about the cage, biting at anything presented to him. At night I observed him vividly green at first; a token, as I presumed, that he had in some measure recovered his equanimity.

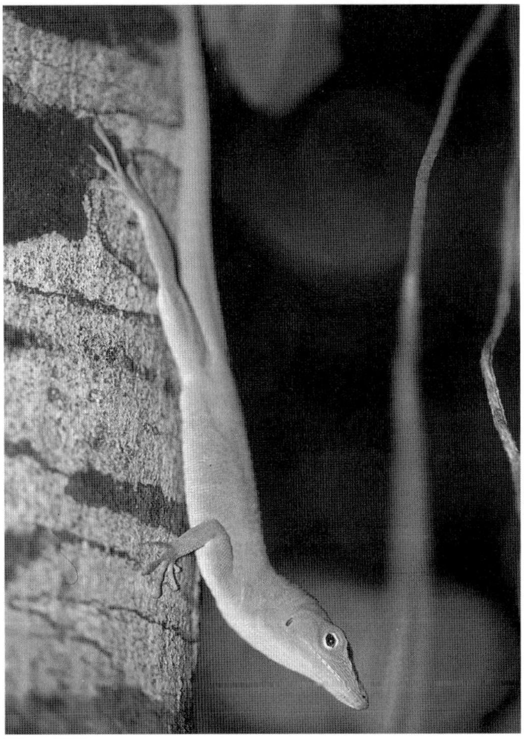

FIGURE 5. A "Venus lizard," *Anolis garmani*, from the Discovery Bay Marine Laboratory in Jamaica (photograph by Richard E. Glor).

"The next day he continued very fierce. I hung the cage out in the sun; two or three times in the course of the day I observed him green; but for the most part he was black. The changes were rather quickly accomplished."

Nocturnal Forest Sounds

"Various and strange are the sounds which strike the ear of one benighted in the forests of Jamaica. Some of these are the voices of night-birds . . . and some are insect sounds. But, besides these, there are some that are certainly produced by Reptiles, though it is difficult to identify them. Nearly every night, at certain seasons, there ascends from the woods around Content a continual snoring of various tones, the voices of numberless Tree-frogs, or, as they are called here, Toads. They are said to reside in the large ventricose leaves of the greater Wild-pines, especially the fine one, *Tillandsia lingulata*, which, about the end of July, sends up a magnificent flower, somewhat like a huge carnation, with broad outer petals of a rich crimson hue and polished surface, and a cluster of smaller interior ones of pale yellow. In the coolness and moisture of these natural reservoirs, always half full of water collected from rains and dews, the Tree-frogs delight to lie, finding in them circumstances eminently congenial for the maintenance of cutaneous humidity, so essential in these reptiles to respiration. They are very rarely seen, and, but for their vocal powers by night, we should scarcely be aware of their existence; the number and universality of these sounds, however, in the mountain-woods, during the hours of darkness, prove that they are abundant. Even when seen by day, their agility in leaping renders it difficult to lay hands on them. The sounds in question bear a strong resemblance to the objurgations of an inveterate snorer, but are much louder; or sometimes reminds one of the groaning and working of a ship's timbers in a heavy gale at sea.

"These are probably the voices of some of the greater *Hylidæ*. But there are other and different noises still. While I am writing this note at Content, — it is a lovely night in June, — all around I am saluted by strange sounds. Now and then comes the singularly harsh and cracked voice of the Gecko, like the note of a child's penny trumpet, or like a stick drawn across the teeth of a comb: — this I am familiar with. But I hear another voice, far more abundant, but quite unknown to me. It is now (about midnight) coming up from every part of the moonlit forest below me, with

FIGURE 6. *Trachycephalus lichenatus* (= *Osteopilus crucialis*), from Mandeville, Jamaica (photograph by S. Blair Hedges).

incessant pertinacity. It is a clear shrill note, so like the voice of a bird, and in particular so like that of the Solitaire, that it might easily be mistaken for it, but for the inappropriate hour, and the locality. Like that, it is beautifully trilled or shaken, and, like it, the individual voices are not in the same key. As I now listen to the mingling sounds, I distinguish two particularly prominent, which seem to answer each other in quick but regular alteration; and between their notes there is the difference of exactly a musical tone. I have little doubt that this is the sexual call of some Tree-frog. The groanings and snorings, which are sometimes so incessant, I do not now hear, except one such sound now and then in the course of an evening.

"Some three or four species of *Anourous* Reptiles are all that Jamaica has produced to my researches, and all of these are of the family *Hylidæ*, whose dilated sucking-disks at the tips of their toes, enable them to cling about the foliage of trees and plants, and to leap from leaf to leaf with security and precision.

"One of these was taken in a bedroom at Savanna-le-Mar, one night in October, having probably hopped in the open window from the branches of a mango-tree only a few feet distant. I was surprised at its changes of colour, in this respect resembling the Chameleons and Anoles, or, still nearer, the Geckos. When I obtained it, the whole upper parts were of a rich deep umber-brown, with indistinct black bands. On looking at it at night, to my surprise I saw a great alteration of hue: it was paler on the head and back, though least altered there; on the rump, and on the fore and hind legs, it was become a sort of semi-pellucid drab, marked with minute close-set dark specks. When disturbed, it presently became slightly paler still, but in a few minutes it had recovered its original depth of tint. In the course of half an hour it displayed again the speckled drab hue, and now uniformly so, save a black irregular patch or two on the back, and a dark patch between the mouth and each eye. The belly, which was very regularly shagreened, was of a dull buff, not susceptible of change. The eyes retained their proverbial beauty, for the irides were of a golden brown tint, like sun-rays shining through tortoiseshell....

"A much finer species, as to form, colour, and size, was brought to me, in May, from the summit of Bluefields Mountain. It was of that curious genus named by MM. Duméril and Bibron *Trachycephalus*, having a distinct neck, a triangular trunk terminating in a point, and an enormous flat head studded with irregular bony ridges.... It is a finer, though more uncouth, species than any yet described. I therefore propose to call it, from its resemblance to the moss-grown bark of a tree, the Lichened Tree-toad (*Trachycephalus lichenatus*) [Figure 6].

"During the short period that this fine reptile remained under my observation, it was impatient of confinement, leaping vigorously. Now and then it inflated its body to a considerable degree, which, as in the former species, was probably an expression of anger, or an attempt to intimidate. A moisture exuded from its skin, which took the form of a very fine froth, even while on the body, and when touched appeared of a gummy nature, adhering to the fingers, and stiffening them as it dried. It left shining marks on the table, too, like the trail of a snail...."

Gosse believed that a second species of *Trachycephalus* might exist "in the western districts of Jamaica," based on an unpublished description and drawing by surgeon and botanist Dr. Anthony Robinson. Gosse gave it the name *T. anochloros* (= also *Osteopilus crucialis*), and provided this information regarding its behavior:

"All that Robinson has recorded of its manners is contained in the following brief note: 'It was given me by Dr. Anderson, of Hanover parish, who informed me that he held it in his hand almost half an hour, after which, having occasion to rub his eyes with the fingers of that hand, which was besmeared with a mucus left by the animal, he was seized with a violent pain and smarting that lasted nearly half an hour, but subsided on washing his hands . . . in a little spring water. The learned Doctor compares this quality to the stinging of the Nettlefish, by which he probably means the *Physalia*; but it seems to me much more analogous to the acrid secretion which exist in the cutaneous follicles of the true Toads; if so, it justifies the application of the term Tree-toads to these animals, to which they are provincially distinguished, rather than Tree-frogs."

Frederick A. Ober

In 1876, under the auspices of the Smithsonian Institution, F. A. Ober undertook the exploration of the Lesser Antilles. He remained in the islands for nearly two years and, although primarily interested in birds (he is immortalized in the name of the Lesser Antillean Flycatcher, *Myiarchus oberi*), he recorded information on many aspects of natural history, including observations on amphibians and reptiles. The excerpts presented here are from *Camps in the Caribbees: The Adventures of a Naturalist in the Lesser Antilles*, first published in 1879.

Dominica

"Late in the afternoon . . . we descried a speck dancing on the waves, which speck was, of course, a boat; and in that boat, when it reached us, I engaged passage for the shore, my unhappy companions drifting about until the next afternoon, sometimes in sight, sometimes lost of view for a long time. As we neared shore I had time to examine the character of the scenery of the western coast, as one object after another was unfolded, and the mass of green and blue resolved itself into wooded hills, narrow valleys, and misty mountain-tops that reached the clouds. A planter's house gleamed white in a valley; a pebbly beach stretched between high bluffs, with a grove of cocoa palms half hiding a village of rude cabins along its border.

"I was approaching an island of historic interest and scenic beauty, of which the events of one and the elements of the other are little known to the world at large. It is the first island upon which Columbus landed on his second voyage. Having been first seen on Sunday, it was called by him *Dominica*, and this event dates from the 3d of November, 1493. Blest isle of the Sabbath day! Many changes has it known since the great navigator first saw its blue mountains and landed upon its fragrant strand.

"Does it read like a fairy tale, this second voyage of Columbus? With three ships and fourteen caravels, containing fifteen hundred persons, he set sail from Cadiz, touched at the Canary Isles, and then shaped his course for the islands of the Caribs, of whose prowess and fierce nature he had heard many stories from the mild people of Hispaniola. 'At the dawn of day, November 3d, a lofty island was descried to the west. As the ships moved gently onward, other islands rose to sight, one after another, covered with forests and enlivened by flights of parrots and other tropical birds, while the whole air was sweetened by the fragrance of the breezes which passed over them.'"

FIGURE 7. A Green Iguana (*Iguana iguana*) from Saba in the Lesser Antilles (photograph by Glenn Gerber).

FIGURE 8. The "Lance-head snake," *Bothrops lanceolatus*, from Martinique (photograph by H. E. A. Boos). *Bothrops* from St. Lucia and the South American mainland are now considered to be species distinct from that on Martinique.

Grenada and the Grenadines

"Skirting the belt of mangroves bordering the lagoon, one morning in March, I anxiously searched the intertwined branches for iguanas [Figure 7]. Grenada is celebrated as being the home of great numbers of these reptiles, which may often be found basking on old walls within the limits of the town. My boatman was a negro, who, accustomed to the appearance of the iguana in the trees, discovered one long before I could distinguish the difference between the reptile and green leaves. Even after it had been pointed out, I had difficulty in recognizing it, so nearly did its colors harmonize with those of the tree in which it was feeding.

"It lay quite still, stretched flat upon a branch, its tail hanging down like that of a snake. Though it was evidently suspicious of our intentions, its quiet was not due to that alone, for it is normally a sluggish animal. Yet, when once thoroughly aroused, it will dash over the ground at great speed. I fired, yet it still clung tenaciously to the bough, and a second shot did not kill it, for it would have escaped had not my boatman pinned it with an oar, after it had fallen into the mud. From one that we captured that morning, the man with me procured a dozen large, white eggs, which he saved to eat....

"The order *Sauria*, the lizard order, is well represented in the West Indies, though in none of the smaller islands between Porto Rico and Trinidad is to be found that greatest of the saurians, the alligator. The Indians of Dominica, to whom I described the alligator, were greatly amazed to hear of a 'lizard' twelve feet in length, as they had never seen one larger than an iguana, which seldom attains a greater length than five feet, and is as mild in disposition as the alligator is sanguinary. The islands, especially the shores, are teeming with lizards of every color, of every variety of marking, and of all sizes.

"Especially do they love the cliffs, and if you are walking through the bushes at the base of any sunny precipice, or over any rocky tract, you will be startled by the frequent dashes made by these reptiles across your path. In a country where you must keep every sense on the alert, to guard against sudden surprise by serpents or poisonous insects, it is very annoying, often startling, to be so frequently disturbed by these active creatures. In the mountains are fewer species, and they are more sluggish, but in the warm lowlands you must be very active to capture one.... They are not poisonous, though repulsive to many, and though some of them will bite severely, they do not inflict dangerous wounds.

"There are many hideous forms, especially among those of South America, like the Basilisk and the Flying Dragon; but in the West Indies there is none more hideous than the iguana. Never was more harmless creature invested with more frightening aspect.... In the islands where it exists it is eagerly sought as food, and its flesh is palatable and delicate, as I can testify from experience, being white, tender, and nutritious.

"The good father Père Labat (worthy missionary and *bon vivant* withal) compares fricasseed guana to chicken for the whiteness of its flesh and delicacy of its flavor...."

Some Summer Days in Martinique

"Centipedes and scorpions also abound here. Indeed, it seems that nature has bestowed upon this island of Martinique all the pests and scourges known to these islands; for only here and in the adjacent island of St. Lucia is found that most venomous and vengeful of all serpents, the Lance-head snake *Craspedocephalus lanceolatus* [= *Bothrops lanceolatus*; Figure 8]. The isolation of this snake on these two islands, when its nearest habitat is Guiana, is one of the most vexing stumbling-blocks to one studying the distribution of animals. How came it here? Was it introduced, or is it indigenous? Was it wafted here upon some floating tree, or was its home here from the beginning? The correct solution of this problem would, doubtless, throw some light upon that more important and gigantic one, Were these islands once a part of the continents? Certain it is, the adjacent islands of Dominica and St. Vincent, separated from these channels less than thirty miles in width, are free from this scourge. Nay, more; it is recorded that, during the wars between the English and the Caribs, in the last century, the Lance-head was carried to the islands just named, but could not be made to live.

"Annually, during the crop season, many laborers are killed in each island, for this snake has its hiding-places in the canes as well as the forests. It has been so abundant that in this garden the pleasant walks and shady drives are nearly always deserted. A serpent over seven feet in length, killed in the garden, is shown in the Museum. There is, it is said, no antidote for its bite.... The poison is quickly fatal, and decomposition rapidly follows.... The dread of the serpent is universal. It seems to possess a hatred for man; and it is seriously avowed by the natives that it will lie in wait for an opportunity to inflict death. The country people live in continual trepidation, and very few of them will venture from their houses after dark, even in the suburbs of the city."

FIGURE 9. G. Kingsley Noble (photograph courtesy of Kraig Adler).

G. Kingsley Noble

G. Kingsley Noble (Figure 9) was associated with the American Museum of Natural History as a herpetologist and an experimental biologist from 1918 until his untimely death in 1940. In 1922 he traveled to the Dominican Republic to search for iguanas (*Cyclura cornuta*; Figure 10) and *Osteopilus vastus*, arguably the largest species of treefrog in the world. Noble was one of the pioneers of collect-ing amphibians and reptiles at night, and experi-mented with a number of light sources. During the field work described below, Noble had acetylene lamps, electric flashlights, and flash powder for night photography (Myers 2000). The text below was excerpted from an article published in 1923, describing his work in a mountain range known as Quita Espuela in the northern part of the Dominican Republic. Here, Noble provides an account of his nocturnal search for the elusive giant, and the source of a "woeful call, such as only a lonely dog — or a frog — could make."

"In the glare of the headlight the jungle at night is transformed into a different world from that observed during the day. Everything within the magic circle of light stands out in relief. Inky shadows thrust forth long arms. Fallen trees spring up and take ominous proportions. Colors lose their values. Every stub and leaf turns to either black or gold. Drops of moisture reflecting back the rays

FIGURE 10. In 1922 Noble was assisted by pig hunters on Isla Beata (Dominican Republic) in his quest to collect Hispaniolan Rhinoceros Iguanas, *Cyclura cornuta* (photographs by G. K. Noble; courtesy of the Department of Library Services, American Museum of Natural History).

of the light shine out like a thousand jewels. The forest becomes a land of mystic exaggerations.

"It is not the strange sights, however, but the sounds which make the night forest awesome, even fearful. Voices, some soft and whispering, others harsh and grating, greet one on all sides. The monotonous chirping of the cricket we recognize at once, but what are those anvils ringing in the distance? A rustling in near-by bushes and our imagination soars! Land crabs, large tarantulas, and whip scorpions seek their prey after the sun has set. One never loiters in the forest at night. Something always seems to be waiting there in the darkness.

"With a last shot of our acetylene rays toward the tree-top 'dog,' we plunged on once more. Two hundred yards beyond we came to a third stream that was swifter and more torrential than the other two. There was something inviting about the white foam splashing in the light of our lamps. We turned and headed upstream toward the mountain. Hardly had we started when a shrill cry arose high above the swirl of the current, it sounded like a jet of escaping steam or a locomotive whistle stammering under excess pressure. We scrambled upward over slippery rocks, gained a bend in the torrent, and there beheld sitting on a rock in midstream and fairly bursting with exertion, a brilliant golden-green tree frog, flecked with white above and partly concealing with his legs four gaudy patches of brightest gold. I recognized at once that it was new to science, and whispered to Juan that we must capture it at any cost. Most frogs will show no concern toward the acetylene lamp, continuing their calling as tranquilly as if the lamp were only the moon staring down upon them. Not so our handsome new species. Hardly had the light settled squarely upon him when he ceased calling abruptly, edged off to the side of the rock, and jumped — whether intentionally or not I could not determine — into the heart of the swiftest current. The white waters threw up their arms a little higher. Our prize had disappeared.

"We were now far from camp and low in spirit. We resolved to take the shortest way back through the jungle. A few yards and we were swallowed up in vines and lianas; but Juan had his machete and we progressed. Nothing is so black as jungle at night. Our headlights pierced the saturated air with difficulty. We stopped to catch our breath. A 'dog' barked high overhead, another just ahead of us replied. Cautiously we moved forward. Again the 'dog' gave forth a smothered bark. I pushed Juan aside, and, scarcely breathing, parted the lianas. There, two yards ahead, was a great brown tree frog, with a sharp snout and a black stripe through the eye. No, it was not the giant tree frog we were seeking, but it was a species nearly as rare. Even as we looked, the frog puffed up his throat into a white balloon, the size of a golf ball, and gave a mournful *ba wo-ow*! Keeping my light shining on his eyes, I slipped my free hand along the frog's back and seized him quickly. Then a very strange thing happened. The frog opened his mouth and screamed like a frightened pig. When frogs call, they keep their mouths tightly closed. Very few have learned the trick of opening their mouth to cry . . . never before was it known that an *Eleutherodactylus* [Figure 11] had adopted [such] 'terrifying behavior.'

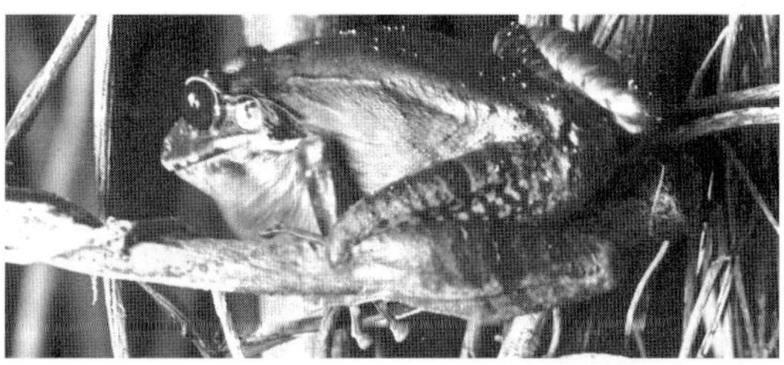

FIGURE 11. A "tree-top dog," *Eleutherodactylus inoptatus* (photograph by G. K. Noble, 1922; courtesy of the Department of Library Services, American Museum of Natural History).

FIGURE 12. The Giant Treefrog (*Osteopilus*, formerly *Hyla*, *vastus*) (photograph courtesy of Louis Porras).

"In my excitement I had noticed nothing but the frog. Before I had him safely bagged, Juan tugged at my arm and pointed out a tree snake asleep on a branch just above my head. The snake [a species of *Uromacer*], nearly four feet long but not half an inch in diameter, so closely resembled the vines that I did not see it at first. As Juan lowered the limb, I recognized the species as a diurnal form that spends the sunlight hours stalking unfortunate lizards. Its mouth is equipped with two long fangs just behind the other teeth. Although these fangs are too small to injure man, they are decidedly effective on smaller prey.

"We moved on joyously to camp. If the giant tree frog did not live in these jungles, at least other interesting creatures did. We gained the ridge at the foot of which lay our camp. A light shone up from the tent door toward the tree tops. Mrs. Noble was signaling to us to approach cautiously. I caught the word *maco grande* — giant frog. I smiled; she, too, had been fooled by the barking frog, I mused, and stepped rather briskly forward. I saw she was holding a flash lamp on something up in the wild plantain in front of the tent door. As I stepped nearer, there took shape against the velvety blackness of the night a tree frog so large that it seemed unreal [Figure 12]. Its four immense feet were flattened out against the plantain stalk but its head, with staring orbs, slowly turned as if contemplating which direction to leap. I thought of my lost species and a chill went down my back. I dropped everything, slipped out of my coat, and stealthily moved nearer. With both hands free I could not miss! Nearer I came. Why not the gun, I thought. Perhaps I could not hold the creature. But it was too late to go back. With both hands I clutched. Something squashy slipped in my fingers. Without daring to look, I dropped the frog into the bag which Juan stretched toward me. In another moment we were inside the tent, with mosquito bar closed, ready to examine our capture. It was then that I noted that my hands were red and swollen. I must have brushed up against a 'poison ivy tree' in the jungle, I thought. We opened the bag cautiously; a penetrating odor like that of burning mustard, though more acrid and sickening, streamed forth. I looked more closely at my hands. To the red swellings was adhering some of the mucus from the frog's skin. In a moment it was clear to me — the skin of the giant frog had badly poisoned me.

"The skin of all frogs and toads contains two kinds of glands: mucus and poison. But the poison of the latter gland was not known hitherto to be injurious to the unprotected hands. Toads do not produce warts, nor can they inflame one's skin in any way. Their glands secrete a poison which only affects mucus membranes, such as those in the nostrils and the eyes. The skin of the giant tree frog, we gradually realized, must be extremely poisonous to burn the hand. We concluded it would be well not to experiment further. The cutting odor alone warned us of the serious results which we might expect if we should brush any of the secretion accidently into our eyes.

"We had captured the king of the tree climbers at our very doorstep. The mysterious voice from the tree tops had come nearer and nearer while Mrs. Noble waited below, and it was only our sudden return that robbed her of the glory of tackling his batrachian majesty single-handed."

Thomas Barbour

Thomas Barbour (Figure 13) was the most influential herpetologist working in the West Indies during the first half of the 20th century. His professional career was spent at Harvard's Museum of Comparative Zoology, first as curator of herpetology and then as director. Barbour traveled extensively in the Antilles, described many new taxa, wrote the first checklists to the herpetofauna, and made major contributions to the herpetology of Cuba and to our knowledge of the diminutive geckos in the predominantly West Indian genus *Sphaerodactylus*. In addition to his technical publications, Barbour wrote several books directed to a popular audience. The

FIGURE 13. Thomas Barbour (photograph courtesy of Kraig Adler).

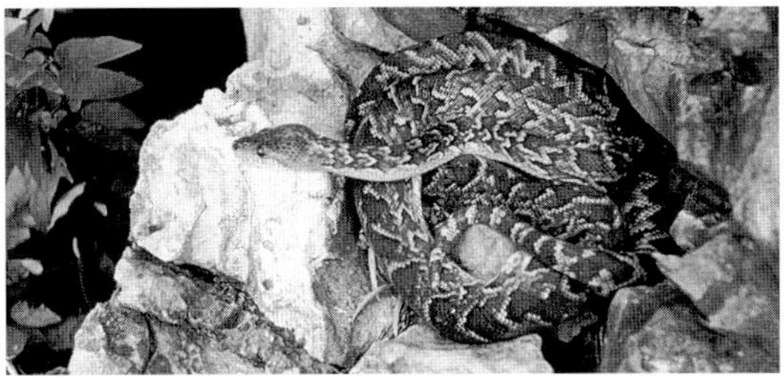

FIGURE 14. A Cuban Boa, *Epicrates angulifer*, from San Vicente, Pinar del Rio Province (photograph by W. H. Gehrmann, courtesy of the Albert Schwartz estate).

selections that follow originally appeared in *Naturalist at Large* (1943, 1950).

Flying Fish and Turtles

"I landed one morning . . . on the Island of Saona, off the coast of Haiti [more accurately off the southeastern coast of the Dominican Republic]. It is a rather flat, uninteresting island and I was not prepared for what I found. I knew that there was a high degree of endemicity on all these islands around the Haitian coast. I knew, also, that Saona had never been visited by anyone in search of reptiles, so I walked around the confines of a small open garden patch, knowing that this was the sort of terrain where one might expect to find Ameiva lizards. Lizards of this genus have a way of splitting up, so novelties may be expected.

"I hunted a long time before I heard a noise in the dead leaves. Ameiva lizards are anteaters and scratch with their paws among the leaves, throwing them about in their search for the insects which may be below them. I approached the sound as stealthily as possible and could scarcely believe my eyes when I saw a perfectly typical Ameiva, and by the same token one utterly unlike any which I had ever seen [*Ameiva taeniura rosamondae*]. I have collected countless numbers of lizards of this genus. I shot this lizard on April 8, 1934. It was lilac gray on the back, washed with fawn colour on the head and turning to pale blue on the tail. A black band, beginning with the eyes, ran along the side of the body and the tail, which was azure blue beneath, while the undersurfaces of the body were glaucous blue, suffused anteriorly with cream colour. The sides of the head were buff yellow. All in all, it was one of the most beautiful and strikingly coloured reptiles which I have ever seen....

"The Haitian peasants are so poor that they will struggle hard to catch lizards, snakes, frogs, and toads — which they do not really like to do — if they can sell them for five cents each, and I mean five cents of a Haitian gourde, which is only worth fifteen cents to start with. We often had as many as a hundred people collecting for us. In this way, on the islands that were populated of course, it was possible to secure in a few days as much material as a single person could have gotten during a long stay, so that while we stopped at innumerable different localities during these voyages . . . and never had very much time at one place, all around Haiti and in the Bahamas we got big collections. You can do this in Jamaica, but not in Cuba.

"We stopped on one occasion at Isle Tortue. I went ashore in the morning and passed word around that we would be back in the latter part of the afternoon prepared to purchase what might be forthcoming, explaining what we wanted. I had a sack of Haitian five-cent pieces. . . . We found that we got much better results from our collectors if we ourselves did not stay where they could watch us. It was so much more fun to stand and stare at strangers than it was to do anything else that the temptation was quite overwhelming, but if we went ashore in the morning and spread the news of what we were prepared to do, then disappeared on board and hauled up the gangway, by the middle of the afternoon we could go ashore and be overwhelmed by a rabble of men and women, boys and girls, with snakes and lizards dangling at the ends of dozens of little lassoes which they fashioned cunningly from shredded palm leaves."

Cuba

"To me there is something particularly appealing about the scenery of the Cuban countryside. . . . The wide sweeping cane fields, their dainty tassels blowing in the breeze, and the giant Ceiba trees, which are so often to be seen, since it is a custom never to cut them down, are a joy to the eye. These great umbrellas, their horizontal limbs each a garden of epiphytic *curujeyes* (plants growing on other plants), as the bromeliads are called in Cuba, are singularly pleasing, especially when the soft green foliage comes out with the first spring rains. Then there are the wide groves of stately royal palms, their pale grey stalks like stone columns surmounted by a section of polished green from which the long, graceful fronds sprout forth. These are used for thatching and the berries are gathered for pigs. There is a law against cutting them down, since the royal palm is the official emblem of the Republic....

"Before long I found myself at Soledad, in Cuba, and I passed out word through the countryside that I was interested in getting a large *Maja*, as boas [*Epicrates*

FIGURE 15. A Cuban Hutia, *Capromys pilorides*; from Guantanamo Naval Base, Guantanamo Province (photograph by Robert Powell).

angulifer; Figure 14] are called locally, and was not in the least interested in small ones. Soledad in Cuba is the site of Harvard's only little ward of Paradise, a lovely botanic garden which I have been privileged to visit for years. While I was sitting on the front porch of Harvard House one hot and sultry afternoon two countrymen came up to the door, politely doffing their hats, and took from their shoulders a pole from which a large sack was suspended. The spokesman of the twain indicated that the sack contained the father of all boas. It wasn't long before we struck a bargain and the sack was mine. I dumped him out of the sack in the dark room, a little detached stucco-and-concrete building adjoining the laboratory. It was obvious at first glance that the snake had dined sumptuously and I was not surprised when looking at him the next day to see that excitement or nervousness had given him indigestion, and a pile of highly aromatic corruption on the floor indicated that not long since he had consumed no fewer than three *hutias* [*Capromys pilorides*; Figure 15]. These savoury rodents abound in the wilder parts of Cuba and each one of these was about the size of an able tomcat. I got the hose and the broom and went to it, slicked the place up, and the buzzards took care of the situation in a few moments."

Archie F. Carr

Archie Carr (Figure 16) was long associated with the University of Florida. He pioneered research on the biology of sea turtles, established a research station at Tortuguero, Costa Rica, and was instrumental in having it designated a national park. Perhaps no other author was as successful as Carr in capturing the flavor of the West Indies in their writing, as demonstrated in the preface to *The Windward Road* (1956, 1979).

"Down in the Caribbean the trade wind blows so honestly that in some islands you rarely hear the cardinal directions used, but people speak instead of living leeward or of going windward to visit an aunt. On all but the smallest or most rugged or least populated islands there are roads that lead to or run along the upwind coasts, and these are known as windward roads. I got to thinking in these terms and liked it, and then it occurred to me that the book I was writing grew mostly out of the hundreds of miles I had walked along the beaches of the Caribbean, where the good beaches are the windward ones built up high and clean by the driven surf. These beaches were the road I walked, and it is a good road. If you are in the tropics and have trouble seeing the good in where you are, work your way to windward where the trade comes in to land.

"Leave your car at Maracas Bay, say, in northwestern Trinidad, and walk off into the marvel of the land-edge there; or travel the heights of north-coast Jamaica, past scenes I never believed in the old stereopticon days. Follow the windward road in Tobago from Scarborough to Speyside, through the West Indies as they used to be, and if the sea is not angry, take a boat out to Little Tobago and look for birds of paradise there. Go west out of Havana, turn into the wind to Baracoa, and stop at the little restaurant by the inlet for *rabirrubias* [= Yellowtail Snapper, *Ocyurus chrysurus*] panbroiled in olive oil, and find the part of town where little houses have clear pools hewn out of the rock of the yards and loggerheads living there, with the people, like confident hens. Get out of Puerto Limón in Costa Rica and climb the stacks and spurs to search out the slips of palm-edged beach set like jewels among the tall rock.

FIGURE 16. Archie Carr (photograph courtesy of Kraig Adler).

"Or on Grand Cayman — if it is July and the broad scrub inside the island claws down the breeze before it gets to Georgetown and looses the mosquitoes in the shade and leaves the open land in the hard grip of the sun, and it is no longer a pleasant place to be, there in the lee of the land — take the road to windward and see the change.

"Take a taxi for it. George Potter has one and there is another driven by a blonde woman named Honeysuckle. Flip a coin and say it comes out George. His wife just got back from the Bay Islands on the *Cacique*, and it was one long storm on the way home and they threw out the cookstove to ease the little ship. George will want to take his wife along — he won't say to keep her in reach, but that is what he means. And then the young will set up a dinning to go, and you wind up sharing the car with a family of six. But they are all cheerful and none of them minds the crowding.

"Now take it easy. You have a goal in mind — to get cool — but keep it in the background till you can do something about it. Forget the solid heat and take the quiet charm of the place so few outsiders ever see. You have to take it — nobody is going to sell it to you. Nobody tries to sell you anything in Grand Cayman.

"So, stop twice before you leave the Iron Shore, as they call the rust-brown fossil reef of the downward edge of the island. Stop first by the graveyard, where the sea sucks and bulges in a little tank cut out of the rock, and fish up the chicken hawksbills they keep there and hold them in the sun and see how you have to turn your eyes away from the five-colored fire from their rayed shells. Stop again to see how the old Cayman glory grew, how the old tradition hangs by a thread in the little shipyard, in the strong slim sculpturing of a mahogany hull. Look hard, because it is the last of something great.

"Then start the seven-mile drive to windward, but keep your pace down, and when you hear the hoot of a blown conch, stop for an image to take away, when a bicycle crunches by on the white rock road, pedalled by a gangling tan man with long bare feet and a wide straw hat, who raises the shell twice a minute and blows one long note to cry the conchs and goatfish in his saddle basket. It is that sort of thing you will see if you are willing....

"It may be that you'll want to stay on [at the little restaurant] for a long time, in the shade, with the coffee, hearing the wind hum over the mimosa thickets, watching the crook-tailed beach lizards [*Leiocephalus carinatus varius*; Figure 17] all about the yard, each on top of its own log or boulder or coco stump, seeing how the two worlds can be on two sides of one little island. . . . ask the girl [at the restaurant] about the path through the thorn to a slant of different shore where Cayman Brac lies off in the unseen distance and no reef drags down the long waves, and the tawny sand stretches straight and smooth, as the green turtles [*Chelonia mydas*; Figure 18] want it. Go there as I did, for no good, practical reason, but full of a vague excitement, full of the wonder of the old travellers who sailed through the hordes of greens massed there from everywhere in Julys four centuries gone. Look off down the heat-blurred lane of untrailed sand and think of the lost fleets and the changing world, and then go back to the shade and wait in the wind...."

FIGURE 17. A curlytail lizard, *Leiocephalus carinatus varius*, from Rogers Wreck Point, East End, Grand Cayman Island (photograph by Robert Powell).

FIGURE 18. A Green Sea Turtle, *Chelonia mydas*, from Tortuguero, Costa Rica (photograph by John C. Murphy).

Suggested Reading

Barbour, T. 1943 (1950 reprint). *Naturalist at Large*. Robert Hale Ltd., London.

Carr, A. 1956 (1979 reissue). *The Windward Road*. Univ. Presses of Florida, Gainesville.

Crombie, R. I. 1999. Jamaica, p. 63–92. *In* B. I. Crother (ed.), *Caribbean Amphibians and Reptiles*. Academic Press, San Diego, California.

Gosse, P. H. 1851. *A Naturalist's Sojourn in Jamaica*. Longman, Brown, Green, and Longmans, London.

Myers, C. W. 2000. A history of herpetology at the American Museum of Natural History. *Bull. Amer. Mus. Nat. Hist.* 252: 1–232.

Noble, G. K. 1923. In pursuit of the giant tree frog. *Nat. Hist.* 23:105–116.

Ober, F. A. 1879 (1899 reprint). *Camps in the Caribees*. Lee and Shepard, Boston, Massachusetts.

Powell, R. and R. W. Henderson. 2003. Some historical perspectives, p. 9–20. *In* R. W. Henderson and R. Powell (eds.), Islands and the Sea: Essays on Herpetological Exploration in the West Indies. Society for the Study of Amphibians and Reptiles, Ithaca (New York). Contributions to Herpetology, volume 20.

The Bahamas

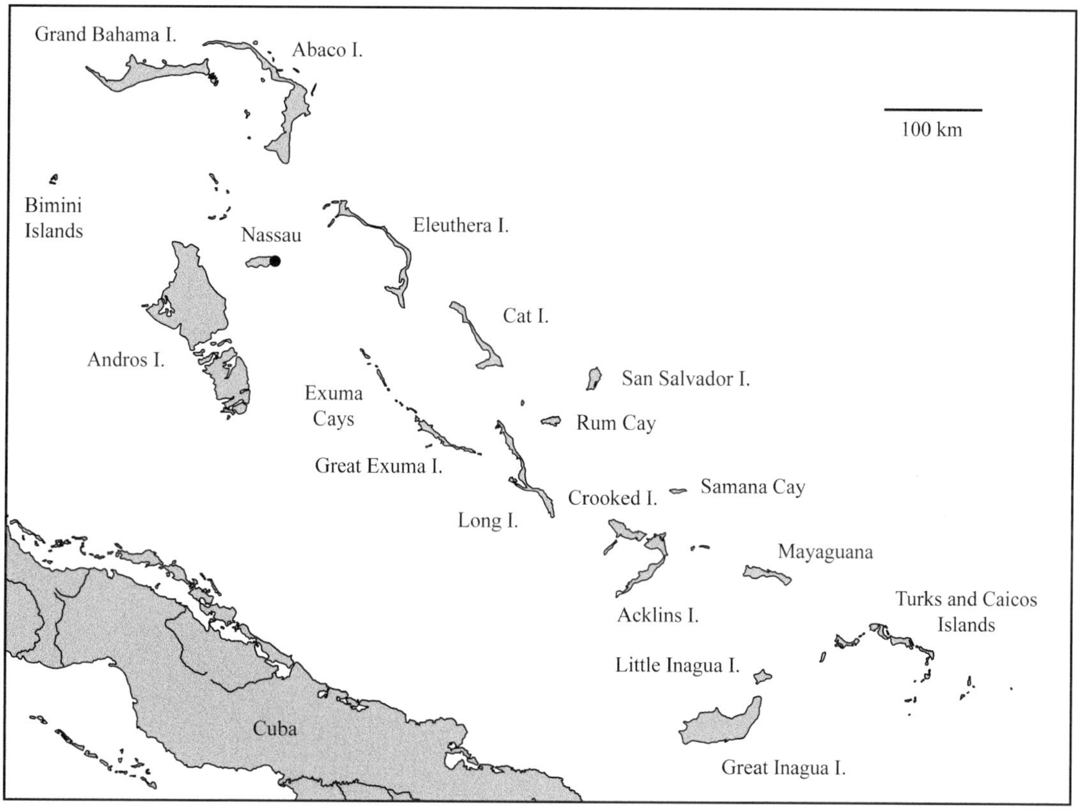
Grand Bahama I.
Abaco I.
100 km
Bimini Islands
Nassau
Eleuthera I.
Cat I.
Andros I.
San Salvador I.
Exuma Cays
Rum Cay
Great Exuma I.
Crooked I.
Samana Cay
Long I.
Mayaguana
Acklins I.
Turks and Caicos Islands
Little Inagua I.
Cuba
Great Inagua I.

"You Study Iguanas in the Bahamas?"

Charles R. Knapp[1]

"You study iguanas in the Bahamas? You poor man."

That's a typical response, laden with good-natured sarcasm, when I talk about my work. At first, I rejoined with descriptions of the logistical difficulties and hardships of island fieldwork. I mentioned the merciless mosquitoes and horseflies leaving their calling cards of itching bumps and rashes. I alluded to sucking mangrove mud, second-degree sunburns, rakes and scratches, aching back muscles, and close calls weltering at sea in small boats with engine trouble. Now I surrender, just acknowledge my good fortune, and move on with the conversation.

Figure 1. Aerial view of the Exumas, Bahamas.

The "Hook"

On my first trip to the Bahamas in 1993, I assisted the International Iguana Society with monitoring selected populations of Exuma Island Rock Iguanas (*Cyclura cychlura figginsi*). If a warm climate, idyllic beaches, and nonstop sunshine are the lure to the Bahamas, then the Exuma Island chain is the hook (Figure 1). The color and clarity of the water are unparalleled and unforgettable. The myriad shades of blue and turquoise blend and swirl as if finger-painted, while the small, mostly uninhabited islands offer the opportunity to walk where few people are fortunate enough to venture.

I spent two weeks on a 32-foot trawler cruising the islands looking for iguana populations. At the time, I was a diffident college graduate new to herpetology, so searching for iguanas was a novel adventure. In 1991, as an aquarist at the John G. Shedd Aquarium in Chicago, I'd become fascinated with the *Cyclura* iguanas under my care. The 1993 trip was my first chance to see these animals in the wild.

The spectacle of an Allen's Cay iguana (*C. cychlura inornata*) lumbering out from the protective shade of a thatch palm left me transfixed. I realized then that I wanted to work with the iguanas in the Bahamas — but at that time, I wasn't sure exactly what I wanted to accomplish.

Retracing History

A few days later, we dropped anchor on the lee side of Bitter Guana Cay. I was aware of this uninhabited, 88-hectare cay for its historical significance, and was anxious to set foot in the sand and search for iguanas. I had read a paper that described Alfred Bailey's 1923 collecting expedition to the Exuma Islands. Bailey had added Bitter Guana Cay to his itinerary after hearing stories from the local people about "guanas" living there. Before visiting the cay, he ventured to a picturesque settlement called Black Point. While at the settlement, he noted that the whole town offered to show him the "guanas" on Bitter Guana. After selecting two men, he was on his way.

Bailey had scarcely reached shore when he saw his first iguana. He noticed that the animals were not wary, and he had no difficulty in securing a few

[1] Department of Conservation, John G. Shedd Aquarium, 1200 South Lake Shore Drive, Chicago, Illinois 60605, USA.

FIGURE 2. The Exuma Island Rock Iguana (*Cyclura cychlura figginsi*).

specimens for his collection. The specimens were eventually identified by Thomas Barbour as a new species, *Cyclura figginsi*, and later reclassified as *C. cychlura figginsi* (Figure 2). Bailey noted that the local people valued the iguanas for food. They told him how easily the lizards could be captured in baited crayfish pots. The local people apparently had been killing and eating the iguanas for years. Since then, many observers have warned of the imminent extinction of the iguanas living on Bitter Guana Cay because of the intense hunting pressure.

Seventy years later, I trudged ashore anxiously anticipating the same welcome Bailey had received. Many of Bailey's descriptions of Bitter Guana Cay hold true today. The towering 10-foot cacti, their branching arms embedded with rigid needles, are still present. A shimmering powder-white cliff still overlooks a long sandy beach crosshatched with purple-flowered railroad vines. However, as I walked across the island, I was saddened and concerned by the lack of iguanas. The few sets of tracks I saw in no way compared with the earlier accounts of abundance. The description of iguanas being taken for food kept running through my mind. After many hours of careful searching, I counted only three iguanas. Bailey had noted taking 19 in a single hour!

Shedd Aquarium Expeditions

Witnessing the fragility of these island iguana populations, I realized my work would be to help these lizards persist in the wild. Two years later, I initiated and began leading field excursions to the Exumas using the Shedd Aquarium's 80-foot research vessel, the RV *Coral Reef II*, as a floating field station. Since 1995, the Shedd Aquarium has offered opportunities for the public to help staff members survey populations of the Exuma Island Iguana (*C. c. figginsi*), Allen's Cay Iguana (*C. c. inornata*), and, most recently, the Andros Iguana (*C. c. cychlura*). Through this unique program, more than 70 people, most without scientific backgrounds, have become involved in iguana conservation. Participants pay a nominal fee to cover their costs and receive basic training in field biology techniques. Thanks to the research assistance they have provided, we have collected a tremendous amount of data in a relatively short time. The participants also have gained knowledge of and appreciation for conservation science in general and Bahamian natural history in particular.

Selected portions of our work to date include determining iguana distributions and surveying populations, performing population biology studies of a translocated colony, collecting blood samples for genetic studies, discovering a new iguana population, identifying destructive development practices, posting educational signs on the islands, and presenting conservation seminars at Bahamian schools.

Although the Shedd trips have been successful, I wanted to expand my ecological knowledge and field techniques. I returned to graduate school, earning my Master's degree and am making progress toward my doctorate at the University of Florida. During this time, I am continuing my research through the Shedd expeditions, and I'm grateful for the companionship and assistance in the field. However, some of my most memorable and satisfying moments have been during trips when I've visited the islands by myself.

Never Alone

While walking – often crawling – through the tropical dry scrub or pine barrens on these remote islands, I am not alone. In my search for iguanas, I can't help but notice tiny Dwarf Geckos (*Sphaerodactylus* spp.) wriggling between fallen palm fronds. During breaks, pugnacious and bold Curlytail Lizards (*Leiocephalus* spp.) scamper over my shoes in search of morsels of cracker or bread. From time to time, hummingbirds whir and hover around my head, zeroing in on my bright red field cap.

When I finally quietly come upon my quest — a large iguana basking in the blazing sun — I never fail to feel a thrill. These lizards blend seamlessly with their dry, rocky scrub habitats, yet once you pick them out, you can't help but be impressed by their conspicuous power and raw beauty. At times,

FIGURE 3. Andros landscape.

I forget my scientific motives and simply watch the animal watching me. Usually with a rolling head bob, the iguana ambles off, with me in calm pursuit of some new ecological question.

An Unexplored Wilderness

I am currently conducting most of my research on Andros, the largest, but one of the least populated islands in the Bahamas chain. Andros is actually three islands separated by surging channels of brine. These "bights" curve between mangrove and pine-forested banks. Small, low-profile cays protrude above the water in the bights, amalgamating the horizon, and making navigation difficult. Although Andros lies roughly 200 km from the eastern coast of Florida, it is extremely isolated and has been described as the largest tract of unexplored land in the Western Hemisphere — perhaps with good reason. The jagged xeric limestone landscape, with its thin veneer of muddy mangroves and dry pine barrens, is punctuated frequently by sinkholes and offers few attractions except to the iguanas and those who study them (Figure 3). Human settlements are small and mostly clustered on the eastern side, adjacent to and protected by the third-largest barrier reef in the world.

Despite this blissful isolation, iguanas are hard to come by. Their apparent low population densities and life history strategies cannot stand up to the pervasive problems of habitat loss, illegal hunting, and depredations by feral animals, including dogs, cats, and pigs, which both compete with and prey on the lizards. Low density and elevated iguana wariness, because of anthropocentric pressures, make capturing the lizards for morphometric and demographic data more difficult, yet all the more exciting.

Iguana Catching

Harmlessly capturing iguanas typically requires a team effort. I deploy teams outfitted with nets, bags, hats, sunscreen, whistles, and water at varying intervals in the seemingly endless expanse of pine yards and mangrove stands. While walking, all is silent except for the perpetual wind whistling through pine needles or the lapping of waves against limestone shores. Everyone is vigilant despite the hazardous footing, biting insects, barricades of thorny vines, and threats of contact with poisonwood. However, one can easily get lost in one's own thoughts while traversing this mystifying landscape.

Daydreams are shattered with the shrill sound of a whistle — the signal that an iguana has been spotted (Figure 4). With hearts pumping feverishly, people begin to scramble over and under branches in the general direction of the whistle, often getting snagged in love vine or shredded by *Smilax* thorns. Our object is to surround the iguana and slowly constrict the ring of people until someone can place a net over the animal. I stress the importance of patience to new research assistants, but with adrenaline surging, the first reaction of many is to charge — causing the lizard to panic and flee through the impenetrable scrub. After a few unsuccessful attempts, assistants learn the importance of teamwork.

Although simple in principle, this technique is often impossible in application. Solution holes concealed by pine needles, jagged dog-tooth limestone terrain, and dense scrub make iguana capturing a hazardous exercise. The group assembles in a horseshoe around the animal.

"We need someone behind the iguana!"

"I'll try, but it's full of poisonwood!"

"Hurry, but don't scare him!"

Hands clench net handles, sweat dribbles off nose tips, yet everyone's focus is on the iguana. People

FIGURE 4. Andros Iguana (*Cyclura cychlura cychlura*).

with an unobstructed view of the animal bark more orders: "Close the gap! Sonja, you have to move to your right! He's moving to his left! Move right! Stop! Stay back!" From others, these spontaneous orders are punctuated with oft-repeated, "Where is he?" "Which way is he facing?" Often we are forced to crawl under branches, snagging nets in the process, to get within capture distance. Experience plays a major role in determining when to hesitate and when to act. A wrong decision can cause the animal to flee — causing immeasurable frustration within the group.

"He's coming your way!" The iguana makes a last ditch effort to "shoot the gap," but a research assistant drops a net on him. When an iguana is in the net, the ensuing action is swift. Members move in rapidly to restrain the lizard. If this is not done quickly, the iguana can break through the sturdy nylon net as if it was gossamer cotton candy, leaving us panting in disbelief. Restraining these animals is not an easy task, nor is it one to be taken lightly. These lizards can attain weights of over 8 kg, and their normal placid dispositions transform to retaliatory aggression when confined to a net. Luckily, most iguanas calm after they are restrained, and data can be gathered without undue risk.

The adrenaline-pumping experience is so overwhelming that we are able to play practical jokes on assistants by placing plastic iguanas, painted to resemble the Andros Iguana, under cover of scrub. We reenact a similar exciting capture scenario and always get an unsuspecting assistant to jump for the decoy iguana, which invariably results in unchecked laughter and hilarious stories around the dinner table that night. Although "real" captures are a rush, so are the releases. As the lizards lumber away, we all feel an overwhelming sense of relief that these magnificent animals still roam the interior of the island. We also feel a sense of satisfaction that the work we do will help them survive into the future (Figure 5).

FIGURE 5. Iguana assistants resting from a hard day in the field on Andros.

Like a Mother Iguana

On these research trips, my questions are numerous and my determination unrelenting. First and foremost, how many Andros Iguanas exist in the wild? Nothing definitive is known about population numbers or ranges. Currently, the species is listed as vulnerable by the World Conservation Union (IUCN), with an arbitrary estimate of 2,500–5,000 animals still inhabiting the interior pine barrens. That these numbers represent only the crudest guesses is reinforced by the fact that recent observations of iguanas by other people have been restricted to animals slated for human consumption — in cages, cooking pots, or caches in the field. Caching iguanas in the field is particularly cruel. Their back and fore legs are trussed behind their backs and their mouths sewn shut until they are retrieved. In this way, animals are kept alive and the meat "fresh" until retrieval.

Detailed life-history data, a crucial component in the development of any conservation program, are not available for this species. Walter Auffenberg conducted preliminary research on the Andros Iguana in the early 1970s, and I intend to continue this work. One objective of my fieldwork has been to locate active termite mounds that the females might use as nesting sites. Auffenberg first documented this behavior, but no quantifiable data are known. Over the years, I've found a number of mounds with empty excavated burrows, but my timing had precluded the location of recently deposited eggs — until the 2001 trip to Andros with a Shedd Aquarium group yielded results. A volunteer spotted a mound that looked suspiciously fresh, with pine needles buttressed against one side. As my heart raced, my hand slowly dug into the mound. Probing, I touched something leathery that gave me a jolt, even though I knew that it had to be an egg. We had found an active nest! With termites swarming over my hand, I pulled egg after egg from the nest for measuring. Then I gently replaced the eggs and carefully refilled the cavity, trying to arrange the pine needles over the disturbed area with as much concern as the mother iguana herself.

The Andros Iguana is the only species of *Cyclura* for which nesting in termite mounds has been documented. Why? Lack of other suitable nesting habitat? Improved environmental conditions for incubation and subsequent hatching success? I don't know — but I intend to find out.

Children of Black Point Meet the Iguanas

In addition to ecological studies, local people must be involved in conservation programs to ensure the long-term survival of these indigenous lizards. From my conversations with many residents of Andros, I've found that perceptions vary widely, from denial about the need to conserve the iguana to complete agreement and understanding. As I continue my research on the island, I also will work to cultivate community pride in — and protection for — the native iguanas.

In June 2001, I had the serendipitous opportunity to advance that goal with the entire population of schoolchildren in Black Point, Exumas — the same settlement that Alfred Bailey had visited en route to Bitter Guana Cay. At the time, I was leading a fieldwork expedition with a group of Chicago-area university students on a small, uninhabited cay directly south of Bitter Guana Cay. We were there to capture as many Exuma Island Iguanas as possible in order to collect measurements, blood samples, and other data for the Shedd Aquarium's ongoing *C. c. figginsi* monitoring program.

Standing on the beach, I reminded the students about proper capture and restraining techniques before sending them into the scrub. Just as they began to splinter off and traverse the island, we heard the abrasive roar of outboard engines. Looking in the direction of Black Point, I saw a boat speeding our way. My curiosity turned to concern as I spotted a second and then a third boat on a direct course for our spot on the beach. I was uncertain of their intentions and felt apprehensive until I caught sight of children in the lead boat. As the boat slowed, the children began spilling over the gunwales and splashing ashore. Trepidation turned to panic as boat after boat unloaded groups of merry, exuberant children onto the beach.

It was a Bahamian Labor Day celebration, and the children were on the island for a picnic. Figuring that the kids would scare away all of the iguanas, my hopes for a productive workday sank — but I could not have been more misguided. I had spoken about iguanas at the Black Point school a few years earlier, and some of the teachers recognized us and our boat. The children and teachers were eager to watch us work and asked dozens of questions. While some of the adults helped us catch iguanas, the children helped measure animals, hold instruments, and record data. Their initial squeals of fear changed into a buzz of curiosity. Never before had I involved such a large group of people so intimately in iguana fieldwork. Encountering the local lizards as the quest of a scientific investigation instead of the quarry of a hunt, the children and adults listened with interest as I explained what I was doing and why it was so important to protect the iguana populations and their habitats. The day turned out to be one of my most rewarding field experiences ever.

Nowhere to Go but Extinct

I continue to visit Andros and the handful of iguana-inhabited islands in the Exumas, but each year I see more denuded and charred acreage among Andros's dry pine forests and more roving yachtsmen cruising the Exumas. People bring potential and distinctly deleterious consequences to the local flora and fauna. What's worse, islands in the Exumas are being sold like Hollywood mansions, and the fragile insular environments are becoming more susceptible to harmful perturbations. Some iguana populations in the Bahamas may appear safe, but they face a slew of potential and real man-made threats from hunting, habitat destruction, and predation or competition by domestic and feral animals. Shrinking or fragmented iguana populations suffer the compounding problems of lack of breeding opportunities and diminishing genetic diversity. The outlook for survival becomes more tenuous in a region renowned for cataclysmic tropical storms. On their small islands, the iguanas may have nowhere to go but extinct.

People have the capacity to push these fragile populations into extinction, but they also have the ability to set aside adequately large protected habitats and share environments with native wildlife. I hope that my work helps these impressive animals persist in the wild. I am in the Bahamian iguana conservation program for the long haul, and I have bright expectations for the future.

Epilogue

In those casual conversations about my work, after I've failed to convey the hardships of island fieldwork, the next question invariably is, "Why iguanas?" I really don't understand the attractive force that lured me to the iguanas of the West Indies, nor do I care. What is important to me is that after all these years, I still feel the same fervor when I see my first and last iguanas of the field season. I haven't found the words to describe this passion, so I just shake my head with a smile and answer pensively, "I don't know."

Acknowledgments
I thank the Bahamas Department of Agriculture and the National Trust for their support and permission to conduct research throughout the years. The Shedd Aquarium, Sandra Buckner, C. Kenneth Dodd, Jr., Lou Roth, and a long list of others have provided monumental support during my work for which I will be forever grateful. Karen Furnweger is still teaching me to write creatively and has offered valuable suggestions on this essay.

Suggested Reading

Alberts, A. C. (ed.). 2000. *West Indian Iguanas: Status Survey and Conservation Action Plan*. IUCN–The World Conservation Union, Gland, Switzerland.

Bailey, A. M. 1925. The iguanas of Bitter Guana Cay. *Nat. Hist.* 25:390–393.

Campbell, D. G. 1978. *The Ephemeral Islands: A Natural History of the Bahamas*. Macmillan Education, Ltd., London.

Knapp, C. [R.] 1998. Vanishing iguanas. *Iguana Times* (*Jour. Internatl. Iguana Soc.*) 7:27–30.

Knapp, C. R. 2000. Home range and intraspecific interactions of a translocated iguana population (*Cyclura cychlura inornata* Barbour and Noble). *Carib. Jour. Sci.* 36:250–257.

Knapp, C. R. 2001. Status of a translocated *Cyclura* iguana colony in the Bahamas. *Jour. Herpetol.* 35:239–248.

Knapp, C. [R.], S. Buckner, A. Feldman, and L. Roth. 1999. Status update and empirical field observations of the Andros Rock Iguana, *Cyclura cychlura cychlura*. *Bahamas Jour. Sci.* 7:2–5.

Knapp, C. [R.] and K. Furnweger. 1999. Conservation spotlight: West Indian rock iguanas. Amer. Zoo Aquar. Assoc. Commun., July:7.

Knapp, C. R. 2003. "You study iguanas in the Bahamas?" p. 23–28. *In* R. W. Henderson and R. Powell (eds.), Islands and the Sea: Essays on Herpetological Exploration in the West Indies. Society for the Study of Amphibians and Reptiles, Ithaca (New York). Contributions to Herpetology, volume 20.

Turquoise Water, Silver Palms, and Fluorescent Green *Leiocephalus*

C. KENNETH DODD, JR.[1]

I made my first journeys to the Caribbean to evaluate critical habitats for endangered reptiles in the late 1970s and early 1980s. Although the visits were not of long duration, I had memorable and sometimes exciting experiences: walking through Luquillo Forest in Puerto Rico listening for small brown coquis while being distracted by Puerto Rican Parrots, a critically endangered species, flying overhead; trudging over the white sandy beaches of Sandy Point on western St. Croix at what was to become a national wildlife refuge for Leatherback Turtles; exploring caves and sinkholes on Isla Mona and being startled by a giant Mona Iguana; and being not so gently lowered from a Coast Guard helicopter amid thousands of screaming boobies onto Isla Monito in order to chase the tiny gecko endemic to that block of limestone in the Mona Passage. I never really dreamed of conducting research in the Caribbean, however, as I seemed forever fated to be chained to a desk in the gray concrete canyons of Washington, D.C.

In the mid-1980s the political climate changed dramatically in Washington, and I suddenly found myself in a research position in Florida. Immediately, I was captivated by the lush vegetation, the new habitat types, the strong light and vibrant insect life, and most of all by the diversity of amphibians and reptiles that I previously had only read about and dreamed of finding. I spent a few years working mostly in the sandhills of north central Florida, an edaphic desert in a subtropical setting, where I found much to keep me occupied. Although the Caribbean was not far distant, I only flirted with the tropics as, on one research project, I used satellite transmitters to track post-nesting Loggerhead Sea Turtles through the Bahamas; exotic place names — Abaco, Andros, Exuma, and "Tongue of the Ocean" — suddenly became familiar.

FIGURE 1. The Exuma Cays.

In the early 1990s, Dick Franz of the Florida Museum of Natural History invited me to join him in a survey of the Exuma Cays Land and Sea Park in the central Exuma Island chain. In order to manage the park, the Bahamas National Trust needed to know what species occurred on which islands, and what factors affected their distribution and long-term survival. Over the next three years, Dick and I, either with Dick's wife Shelly or colleagues from other institutions (David Lee and Mary Clark of the North Carolina Museum, Sandy Buckner of the Bahamas National Trust, and Don Buden), visited nearly every island in the park looking for living and fossil amphibians and reptiles. I remember my first impressions flying over the low, gray-and-green limestone islands often fringed with white sugar sand and surrounded by the most beautiful turquoise water — paradise (Figure 1). I could not feel the noseeums or the microwave white

[1] United States Geological Survey, 7920 N.W. 71st Street, Gainesville, Florida 32653, USA.

heat of mid-afternoon, nor would I yet know the constant wind blowing through the silver palms, the treacherous dogtooth limestone, or the incredible undersea life which we enjoyed at midday when the lizards had enough sense to seek cover underground — but it sure was beautiful from a small fixed-wing airplane.

Being helpless on water (neither Dick nor I care much for boats), we were usually transported to the various islands in the morning by Peggy Hall, then the park's only warden. Peggy must have been in her mid to late 60s. She was wiry and tanned from a lifetime spent in the islands, and she was used to doing things for herself and her old dog, Powerful. She lived on a converted tugboat, named Moby, anchored off the warden's office at Waderick Wells Cay, always keeping a wary eye out for poachers. After patrolling in the afternoon, Peggy would pick us up and bring us back to Waderick Wells. On days when gas was low or when bad weather threatened inter-island travel, we concentrated our efforts on Waderick Wells, where we were conducting basic life history studies on the lizards, especially the curlytails (*Leiocephalus carinatus*).

Curlytails are handsome and robust Caribbean lizards. They get their common name from their long tails that are conspicuously curled up and forward over their backs. We painted dots on the lizards with bright red fingernail polish for individual recognition in our mark-recapture study, and we coated certain animals with bright green fluorescent powder in order to track their movements. At night, we crawled through the dense undergrowth with a black light searching for the glowing fluorescent powder trails. In this way, we were able to map habitat use and tell where the lizards wandered during the day and slept at night. All the curlys on the northern end of the island eventually had either red dots or

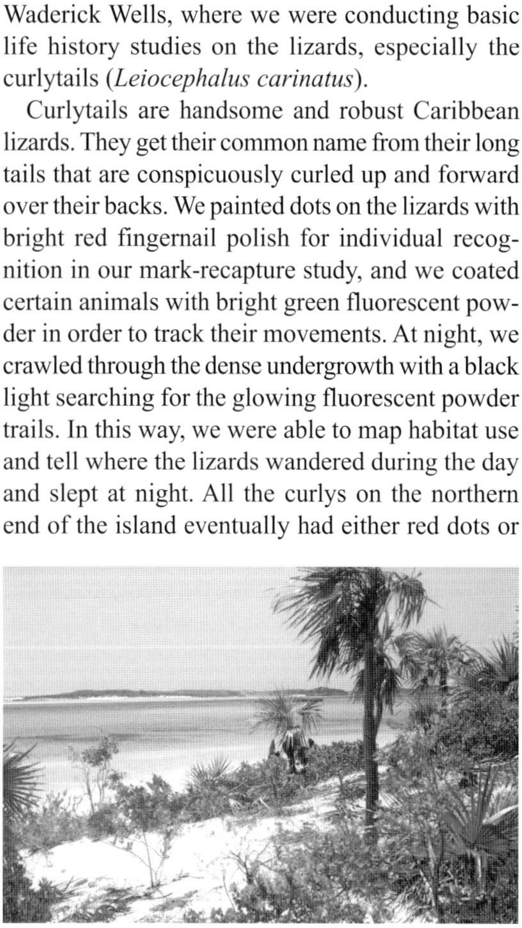

FIGURE 2. Habitat in the Exuma Cays.

green powder on them, a phenomenon that often puzzled the confused visitors from passing yachts. At that time, the office had no electricity, so when darkness came and the mosquitoes began their nightly quest for blood, we had little to do after tracking lizards except to sit on the porch, hope for a breeze, drink rum, and watch the moon rise over the ocean.

A narrative is a difficult way to convey the thoughts and feelings of working on the low limestone islands of the Bahamas. Fortunately, I kept a journal while I was there in order to recollect the magic of those trips. Rather than simply recite where we went and how beautiful it all was, the words penned at that time now provide a much better synopsis of what it is like to be a herpetologist in the Caribbean, of why we go there, and of what it makes of us upon return.

21 May 1991

Gorgeous setting here on the north edge of Waderick Wells — the nature center/office is perched high on a hill overlooking the protected harbor. The office has two main rooms separated by a breezeway, and a covered porch wraps around the entire building, which is perfect for sitting and gazing at the ocean in the distance. The building has an attic where we will sleep, and even a hatch in the roof leading to a platform that enables us to look far off at the distant sea and islands. There is no electricity, and all the water comes from a cistern built into the limestone under the main part of the building. At night, it is truly dark and quiet, except for the wind blowing through the palm fronds. When the moon is bright, we won't need additional lights out here. Giant land iguanas, Ping and Ling [semi-tame *Cyclura* then living near the warden's office] live around the house, where they are fed along with the hutias (*Geocapromys ingrahami*) [an introduced rodent on Waderick Wells, but endemic elsewhere in the Bahamas]. Bananaquits and Bahamas Mockingbirds steal sugar and crumbs.

The high winds of yesterday and today have made it quite delightful (Figure 2). No bugs, and the temperature has been very pleasant. Not much of a sunset this first night because clouds completely cover the sky. No rain in a long time, however. Constant strong wind. After arriving, we immediately set up our main transect, which we will run at least twice a day, from the warden's office down to a small causeway in a shallow channel that bisects

Waderick Wells. We are mainly interested in curly-tails, in terms of lizard ecology, but we are recording all amphibians and reptiles, including taking voucher specimens from the various islands.

24 May 1991

Breezy morning. The cloud patterns hint of rain, but here it might do anything. While sitting in the nearly full moonlight last night, we heard the haunting calls of many Chuckwill's Widows echoing over the otherwise silent island. Nice sheen on the calm water. Already (7:30 AM) it is showing turquoise. Bananaquits hop everywhere — bright yellow and black (or gray-backed) birds, with light gray throats and masks. They have an insatiable craving for sugar. We put sugar out on the porch railing and soon have 30 birds scrambling and making a fuss. I held four in the palm of my hand as they searched for sugar grains. They sport a splash of brilliant red at the base of the bill, and an incredibly fast little wormy tongue. It's hard to believe how rapidly they use it to gather the sugar.

26 May 1991

Another beautiful and quite hot day. Took a run over to Bell's Cay. Once inhabited, we saw the many stone-walled ruins dating from the Loyalist Period (late 1700s). Gorgeous setting, but it must have been a hard life with the heat, lack of water, and biting insects. Found a plot with sea island cotton still growing; perhaps it dates from that early period. More than 12 pits and caves. Some require a ladder for access, but we managed to crawl into a number of them. Nice roots going down into a few helped as handholds. One in particular had a solid fern wall with a kind of backhole entrance. From down below, the surface is a green blanket of low trees, vines, drapery and columnar cactus, and orchids. Found tiny *Eleutherodactylus*, but could not catch the adults. Also spotted a whip scorpion.

27 May 1991

Dick and I cornered Peggy into taking us to White Bay Cay. Very hot in the direct sun. We both worked rather hard, but reptiles were scarce. We caught only one *Leiocephalus*, six *Anolis sagrei*, one *Sphaerodactylus nigropunctatus*, and one *Eleutherodactylus* while searching through the litter. The forest is very thick, but there are some large trees and some open areas. No ameivas. Saw large bromeliads and a huge columnar cactus. I walked around the entire island — it was bigger than I had thought it would be — only to find Dick splashing naked in the water. Photo op! Voucher specimen! I am struck at how different even nearby islands are. Why no ameivas? Why are ground geckos so rare (nearby Sandy Cay is full of *Sphaerodactylus*)? Why were the *Leiocephalus* so wary? What are the selection pressures operating here? Back on Waderick Wells. Pleasant evening. Nice breeze. Blowing silver palm fronds. Rum. Good food. Watching the island and the waves. Waiting for sunset.

31 May 1991

Watched sunrise this morning. The ocean and inlet are dead calm. Very pretty with the sun on the white yachts in the harbor. The single barracuda was out patrolling again. Last night the stars were fantastic as we sat on the dock. No shooting stars, but a number of satellites. Billions of stars! A large number of green fluorescent globs floated by. We at first thought they were ctenophores, but Peggy says they were squid. Really special to see them floating in the calm sea.

12 March 1992

Waderick Wells, Exumas. "Home" again. Fourteen yachts in the harbor. Pleasant breeze across the house, but the mosquitoes were bad last night. I slept in the loft and put my tent up on the bed — that kept the bugs out even if it was a bit unusual looking. Caught *Anolis distichus* with a broken tail on the roof beam, the only one seen thus far. Billions of stars were out at 2 am, but so were the "mossies." Blue turquoise water, with hardly any chop. No Bananaquits.

13 March 1992

Friday the 13th. Clouds rolling in from the southeast; they are ever changing. The wind full but not strong enough to keep the sand flies away. The mosquitoes were intense last night. We sat talking until dark and were badly eaten — the beautiful night could not be enjoyed. Bedtime came by 7:30 because of the relentless attack. Yesterday we worked the house region and then on down Julie's Trail [the trail from the warden's office to the causeway over the shallow inlet that divides Waderick Wells] to the causeway. This will be our main sampling transect. We caught 19 *Leiocephalus* — they were generally much more skittish and wary than last year, and many of our lizards proved to be recaptures (Figure 3). They

FIGURE 3. Dick Franz attempting to noose a Curlytail Lizard, *Leiocephalus carinatus*.

remember us! Warm but overcast in the morning. Getting our catching (noosing) skills sharpened.

After lunch I snorkeled around the dock and rock wall. Water cool but pleasant. Saw a live sea slug, but it was not very bright. Dull yellowish with two dorsolateral flapping folds. Wavy. I spent some time floating in the water watching it. It would go along churning the bottom, then a current would roll it over so that it would have to right itself. Again and again. Saw many beautiful reef fish, several conchs, little brain corals, and the green calcareous algae. In the late afternoon, we repeated our surveys along Julie's Trail on the way to the beach. Sunny, beautiful late afternoon swim. Calm clear water. Perfect setting for the evening.

14 March 1992

The Chuckwill's Widows were calling again last night over on Narrow Water Cay. Very strong winds, at least through 11 PM. Surprising to have to think of the wind chill factor in the Bahamas — glad I brought a sweatshirt. Woke up at 4 AM — it was completely jet black with a fine mist in the air. I think we are in trouble, weather wise. Very still.

8 AM. Heavy squalls blowing through, driving rain, and cool. At other times, a strong mist. Took pictures of the rain. This is supposedly the driest season of the year, but you would never know it today. Yesterday morning we worked Julie's Trail, but did not get many new lizards. Only four *Leiocephalus* and two *Ameiva*. Saw a few of the recaptured lizards. The spots are easy to see. Many lizards were sitting in the low branches of the trees, so we recorded perch heights, widths, and ambient temperatures. Starting to feel a little sunburned. The Tropicbirds are flying around BooBoo Hill despite the wind and rain. No sounds but the wind whistling through the beams of the office and across water and vegetation. Saw a very large ray working the area from the dock to the boat anchorage.

15 March 1992

After a rather chilly night, we hiked down to Gecko Cave. Nice weather — not too hot but plenty of sun. Relocated the cave, climbed down in and saw geckos (*Tarentola americana*) everywhere. I crawled in with the camera and tried to take photographs, but I don't know how much reflectance I'll get from the light tan walls. We caught 13, and processed them in the mosquito-filled entrance sink (Figure 4). Coming back, we went down into two of the pits by the Loyalist ruins. Both were direct shafts of about 7 and 5 meters, respectively. No passageways and no geckos. There was a great deal of deep fine sediments in the bottom, but no bones in evidence. Sifted through the sediments, but found only shell pieces. The sediments were quite deep, however, and maybe something would show up in screens by trenching. Very warm and sweaty at the bottom. Mosquitoes prevalent.

16 March 1992

Another cool night. Not a cloud in the beautiful morning sky — going to be a great day. Headed over to Alligator Cay and spent three hours combing it for lizards. Found *Leiocephalus*, *A*[*nolis*]. *sagrei*, *Sphaerodactylus notatus*, and seven *Cyclura*. Saw *Cyclura* burrows, hideouts under rocks, a small tail drag, and got very close to an adult. In fact, so close that I could not get him in focus with a 150 mm lens. He was polite — let me change lenses and shoot lots of pictures. Nice patch of forest, all to myself. In leaving, we had to wade way out to the skiff because the tide was out — we had to watch our step because

FIGURE 4. The author emerging from Gecko Cave with a bag of *Tarentola americana*.

of the urchins, rays, and other spiny creatures. Spent the afternoon catching lizards over on the east side of Narrow Water Cay. *Ameiva* trails everywhere, but no animals seen. Dick managed to dig one out — digging from both burrow entrance tunnels toward the center. As he got close, the lizard came bursting through the sand at full throttle. The trick of this method is, of course, to anticipate the exact spot from where the lizard will catapult in order to pounce effectively before he runs off. Hot and bright on the beach — we caught a Brown Runner [a snake, *Alsophis vudii*] and marked nine *Leiocephalus*.

17 March 1992

Fuel has become a problem. Can't go far until deliveries are made into the central Exumas. We went out to the Malabar Cays for the morning. Made a beeline for a tree with a termite mound in it and sure enough found the *S. nigropunctatus* that I thought should be there. Got a good series of specimens, in fact, but no *Tarentola*. Naturally we saw *A. sagrei*; they are found on even the smallest islands as long as there is some vegetation available. The Malabars are incredibly rocky and difficult to work — low trees, dogtooth limestone. Found a small pit in the saddle and climbed down into a tight chamber with another small chamber going off it. A number of other passages led off, but I couldn't see down them. No geckos, but in the dirt scraped from the floor, I found a small bone. Took three bags out to sift later for fossils. [Note: The dirt was filled with small fossil bones, i.e., vertebrae, ribs, limbs, and jaws of lizards and birds. We even discovered the vertebrae of a boa which has been long extirpated from this tiny rocky island]. Turned over a rock and found a Brown Noddy sitting on an egg. She took off, but feigned injury to lure me away from the nest.

18 March 1992

Peggy dropped me off at Sandy Cay before going on poacher patrol. I am alone on a tropical Bahamian island. Sandy Cay at low tide. Worked the palm forest and immediately started finding *A. sagrei* (5) and *S. nigropunctatus* (13). I stopped then; it was too easy. Saw many large silverfish and roaches. Once I even grabbed a lizard and a scorpion at the same time, but fortunately the arachnid didn't get me. Serious sampling took about 45 minutes. Spent the rest of the time walking the stony west beach. Many conchs, all alive, mostly small. An Osprey at her nest on the cliff watched me intently and occasionally called. Very hot white sand; a few shells. No *Ameiva* trails or burrows, and no *Leiocephalus*. Bahamas Mockingbirds work the palm thatch in the area I just surveyed, but no sea birds are nesting at this time of year. The large sandspit to the west was exposed, and it was possible to walk far out onto the strip of sand, far from the main portion of the island, seemingly into the sea. The sandspit was very broad and covered at scattered intervals with beautiful orangish conch shells. Some were alive and partially buried in the wet sand, while others were dead and on their way to becoming fossils. Wind, the sea, bright sun, and me on the deserted island. Peaceful and beautiful. Later . . . full bright moon tonight. The low tides have been very low, exposing a sandbar in the Waderick Wells channel. The sandbar looks silver in the bright moonlight.

23 March 1992

Sandy Buckner of the Bahamas National Trust arrived today in a seaplane just before our prenoon lizard survey. She went out with us into the bright sun, took lots of pictures, and asked lots of questions about lizards — Sandy has been totally enamored with lizards ever since she began studying them in her yard in Nassau. After lunch, we did another lizard transect, and powdered a *Leiocephalus*. Then, Sandy and I went out so I could teach her our "high tech" lizard noosing method [a noose made of unwaxed dental floss tied to the end of a handy stick]. Chased lizard #16 all over the place. As I was again getting close, and as he carefully eyed me over his shoulder just out of reach, I stepped into a solution tube filled with sulfur water and sank down to above my knee. Almost couldn't get out. Badly skinned my shins, but of course we still had lizards to chase, blood and bruise or not. Sandy tried catching a juvenile, but to no avail, but I was able to grab a good sized female *Alsophis*. We eventually found a big patient male curlytail who let her get a noose around him. Of course the tie broke and the lizard ran free, but I eventually caught him. After dinner, a tremendous thunderstorm set in. Very high winds and sheets of rain. Lightning! The worst storm I have seen in the Bahamas. Went to sleep, tired, in the roaring wind.

24 March 1992

About 10:30 AM, Peggy announced that she was going to make a fuel run to Sampson Cay, and that it was calm enough to drop us off at Rocky Dundas. Gray and threatening skies on the cool run down there, but no landing problems. We saw the entrances to the sea caves on the east side, but the tide would not let us in. Landing can only be tried at low tide on a calm day. The bulk of the island is a tableland with high cliffs; it is very difficult to get around, and it is covered with poisonwood. The big sea cave is most impressive; one opening is located on the top of the island, and plunges vertically far down to the sea which enters through an opening to the east. Easy to fall into. I could not tell the dimensions of the room, but perhaps it dropped 25 meters to the sea, and was equally large in diameter. The wall on the island side has sheered away creating huge boulders. Behind the rock boulders, a very nice tropical forest is found — large trees, palms, bromeliads. There are many species of trees with diameters of up to 54 cm. Very rocky. In the cliff face, cave shelters, some with formations, are common, but none seem to go anywhere. No fossil bones.

The forest to the southwest is open with many trees, and is meadow-like in some areas. I found a Thick-billed Vireo nesting and saw a number of other birds, including a White-crowned Pigeon. Saw a single *S. nigropunctatus* but it escaped down a hole. Caught three *S. notatus* under palm debris and rocks on the wall. *Anolis sagrei* is very abundant. Centi-pedes, scorpions, arachnids. Interesting forest. Rained while we were there off and on for an hour and a half — everything was wet and humid. Dripping. Couldn't get my glasses cleaned. Around the fringe of the island, many piles (more like fences) of old conch shells, weathering away. There were two principal fences, each about 1.75 meters high and maybe a hundred meters long. Some may date to Arawak times. There is also a nice palm forest, but I found no ground geckos there. Wet soils beneath the vegetation — maybe that inhibits them. I walked around the periphery of the island taking the gray day in, looking for reptiles, and sometimes just looking. The sun finally came out, and I realized how bloody tired I was. Played in the rocky pools until our ride arrived to pick us up. Sandy looked bedraggled — I think we wore her out. Nice fast trip back through the turquoise waters. That night, very heavy winds rocked the warden's compound all night. It was like being on a vibrating bed.

25 March 1992

We got up early and began packing. The wind had roared all night and I guess we were just ready to leave. Ping and Ling were basking in the sunshine, but it felt cool in the wind. About 10:30, we started loading and at 10:45 we pulled out in the Red Devil [the Red Devil was the nickname of the outboard runabout that Peggy used to patrol the park; it was about 5 meters long, and known for its idiosyncrasies]. Peggy, driving; the rest of us, in rain slickers with our bags tight against the floor. Me next to Peggy, and Dick and Shelly huddled in the back. No one was looking forward to the high seas and the waves. A little wet but no problem until Conch Cut [a cut is an opening between islands that allows a strong current to flow east to west through the island chain], as expected. Rolling waves washed over the side, and several times I was doused. Peggy stood up in the cuts gripping the steering wheel, while staring intently ahead. Actually, Conch Cut wasn't as bad as she thought it was going to be, and we became a bit complacent. Then we hit Pipe Line Cut.

Holy s—! Four to six foot waves had us rolling, slamming, and soaked. Our bow would go way up in the air, and I feared a deep trough on the other side which would plunge us straight to the bottom. Fortunately we managed to stay on top of the crest of the waves, thanks to Peggy's driving skills. I only glanced back once, almost straight down, at Dick and Shelly shivering and trying to hold on. Several times we rolled badly, laterally, and at one point a grim-faced Peggy looked over at me and remarked in calm seriousness, "We've got to get out of here." I'm willing, I thought, if I survive. Wave after wave, drenching salt water, windy, cool. The trouble was I could not see the damn waves coming (my glasses were wet and covered with salt spray), and quite frankly I didn't like it one bit. Bucking, sliding through wave crests, pounding over the top. Too tense to be scared — not enough time to panic properly. Finally, we scooted through, but we bounced hard all the way to Staniel Cay. *No more water!* Headlines: "Biologists Drown in the Bahamas."

We parked ourselves at the Happy People Bar on Staniel Cay to wait for the 3:30 airplane, calming our shaking nerves and knees, and drying off. And wait, and wait. Thought the plane was lost, but it showed up about 5:30. High winds slowed him down, we were told (for two hours? — some winds). The plane was a six-seat charter from good old Safe Air. The pilot kept looking around for something before we took off; he spent about 10 minutes rummaging through the cockpit, under the seat and on the ground, but apparently he didn't find what he was looking for. An easy flight to Nassau for 40 minutes to clear customs. Maybe we'll make it yet.

About 7 PM, we finally left Nassau. Dark. The pilot finally found what he'd been looking for — his glasses! He also had hearing aids in both ears, which caused us a bit more disquiet. As we flew off into the dark, it became clear that he had another problem — no dash lights. But he did carry a small pen light which he turned on occasionally to read the instrument panel. Shelly had a death grip on Dick's arm. The roar of engines in our ears, but thoughts of being hopelessly lost on a stormy night in the Bermuda Triangle occupied our brains. Is he going in the right direction, or will we just run out of fuel somewhere over the black Atlantic? We flew through the dark at the top of the storm and turbulence. At times I could look up and see stars, but since we were at the top of the clouds we bounced vigorously. Nothing like being in a dark tube bouncing through the black sky, guided by a pen light. Time stands still, and position is unknown. More headlines: "Biologists Killed in Bahamas Plane Crash." Finally the lights of the Florida coast. I was actually happy to see Ft. Lauderdale in the pouring rain. Found a motel in Boynton Beach and crashed (figuratively, of course).

26 March 1992

Home again, same aftereffects of culture shock; Florida has too much noise, too many people, too much confusion, too many reports of bad news. Feeling unsettled and out of place. And very, very tired.

15 July 1993

Back to the Bahamas. Hot bright afternoon but with a nice breeze. The light is so intense it almost hurts. Mostly blue sky, but I suspect we could see a few scattered thunderstorms later this afternoon. This morning I retrieved my lizard noose which I had left down by the causeway beach. We then went over to Alligator Cay for a while. Captured eight *Leiocephalus*; saw both *S. notatus* and *A. sagrei*. Many *Cyclura cychlura* all over the island. Saw a pretty good-sized male, and what appear to be small juveniles, about 30 cm or so. Looks like reproduction is occurring. Found an old large helmet shell. Not enough time! Very serene listening to the wind over the palm fronds and island vegetation. Single note chirp of the Bananaquits, the song of an occasional mockingbird. Bright yellow-green vegetation in the hot bright sun. Turquoise and white water.

By 3 PM, the sky has cleared completely, and has turned into liquid bluish-white heat. The sun is so intense it is like being in a microwave oven. The breeze on the porch is quite pleasant, however. I'd hate to be working out in the bush today.

16 July 1993

Yesterday at 4 PM, we headed north back to Shroud Cay. We were left off on the northernmost part of the island, and eventually worked around toward several of the southern beaches. Hot and sweaty. I crossed a small ridge to the lagoon. Big web spiders, many mosquitoes — to the lair of the giant land crab. Holes everywhere. I then turned right and went through an incredible tangle of vines, low bushes, and cracked and broken limestone. Very difficult to maneuver. Looks like *Tarentola* country; indeed everything should be here. Giant agave — 1.5 meters diameter at the base and taller than me. Lots of good-sized bromeliads. I finally managed to find the coast and walked along the dogtooth limestone to check several more beaches. Found *Leiocephalus* (very rare here), *Ameiva auberi*, *S. nigropunctatus*, *S. notatus*, *Anolis sagrei*, and *A. smaragdinus* (in a big poisonwood tree). Sweaty and hot till about 6:30. By the time we were picked up, the water was dead calm. Very large setting sun directly into the ocean. Tonight we even had ice with our rum and Kool-Ade. Millions of stars, but no moon.

17 July 1993

Yesterday afternoon we ran another lizard transect. Found a few lizards but not many. Seems to be a small window of activity of about 20 minutes in late afternoons on hot clear sunny days. As yet we have seen no small *Ameiva*, so hatching is much later than for *Leiocephalus*. I sat in the shade watching the nearly calm water and very beautiful golden late afternoon light. The high tide in the channel was about the highest we have ever seen it. The little nighthawk has finally laid her single egg — right in the trail. Saw one egret on the hike back. Standing, looking at the mangroves, rocks and vegetation, one can become lost in time. It could be any time; no way to tell.

This morning (after a rather unpleasant night when the wind dropped to calm), we went over to Narrow

Water Cay to chase our marked lizards. We found 15 curlytails, including four of the nine marked last year. Good morning's work. Saw a Horse Conch in the shallow water. After lizard catching, Dick and I went swimming out over the far reef. I don't think that it is possible to describe the colorful fishes, corals, sponges, sea worms, and other creatures of a coral reef. What a wonderful sight! Calm, peaceful. Strong tidal surge — we could swim into it and drift right on back. Sitting in the shallow warm crystal clear water — blue sky, gray-green islands with white sandy beaches, the multi-shaded turquoise waters. Incredible to simply be in a place like this. It's hard to imagine the other world out there as we watch indigo fish swim back and forth and follow sea fans floating in the crystalline waters. Beautiful — beautiful. Fishes: Squirrelfish, Flamingo's Tongue (a mollusk), Nassau Grouper, Queen and French angelfish, multi-colored wrasses, Queen Triggerfish, Blue Tang, Rock Beauty, Spotfin Butterfly Fish, among many others.

18 July 1993

Up at 6:15 to a sunrise in a perfectly clear sky. The sand flies are bad — kept us in fits all night. A million stars with no moon, and three shooters [= shooting stars]. Everyone went to sleep early except Dick and me. Calm, peaceful. The calm morning has done nothing to keep the buggers away.

Yesterday afternoon was another hot and lazy afternoon. We just mark time until it is cool enough to do the afternoon transect. Not many lizards, as usual these hot clear days, so I photographed the transect. The late afternoon tide was very high again when Dick and I took a quick dip in the calm sea. This morning we hiked back down to Gecko Cave. Saw 11 *Tarentola* but caught only five, including lizard #14, a recapture. The cave had some water in the entrance, but otherwise it looked the same. There was even a trace of fluorescent green powder on the wall from our visit last year. Mosquitoes. Then a good soaking on Beryl's Beach. No orchids in bloom, but the joeweed is an extremely fragrant flowering bush. We have looked but have not seen anything that pollinates it. Very strong and sweet aroma.

Dick and I then went up to the northernmost end of the island [Waderick Wells]. Slight breeze but burning hot bright sun. Became very thirsty in the intense radiation. Walked back along the eastern ridge; saw many hutia tracks and fecal pellets, an iguana tail drag, and a large Manta Ray working along the wall. A fledgling Tropicbird squawked at me. It's pretty up here — we can look into the multicolored water and see the fish, sponges, and fans. Glad to make it back to the house though. Conch chowder over grits for lunch (yum). Breeze better now. Dick, asleep in the hammock. Bananaquits, drinking orange Gatorade.

After dinner, we did a transect on Powerful Beach, but only saw two of our marked lizards. We then measured the transect from the house to the causeway. Beautiful night with floating iridescent ctenophores. Sand flies were a bother. Many hutias off the porch — calling their soft vocalizations. They are even climbing the trees.

19 July 1993

Up early (6:20) to a perfectly still and calm morning. The ocean is like glass. Of course, the sand flies are most unpleasant. Dick and I cleaned up the kitchen, swatted sand flies, and waited for Sandy Buckner to arrive in the seaplane. Then, Sandy, Dick, and I went over to Alligator Cay. This time we had no trouble finding plenty to photograph. Adult *Cyclura* and possibly three size classes of juveniles. Sandy lures them in with bright red maraschino cherries. The funniest thing was to watch the curlytails swoop in and grab a cherry, then take off through the forest carrying their huge bright red treat. Later, I spent more than an hour in the water, fish watching: giant crawfish [lobster], damselfish — black body, yellow fins, with brilliant iridescent blue dots on the head.

20 July 1993

Last night was like sleeping in a sauna; the heat was almost overwhelming. This morning the sand flies are intense and the Bahamas oven has begun to bake now that the sun is up. Oatmeal, tea, and sand flies. Dead calm pervades the islands.

This morning, we went north to Bush Hill Cay. Calm water, the rule; we could see the bottom along the way. In the boat's breeze, the trip was most pleasant. We found a large boat buoy, and a little farther along we found a freshly baited fish trap. Hauled it in as a trophy [fishing is not allowed in this marine park]. Ray [the warden who replaced Peggy Hall] thinks it was set by someone staying on Pigeon Cay, so he wants to stake it out. We hit Bush Hill with no further adventures. The iguanas [*Cyclura cychlura*] are everywhere and of all size classes. They are colorful and most tame. We caught two to count scales (hey, try [catching an iguana] with dental floss!) in order to verify their place of origin. However, one took a good chomp on Dick's finger — he was not amused, although the iguana had a point. Many chases, lots of head-bobbing. They are most curious and easily lured in with red fruit, including tomatoes and orange peels. Found one opened clutch with 10 dry eggshells scattered nearby. Saw several nest diggings, but no open burrows. Very hot in the island's interior. The heat seems to bother me — or maybe it is just the intensity of the sun. I sweat; I drink. But mostly I want to crawl into the shade to find a cool, breezy (ha!) spot. (3 PM) Blazing intense white light and sun.

In the afternoon, Sandy, Dick, and I got a late boat ride down to Rendezvous Beach. Hiked up to the Loyalist ruins to photograph the sailboat carved into the plaster on one of the ruin's walls. Then we went to Gecko Cave, but only got one *Tarentola*; it escaped. Hot and very humid down in the hole. Long, humid, sweaty hike back to the warden's house. Glad we were able to get pictures of *A. distichus* (at Gecko Cave). Drenched and tired upon return. After dinner, I just basically collapsed.

21 July 1993

Up very early after a very sound sleep. The bugs didn't bother me, and it was not as hot as before. We headed out to Hog Cay to look for hutia tracks. Worked the entire island — saw *Leiocephalus* and *Ameiva* burrows, but no definite "hooter" tracks. Took pictures of several tracks that might be *Leiocephalus* for comparison. Trash and debris all over the east (windward) side: much oil tar, a caustic chemical drum, a U.S. Coast Guard life jacket, bottles, light bulbs, and just plain trash. For such a beautiful place, the trash that accumulates on these islands is disheartening to see. Went back the Exuma Sound side of the island, thus circumnavigating Waderick Wells. How did the hutias come to Hog Cay if they are still there? Said good-byes about 3 PM. Took one last walk down the trail and found the little female [one of the very first curlytails we caught in 1991] perched in her tree. Last sighting — when I came back, she was gone. Smooth boat ride down to Staniel Cay. How much different than last time! Warm sun on breezy water (yes, it finally arrived).

22 July 1993

The plane arrived exactly on time. I had spent about an hour and a half watching the ocean and an *A. smaragdinus* crawl around a tree four and a half meters off the ground. Hard to see — mottled green and brown. Easy pleasant flight right along the entire Exuma chain. Perfect weather. Photographed "our" islands: Little Bells, Soldier, Waderick Wells, Hawksbill, Little Wax, Bush Hill, and Shroud. At Ft. Lauderdale we dodged a terrific thunderstorm; ferocious lightning, rain, and thunder. Back to Florida, home of the lavender strip malls. We are still on island time.

The result of these visits was a series of scientific publications on the biogeography and conservation biology of the amphibians and reptiles of the Exuma Island chain. Our (Dick's and mine) preliminary observations have led to more detailed studies by our students, Chuck Knapp, on the *Cyclura* of Alligator Cay, and Celeste Shitama, on the life history of the curlytails on Waderick Wells. The reports map species ranges, discuss how the animals arrived at their present distribution, and detail the historical and current factors affecting their long-term survival. But working in the Bahamas or elsewhere in the wider Caribbean is not confined to dry scholarly presentations; it is also the sun and sand, the jagged limestone, the interesting, new, and wondrous herpetofauna, the rum and smells of island cooking, and — most importantly — it is the colleagues and friends who chase the critters and share the adventure with you. It is hard work, sometimes hot, wet, buggy, and extremely unpleasant, but, at other times, it is cool and dry. You cannot imagine a more beautiful or ideal place to engage in biology, especially while watching a huge full moon rise over a green-black island nestled in a silver sea.

The out islands, away from the tourist resorts, are truly special. They are also amazingly fragile, both in their human sociology and in their island ecology. What has often amazed me is that so little is known about the Bahamian herpetofauna, which lies so close to the teeming United States mainland with its internationally renowned universities and resources. The islands offer a wealth of opportunity for life history and paleontological studies in a place of beauty and hospitality. Although my present position does not allow for much international work, I hope someday to get back to the islands and to renew old relationships with the curlytails and Lion Lizards, with Brown Runners and Reef Geckos. And the sun and turquoise water, the heat and the mosquitoes, the reef fish, and the wind blowing through the silver palms. Maybe I'll even see the little green fluor-escent female *Leiocephalus* perched in her tree at Waderick Wells.

Acknowledgments

I thank Dick Franz for inviting me to help with the Exuma Cays survey; Peggy Hall, Ray Darville, and Sandy Buckner for taking us to the various islands and being most enjoyable companions in the field; and Marian Griffey for her comments on the manuscript.

Suggested Reading

Dodd, C. K., Jr. and R. Franz. 1996. Species richness and biogeography of the herpetofauna in the Exuma Cays Land and Sea Park, Bahamas, p. 359–369. *In* R. Powell and R.W. Henderson (eds.), *Contributions to West Indian Herpetology: A Tribute to Albert Schwartz*. Society for the Study of Amphibians and Reptiles, Ithaca, New York. Contributions to Herpetology, vol. 12.

Franz, R., C. K. Dodd, Jr., and D. W. Buden. 1993. Distributional records of amphibians and reptiles from the Exuma Islands, Bahamas, including the first reports of a freshwater turtle and an introduced gecko. *Carib. Jour. Sci.* 29:165–173.

Franz, R., C. K. Dodd, Jr., and S. D. Buckner. 1996. A review of the herpetology of the Bahamian Archipelago. *Bahamas Jour. Sci.* 3:22–30.

Appendix. Common and scientific names of Exuma Island amphibians and reptiles mentioned in the text.

Amphibians	
Eleutherodactylus planirostris rogersi	Greenhouse Frog
Reptiles	
Alsophis vudii vudii	Bahamian Brown Runner
Ameiva auberi obsoleta	Exuma Islands Ameiva (or Lion Lizard)
Anolis distichus distichus	Bark Anole
Anolis sagrei ordinatus	Bahamian Brown Anole
Anolis smaragdinus smaragdinus	Bahamian Green Anole
Cyclura cychlura inornata	Allen's Cay Iguana
Leiocephalus carinatus virescens	Exuma Islands Curlytail Lizard
Sphaerodactylus nigropunctatus gibbus	Exuma Islands Black-spotted Ground Gecko
Sphaerodactylus notatus amaurus	Great Bahama Bank Reef Gecko
Tarentola americana warreni	Bahamian Wall Gecko

Dodd, C. K., Jr. 2003. Turquoise water, silver palms, and fluorescent green *Leiocephalus*, p. 29–38. *In* R. W. Henderson and R. Powell (eds.), Islands and the Sea: Essays on Herpetological Exploration in the West Indies. Society for the Study of Amphibians and Reptiles, Ithaca (New York). Contributions to Herpetology, volume 20.

Adventures of the Iguana Patrol

JENNIFER M. VALIULIS,[1] KIRSTEN N. HINES,[2] AND JOHN B. IVERSON[3]

Guarding a Studied Nest

Golden rays cut the gray of dawn. Gems sparkle across gently rippling waves.
A solitaire crab scurries under jagged black rock. The odor of sunscreen drifts across white sands,
Three humans emerge from beneath palm fronds.

A female iguana warily eyes the unwelcome visitors. The scaled head bobs warning at the oncomers.
The head bobbing becomes more adamant as they approach her nest.
The iguana slowly backs up to a heap of sand, head-bobbing persistently.

The humans pause momentarily to observe this defiant reptile.
Decidedly, the researchers converge upon the sand pile. They drop to their knees and push the sand aside.
Sweat streams down bronzed cheeks as the sun beats down.
Dirt and sand cake into wounds ripped open by relentless roots.
The disappearing pile is tediously converted into a winding burrow.
The humans take turns toiling in the sand as the heat overcomes.

Head bobs futile. The lizard paces in the bushes as her precious mound is flattened.
In a moment of bravery, the confused mother tests upturned soil.
Nervousness overcomes and the iguana returns to lurking in the bushes,
Repeatedly bobbing her head sagely.

Excited joy lights up a digger's face. A side chamber is discovered.
The chamber proves empty. Joy is replaced by exhaustion.
An exasperated human sigh, but continued digging.
Finally, a peak in excitement as an ecstatic laugh replaces the sigh.
Probing fingers break into an open chamber.
At the touch of smooth, leathery eggs, stinging eyes and throbbing wounds are forgotten.

The mother iguana tears after an intruding lizard. Just as suddenly, she returns.
Curiously, she eyes the white ovals lying in the shade by her burrow.
She approaches, but quickly retreats, bobbing her head menacingly.

Having replaced the eggs in their hidden chamber, the researchers disappear beneath the palm fronds.
The diggers retreat to cool, welcoming waters with unsurpassed grins.

A bewildered mother emerges from the bush. Tentatively she explores the reconstructed area.
Finally, an iguana-style grin.
The lizard puts finishing touches on a remodeled sand pile.

— *KIRSTEN HINES*

Teenagers on a marine biology trip eagerly crowd around a figure on the ground. She is prone, and her arm is buried shoulder-deep in the sand. Intent on finding the treasure, sweat drips into her eyes, and sand creeps into every crack and crevice. A frantic iguana circles the group, bobbing her head in warning. The figure struggles to stretch deeper into the burrow, tearing flesh against underground roots. The relentless sun beats down. Then, her fingers touch something pliable and leathery. "I've

[1] Department of Biology, Colorado State University, Fort Collins, Colorado 80523, USA.

[2] Department of Biological Sciences, Florida International University, North Miami, Florida 33181, USA.

[3] Department of Biology, Earlham College, Richmond, Indiana 47374, USA.

got 'em!" she cries, and the crowd leans in as she triumphantly pulls out a nest full of iguana eggs one by one. "This's what we've been studying for the past month. This is why we're here."

Plans and Preparations

The purpose of our trip was to spend a month studying the nesting ecology of the Allen Cays Rock Iguana (*Cyclura cychlura inornata*), a rather large herbivorous species that, despite its scientific name, is beautifully colored with patterns of gray-black, blue, cream, pink, red, and orange. Only two native breeding populations are known to exist, and these are located on two small islands, Leaf Cay and U Cay, in the northern part of the Exumas island chain of the Bahamas.

In anticipation of one day studying the nesting ecology of these iguanas, John Iverson began a mark and recapture study in 1980. Every year or two, he charters a sailboat for a week or more, and takes about a dozen Earlham College students and alums to the Bahamas for a week of iguana catching. Kirsten Hines and Jen Valiulis are both Earlham alums who have accompanied John on previous trips to the Bahamas. John and his "army" of assistants have marked 1240 iguanas and recaptured 2330 marked individuals for a total of 3570 captures in the course of 22 years. This is the second-longest ongoing field study of any lizard population in the world (a single population of land iguanas in the Galapagos has been subjected to an ongoing study for only one year longer). Through this mark-recapture program, we can fairly accurately predict the age of an individual simply by knowing its length, sex, and capture history. This information gave us the foundation that we needed to effectively study the nesting behavior of this species.

In May 2001, in addition to taking measurements, we used latex paint to mark all captured iguanas with numbers on their sides. This way we would be able to identify individuals when we returned for nesting season without having to disturb any nesting behavior. Male and female iguanas were distinguished by painting the numbers on different sides: males' numbers were painted on their left side, females' numbers on the right. The system was foolproof because we only had to remember that females are always right!

In June, the three of us converged on Nassau from our respective corners of the United States and busily set about collecting gear and supplies. Kirsten and John arrived first, and had the unenviable task of shopping for a month's worth of non-perishable food. Our transport to the island was a chartered sailboat. The couple that ran the charter company generously agreed to take us on the six-hour expedition to Leaf Cay and check in on us every week or so, as well as bring us supplies when we ran low. Our only means of communication with them (or anyone else for that matter) was a temperamental two-way radio on which our code-name was the "Iguana Patrol."

On Patrol

We arrived on Leaf Cay on one of those unbelievably hot days during which the only relief comes from air conditioning or, in our case, frequent dips in the ocean. We soon found that brutal heat was the norm — but we quickly acclimated. After being there for only a week or two, we would start shivering and huddle in our sleeping bags if the temperature dipped below 82°F. As we unloaded the enormous mountain of supplies that would sustain us for the next month, we could hear shuffling sounds in the sea oats around us. Occasionally, an iguana would emerge from the grass, eye us suspiciously, and poke around our ever-expanding pile of gear.

We set up our camp and eagerly did a quick survey of the island to get a sense of whether or not the iguanas were nesting yet. One of the difficulties in planning this trip was the lack of any previous knowledge regarding when and for how long these lizards nest. So, using knowledge about the timing of nesting in similar species and a little bit of guesswork, John decided that we should arrive in mid-June and stay for a month. After all plans were in place, we learned that a single nest of the Andros subspecies of this same species (*C. c. cychlura*) had just been discovered in early May. This information caused us endless anxiety, since we had hoped to catch at least part of the nesting season. However, in our initial survey of the island, we turned up no obvious signs that females were nesting. Of course, we weren't exactly sure what to look for. We knew that iguanas would lay their eggs in burrows and would presumably cover the burrows completely. We also expected some degree of nest-guarding by the females. Other than that, we were just looking for anything out of the ordinary.

Survivors

We soon started making comparisons between our life on the island and the TV show "Survivor." In the first season of that show, 16 people were placed on an island to try and survive by their own wits (or lack thereof). In addition, they voted one person off the island each week. The lone survivor at the end of the show won a million dollars. We didn't have any cash prizes waiting for us at the end of our month on the island, but we soon began facetiously voting each other off the island anyway. Kirsten was the first to go for pulling out her sleeping bag on a "chilly" 85° night. Any similarly unusual behavior or silly remark was punished by threats of banishment from the island.

First Nest!

The first couple of days were spent surveying both islands, catching unmarked iguanas, and monitoring possible nesting sites. To our immeasurable delight, while surveying U Cay on the third day, Kirsten discovered a large female in the act of covering her nest. We stood there absolutely hypnotized by this emaciated iguana as she kicked sand onto a mound over the opening of her burrow. She was obviously exhausted, yet she fiercely defended her nest by sprinting after any other iguana that ventured too close (Figure 1). She seemed to be aware that we were watching, and she would frequently stop digging to circle the mound and vigorously bob her head. We watched her do this for 45 minutes before it occurred to any of us to record anything. We're not sure what made this behavior so mesmerizing. Perhaps it was the realization that we were probably the first people to witness this behavior in this species. Despite the frequent visitation of this island by tourists, most would overlook a skinny lizard working so hard to conceal her eggs. Maybe it was actually watching this behavior after spending so much time speculating about how, when, and where it would occur. Possibly it was amazement at how lucky we were with our timing, because she appeared to be the first iguana of the season to nest. We left her undisturbed to finish covering her nest, with plans to return the next day to excavate the eggs. We reloaded the inflatable kayak that served as our only transportation between islands and paddled back to our camp on Leaf Cay, exhausted, skin tingling from a little too much sun, but happy. We knew that the waiting was over and our quest to study the nesting behavior of the Allen Cays Iguanas was likely to be successful.

The next day, we returned to find her in the cover of the bushes near her nest, intently surveying the area. If any creature ventured too close, she bolted after it from her shady rest until she was satisfied that it would not return. Although she unabashedly chased male iguanas that were twice as large as she was, she simply had no idea what to do with us. We were much bigger than anything she had encountered previously. We boldly approached the nest and actually dared to disturb the mound she had worked so hard to build. Then, we began digging up her burrow and endangering her eggs! As we continued to dig and throw sand around, she frantically circled us, bobbing her head insistently (Figure 2). If we remained motionless, she would even walk right under us, tasting the soil and inspecting the damage we had caused. Finally, after a frustrating wrong turn or two, we found the actual nest chamber and the precious eggs. We carefully pulled them out and admired them as if they were made of pure gold. Their size and number were amazing — no wonder the mother looked so emaciated. The eggs had occupied a significant portion

FIGURE. 1. Female iguana defending her nest site against the approach of another iguana (right).

FIGURE. 2. Female iguana attempting to drive Kirsten Hines away from the iguana's recently completed nest.

of her abdominal cavity and had accounted for nearly 20% of her gravid body weight. Five oblong eggs were each about the width of a golf ball but twice as long. After measuring each one, we returned them to the nest chamber and added a temperature logger, taking care to replace everything exactly as we found it. Then, we filled in the burrow and reconstructed the mound. Our efforts apparently weren't to the mother's liking, as we came back later to find her rearranging the sand. She continued guarding the nest for most of the month we were there (as did a number of other nesting females).

Mother Iguanas' Worst Nightmares

Now that we knew what to look for, we found other nests, although it was increasingly obvious we arrived at the very beginning of nesting season. Within a week, we began to find several nests a day. The search for new nests was always exciting. Each morning we would comb the island searching for the tell-tale mound of sand and an aggressive mother. John had a talent for collapsing in nest chambers simply by walking over them. His "divining toes" became our most reliable tool for locating nests.

We soon became a mother iguana's worst nightmare. With complete disregard for a mother's attempts at nest defense, we would storm in, flatten her carefully constructed mound, excavate her burrow, and pull out her eggs. The best moment of any day was when, after digging up a long burrow, spending long hours sprawled on the hot sand, arms stretched to the fullest, roots tearing at our flesh, excavating numerous side chambers, anxiously probing for the egg chamber, with a mouth full of sand and eyes full of sunscreen and sweat, we found the eggs (Figure 3). More than once our excitement was premature as we broke into a chamber from a previous years' nest only to find a pile of shriveled shells. A close second for the highlight of a day was when, having measured and replaced the eggs, and properly re-covered the burrow, we jumped into the cool ocean to wash off before starting again. Small inshore fishes seemed to share in our enjoyment as they swarmed in and out of the muddy cloud we generated in the shallows several times a day.

Emulating Iguana Behavior

Although iguanas were the main focus of our month in the Bahamas, a number of other events made this such a unique experience. One interesting aspect to doing research in a place such as the Exumas is the constant interaction with tourists. The Allen Cays are a popular destination for chartered boats as well as a stopover for outfitters that run daily trips out of Nassau in high speed cigarette boats.

Tourism obviously has a significant effect on the iguanas. Scores of them will run down to the beach at the sound of an approaching boat. Those residing on or near the beach readily approach humans and literally beg for handouts. Many people comply and feed them all kinds of goodies that iguana guts aren't designed to digest. We witnessed people feeding them anything from raw ground beef to pieces of foam from shipwrecks. For these reasons, we were understandably suspicious of any vacationers that showed up on what we selfishly considered "our" islands. We did our best to avoid significant contact with them. We imagine that the wariness went both ways.

We probably caused a number of nervous moments for the happy vacationers by emerging from the island interior, bloody from working in the bush and being scratched by ornery lizards, often covered in sand from the most recent nest excavation, with pillow cases filled with thrashing lizards, and racing

FIGURE. 3. Jen Valiulis displays the result of a successful nest excavation while John Iverson hides from the sun.

FIGURE. 4. "Show and tell" for the tourists on Leaf Cay.

to the cool embrace of the ocean. Admittedly, bemused by some of the shocked reactions, we may have exaggerated our uncivilized state. Some people gave us looks similar to the glares we received from a mother iguana as we approached her nest. The braver ones asked what we were doing, and we were quite proud to tell them. Once over their initial shock, most were genuinely interested in our work. Many of them appreciated the parallels between "The Iguana Patrol" and "Survivor" when they realized how primitively we were living.

We realized that the impact from tourists wasn't going to disappear if we ignored it, so we did our best to educate people about the life history of the iguanas as well as their natural diet. When the powerboats roared up for their daily visits, we tried to have a recently captured iguana ready for educational "show-and-tell" (Figure 4). In exchange they gave us cold water, an occasional cold beer, and other needed staples. In our constant battle with the heat and sun, we treasured these small things. Even in coolers, ice would melt in a matter of hours, and sodas would explode from the heat. Consequently, we were usually stuck drinking warm water flavored with powdered tea to mask the distinctly sour taste of Nassau water. Any kind of cold liquid was pure ecstasy. Soon, we were imitating our iguana counterparts and running to the beach at the sound of the boat engines, eager for any handouts.

Red in Tooth and Claw

Although we tend to refer to iguanas in the affectionate manner unique to herpetologists, they are not the friendliest critters around. Upon capture, they thrash about trying to slash the captor (and anything else within range) with their claws. Their bites can be equally painful. Razor-sharp teeth, normally intended for tearing leaves, can easily slice through skin. If that isn't enough, iguanas also have long tails covered in sharply pointed scales that, when whipped against unsuspecting skin, can leave painful red scrapes and welts (Figure 5). Anyone who has ever spent time catching these lizards is familiar with the concerned glances from people who, after the iguana catchers return home, look at all the cuts and slashes on wrists — and suggest therapy.

An Important Lesson

When camping for a month with only minimal protection from the elements, you become very familiar with the vagaries of the weather. By and large, we were extremely lucky with daytime weather. We had only one day during the entire month during which inclement weather limited our work. Nights, however, were an entirely different story.

Visions of sleeping on the beach with ocean breezes and the gentle lapping of the waves prompted us to set up our tents and sleeping bags on the sand, away from the protection of the interior vegetation where the rest of our camp was located. For the first few nights, we slept blissfully. In the evenings, before going to bed, we sat on the

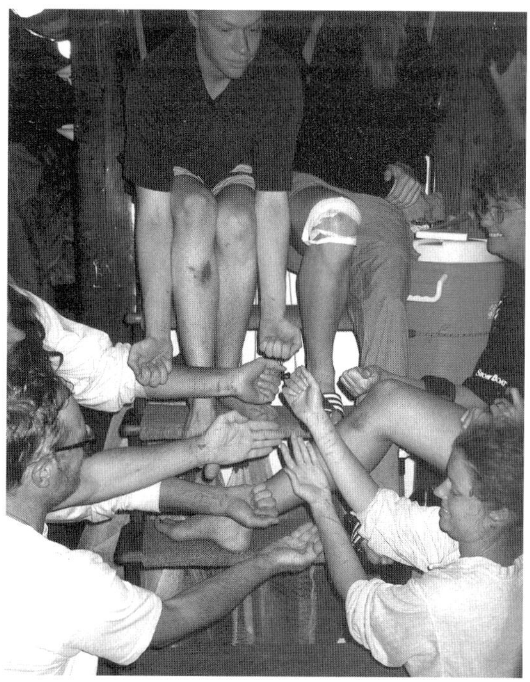

FIGURE. 5. Accumulated injuries from capturing 423 iguanas during five days in March 2000 (John and Kirsten in foreground).

beach remarking on the beauty of the stars overhead and the distant lightning and how some poor soul was getting drenched in the distance. Our luck didn't hold. One night we woke to find water pouring in through the open flaps of our tents. Few things are as uncomfortable as lying in a puddle of water with sheets of rain pelting your body, all the while trying to keep your tent from turning into a kite. Eventually the wind and rain got so bad that we had to pick up our tents and move them to a more protected site in an effort to salvage some soggy sleep. Colin Meeks, another Earlham alum who had shown up the previous day to help us for a week, was immediately voted off the island for bringing the bad weather with him.

The next morning in bright sunshine we remarked on how crazy the storm had been and, happy that it was over, returned our tents to their positions on the open beach. We were convinced that the storm had been an unusual event. We were wrong. The next night was worse. Completely exhausted from the previous night's adventure and a hard day of digging, we retired to our tents early to the sounds of John's assurance that it wouldn't rain tonight. After all, the lightning was well to the south, out of the path of our prevailing winds. Within an hour, we awoke from a dead sleep to gale-force winds, thunder, and lightning all around us — and, just as we crawled out of our bags to haul our tents to the more sheltered inland, a torrential rain began. Somehow, still half asleep, we successfully stumbled up the hill, despite our fears, the painful sting of the rain, the "freezing" 78° temperature, and our efforts to keep our tents from becoming parasails. As we wrestled with the tents in an attempt to stake them down, a bolt of lightning hit nearby. We aren't sure if it was the force of the lightning or the crash of thunder, but we were all knocked to the ground. As Jen crawled back into her tent, she realized that Kirsten, John, and Colin were gone. Not knowing where they had gone, she spent a very long 15 minutes pressed against the side of her tent in an effort to keep it from blowing away, all the time shaking in abject terror.

Between the seconds separating the flashes of lightning, she was in complete darkness and all she could hear was the whipping wind and the crashing thunder. This was perhaps the only time during the whole month that she questioned her decision to make the trip. All of a sudden, a tent flap flew open and Kirsten fell in, soaked to the bone. John stuck his head in the other side of the tent telling her to get in the tent before she got wet. They both looked at him in disbelief — and then laughed. Jen had never been so relieved to see anyone in her life. The storm continued to rage, but with two people in the tent, it seemed more like an adventure than a life-threatening event. With lightning still crashing around them, Kirsten and Jen laughingly admitted to each other that they had each been reassuring themselves with John's mantra that "if you get hit by lightning, it's your time to go." In the interim, Colin had jumped into the other tent to keep it from blowing away, and Jen and John had secured the rest of the camp against further wind and rain damage.

The storm began to lose some of its energy. Still shaking from the cold and the waning adrenaline, we tried to get back to sleep in our soaking tents and sleeping bags. It was not a comfortable night for the Iguana Patrol, but we did learn an important lesson. From then on, we spent our evenings on the beach enjoying the breeze and the lapping of the waves, but, when it came time to sleep, we returned to the shelter of our tents in the interior, despite the heat and lack of breeze. The next morning, John was voted off the island for his erroneous weather predictions.

Giants

Across a small channel from Leaf Cay was the namesake of the archipelago, the large island of Allen Cay. Despite being twice the size of the other islands, its extremely rocky terrain and almost complete lack of sand rendered it fairly inhospitable, even for iguanas. Surveys prior to the 1980s had turned up no signs of their existence. In the mid 80s, our own casual surveys of the island had revealed a small number of subadults that we presumed had been introduced. More recently, we had heard reports that a giant iguana was living on the island — and, in May, our boat captain, a former Navy SEAL, in full camouflage and face paint, tracked it down and caught the four-foot lizard bare-handed, a feat that was impressive even with smaller animals.

We decided to take a day off from digging up nests and go over to Allen Cay in search of this monster. We arrived on the shore and, before we could even get our gear unpacked, we saw him. He was about 15 feet away in the low bushes, staring at us with his red eyes. He was enormous! We stared at him, not believing our luck. In May, hours were spent locating the lizard, and now, a month later, here he was, just sitting there waiting for us to catch him. As we unpacked our gear, we began to wonder how in the world we were going to catch him. Our usual methods of dip-netting or noosing were definitely not going to be enough. Our typical noose is actually an extendable golf ball retriever with a high-strength fishing leader fashioned into a loop attached to the end. This allows us to catch particularly skittish iguanas by standing back from them and slipping the loop around their neck and then reeling them in. It doesn't do any damage to the lizards other than making them really mad. However, this guy was so big that he would undoubtedly break the noose with minimal effort.

We decided that, despite our inadequate gear, we would have to try something soon, before he wandered away. We first tried the noose, so we could stay as far away as possible, and then we used the net for backup. This worked — to a point. As soon as the noose and net went over his head, he tore holes in the nylon mesh as he bent the noose pole to its limit. Colin and Kirsten had no choice but to pounce on him. It took both of them to hold the squirming lizard down. Once under control, he was even more imposing than when we first saw him in the bush. He glared and hissed at Colin and Kirsten and, if his snapping jaws were to hit their mark, someone could easily lose a finger.

We excitedly radioed John, who was on U Cay looking for new nests, to tell him the news. He quickly hitched a ride on a zodiac from one of the boats anchored nearby. We proudly showed off our catch. We had thrown a pillowcase (our standard holding bag) over the lizard's head, and that seemed to calm his thrashing. We measured him and found that, including the tail, he was 1.2 m long (over 4 feet)! Although he exceeded the capacity of our scales, we estimated his weight at over eight kilograms (almost 18 pounds). The largest animals from 22 years of work on the other two islands were just over a meter in total length and only 5 kg in body mass. Even more thrilling was the realization that this was not the same iguana captured in May — so more than one giant iguana lived on this island. After taking pictures of each of us proudly holding the monster, we released him and John returned to U Cay.

Thrilled and a bit cocky from our initial catch, the rest of us decided to search for the other one seen in May. We climbed up to the rocky ridge, one of the highest points around, where we could see for miles over the open ocean to the west and "our" islands to the east. After admiring the view, we started looking around, only to find a rather gruesome site. Headless bird carcasses were strewn about the rocks. Something had killed and eaten most of the meat. Upon closer inspection, we realized that they were Audubon's Shearwaters, a highly endangered species that nests on the island. Slowly, we began to suspect that a lack of competition might not be the only reason for the gigantic sizes of these iguanas. We collected some of the carcasses and uneasily split up to search for the other iguana.

Progress was slow because the dogtooth limestone was jagged and sharp. About 20 minutes later, as Jen was returning to get something from her backpack, she saw some movement on the rocks. Looking closer, she saw an even larger iguana munching on some leaves. It eyed her, unconcerned. Obviously, it did not consider her much of a threat. Jen excitedly called for Kirsten and Colin, who had to pick their way across the rocks. The minutes until they arrived were nervous ones. If he decided to run away, she would not be able to chase him across the rocks without impaling herself. If he charged her, he could definitely cause serious injury. Finally, Kirsten and Colin arrived and, still confident from our last catch, we tried to throw the net over him. He ran right through the mesh netting as though it wasn't even there, tearing a huge hole in the net. He scampered into a nearby hole in the rocks, where we could hear his deep breathing. He seemed to be growling from within his retreat as he exhaled air and wedged himself deeper in the hole. Although we could see his tail, none of us had any desire to stick our hands in there to drag him out.

In the meantime, John, having finished his surveys of U Cay, arrived in the kayak. We regrouped and decided that we couldn't let this one get away. We carefully sealed off all but two potential exits of the honey-combed retreat. Colin and John were assigned to guard the two left open. They covered their forearms in extra pillow cases, pulled on heavy gloves, and began to scheme. How could they chase the iguana out from the cover of this labyrinth? They took turns nervously easing their arms into the burrow, trying to reach and grasp the hindlimbs while making every effort to stay away from the beast's jaws. To John's suggestion that he try to pull it out through another hole, Colin replied, "I think it'll

FIGURE. 6. John wearing the grimace of fatigue after an hour of midday work extracting an iguana from the honeycombed limestone of Allen Cay.

bite my hand off, but if you want me to, I'll do it." Kirsten and Jen stood by uneasily, fully expecting to have to put their first aid skills to use soon. Patiently and cautiously, Colin and John provoked the iguana until it reoriented itself within the rocky retreat in such a way that its head was wedged into a side chamber, rendering it unable to rear back and bite a probing hand. Lying flat on the jagged rock to extend their reach, Colin and John carefully worked their hands up the tail to securely grasp both of the hindlegs. A well synchronized jerk backward allowed John to slide his free hand up to the animal's neck and pin it to the substrate. Apart from the flesh torn by the rocks, final removal required only coordinated control of the writhing animal — Colin trying to restrain the lashing tail and John attempting to control the viciously snapping head (Figure 6). Once the monster was extracted, it took all four of us to process this 1.25 m, 10 kg giant (Figure 7).

Several return trips to Allen Cay yielded a total of 11 monstrous iguanas, although we actually captured only seven of them. We have yet to determine why they're so enormous. Discussions with previous visitors to the island generated several accounts of iguanas eating young shearwaters, but we don't know if this carnivory, highly atypical of Rock Iguanas, occurs frequently enough to fully account for such large body sizes. We also are not sure whether the iguanas can actually breed on the island, or if the colony will perish with the eventual deaths of the few oldtimers that live there now. These iguanas probably arrived as a consequence of relocation by some well meaning tourist who, thinking that they were doing their part to help the species, brought a few over from the other islands. We have never observed these iguanas swimming long distances, and assume that they're either incapable or unwilling to cross the channel between the islands. In the meantime, we are faced with a problem. The huge iguanas obviously are thriving, but their success may be at the expense of the shearwaters. Which of these two endangered species should receive protection priority?

Looking Back

Scientifically, our trip was a resounding success. We excavated 40 iguana nests, more than we had dared to anticipate when we arrived. We documented all of the intricacies of nesting behavior, not to mention nesting preferences (location and substrates) and details of clutch and egg sizes. We also discovered a new colony of giant lizards that presumably feed on birds.

Also, by literally living among the critters of the Allen Cays, we gained insights into how they lived on a daily basis and came to know some of them personally. We had several "regulars" that fre-

FIGURE. 7. Colin Meeks (right) and John Iverson prepare to "bag" a giant iguana on Allen Cay.

quented our "patio." "Tuffy" (Figure 8) was an aggressive little Curly-tailed Lizard that defended his territory against any intruder, no matter how large, and fed on crumbs of our bread and chips. "Stubby" was a large male iguana missing half of his tail. He was always begging for food and even managed to get into our kitchen tent one afternoon. An assortment of mockingbirds also frequented our mealtimes and fed from our fingers. Leaving these animals was like leaving friends.

The morning we left the islands was like any other: warm, breezy, and sunny. The powerboats roared up and the tourists poured out to gawk at the iguanas and feed them grapes. We gave our usual spiel about our research, but this time we joined them as they left the island. As we sped away, the theme song to "Miami Vice" reverberating from the boat speakers, we wistfully watched the Allen Cays grow smaller behind us. We knew that we would be back, but it's impossible to live in such a unique and amazing place without being affected by it. Despite previous trips to the islands, living "in sync" with the natural rhythms drew us so much closer to them, not to mention each other.

Perhaps the uniqueness of the experience of spending a month in the West Indies is best demonstrated by the shell-shock of returning home. Of course, the inevitable backlog of snail- and e-mail awaited, but other, more subtle adjustments seemed to be the hardest to make. When it gets a little too hot, we can no longer jump in the ocean. Even if we could, we couldn't just jump in at any old time, because now we had to wear actual clothes, rather than running around all day in swimsuits. We no longer need to apply sunscreen five times a day in the constant struggle against sunburn — and what is that strange beeping sound emanating from that electronic device on the desk? Most obvious, however, is the fact that the pace of life suddenly speeds up by several orders of magnitude. As clichéd as it sounds, island life runs at a pace of its own. When everything you do is dictated by the schedule of the tides or the weather or even the iguanas, you learn to be patient and not to get frustrated when things don't proceed exactly as planned.

People were very friendly and went out of their way to lend a hand. Every person with whom we interacted demonstrated this in their own way. Our Bahamian friends and colleagues facilitated our permits, helped solve our logistical problems, and provided transport to and from the islands. Powerboat operators brought us food, supplies, and cold drinks, and then refused any kind of payment. The local fishermen, concerned that we weren't eating well enough, cooked us a delicious meal of conch and grouper, and took time away from their livelihoods to give us much-appreciated rides between islands. Even the tourists on vacation, despite our savage appearance, occasionally invited us aboard their boats for drinks or dinner. Each one of these people made our month that much more memorable, and made it that much harder to leave.

FIGURE. 8. Jen becomes attached to "Tuffy," a curlytail on our "patio."

Acknowledgments

This research project has been funded primarily by the participating students (74 total through 2001) and faculty (six total) of Earlham College, who have volunteered their time and shouldered most of their own expenses. Additional financial and logistical support has been provided by Dr. Charles and Sally Test of Indianapolis, Mr. Hugh and Mrs. Sandra (Past President of the Bahamas National Trust) Buckner of Nassau, Nigel Bower and Chris MacDonald of Powerboat Adventures of Nassau, Ron White of Good Fortune Charters of Beaufort, North Carolina, and Bruce Dunham and Sheila Young of 7 C's Charters of Charleston, South Carolina. Research permits were provided by the Bahamian Government and the Bahamas National Trust. Assistance in the field by many other visitors and yachtsmen also is appreciated. Comments on early drafts of this essay were provided by Jon Bossenbroek, Ronald and Lois Hines, Sheila and

Peter Iverson, Ralph Saporito, and George and Elisabeth Valiulis.

Suggested Reading

Alberts, A. (ed.). 2000. *West Indian Iguanas: Status Survey and Conservation Action Plan*. IUCN/SSC West Indian Iguana Specialist Group. IUCN, Gland, Switzerland.

Burghardt, G. M. and A. S. Rand (eds.). 1982. *Iguanas of the World*. Noyes Publ., Park Ridge, New Jersey.

Iverson, J. B. 1978. The impact of feral cats and dogs on populations of the West Indian Rock Iguana, *Cyclura carinata*. *Biol. Conserv*. 14:63–73.

Iverson, J. B. 1979. Behavior and ecology of the Rock Iguana, *Cyclura carinata*. *Bull. Florida St. Mus., Biol. Sci*. 24:175–358.

Iverson, J. B., K. N. Hines, and J. Valiulis (author order yet to be determined). Unpubl. MS. A preliminary study of the nesting biology of the Allen Cays Rock Iguana, *Cyclura cychlura*. Submission expected in 2003.

Iverson, J. B. and M. R. Mamula. 1989. Natural growth in the Bahaman iguana, *Cyclura cychlura*. *Copeia* 1989:502–505.

Iverson, J. B., G. R. Smith, and L. Pieper. In press. Factors affecting long-term growth of the Allen Cays Iguana, *Cyclura cychlura inornata*. *In* A. C. Alberts, R. L. Carter, W. K. Hayes, and E. P. Martins (eds.), *Biology and Conservation of Iguanas*. Univ. California Press, Berkeley.

Knapp, C. R. 2000. Home range and intraspecific interactions of a translocated iguana population (*Cyclura cychlura inornata*). *Carib. Jour. Sci*. 36:250–257.

Knapp, C. R. 2000. Status of a translocated *Cyclura* iguana colony in the Bahamas. *Jour. Herpetol*. 35:239–248.

Valiulis, J. M., K. N. Hines, and J. B. Iverson. 2003. Adventures of the Iguana Patrol, p. 39–48. *In* R. W. Henderson and R. Powell (eds.), Islands and the Sea: Essays on Herpetological Exploration in the West Indies. Society for the Study of Amphibians and Reptiles, Ithaca (New York). Contributions to Herpetology, volume 20.

Cuba

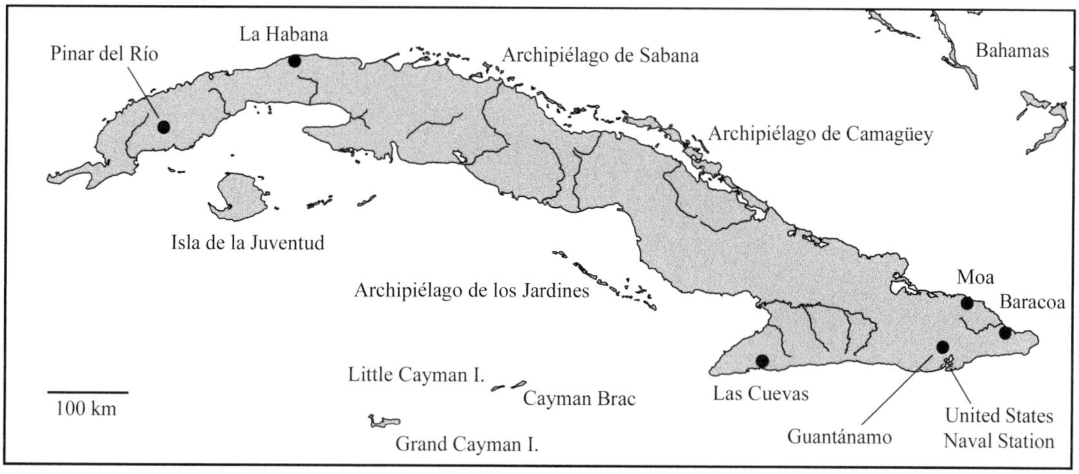

Pinar del Río
La Habana
Archipiélago de Sabana
Bahamas
Archipiélago de Camagüey
Isla de la Juventud
Archipiélago de los Jardines
Moa
Baracoa
Little Cayman I.
Cayman Brac
Las Cuevas
100 km
Grand Cayman I.
Guantánamo
United States
Naval Station

The Frog Detectives of Pico Turquino

S. Blair Hedges[1]

For many Cubans, Oriente is like a different country. It is the large and rugged upland region of eastern Cuba, about 1000 km east of Havana. In 1974, Oriente Province was divided into four smaller provinces, but Cubans still use the older name. In this essay, I recount my first expedition to Oriente in July and August of 1989. It was a special trip because we explored the highest mountain in Cuba, Pico Turquino (Figure 1), and because of the detective work that led to the discovery of two new species of frogs.

Some might ask why I have such a fascination with species previously unknown to science. I've thought about this and have come to the conclusion that it is a combination of things. I certainly enjoy the sense of wonder at new discoveries. I also gain a sense of satisfaction from withstanding the rigors of field work and recognizing as new a creature that no scientist had previously encountered or recognized as unique. In addition, I admit to some pride in the detective work that discoveries sometimes entail, and I enjoy the challenge of coining appropriate new names.

Figure 1. View of Pico Turquino (center, in distance) from the south coast at Las Cuevas, Cuba (© S. Blair Hedges).

Getting There

Getting to Cuba is *always* an adventure. Richard Thomas, Professor of Biology at the University of Puerto Rico, accompanied me on that first trip (and many others). A day before leaving, the Cuban Interests Section in Washington, D.C., at that time located in the Czechoslovakian Embassy, told us that our special scientific visas were postponed indefinitely, for unknown reasons. However, our Cuban colleagues suggested we come anyway and try to obtain the visas after arriving in Havana. Scientific visas were not needed for entering Cuba, but were required for conducting research. Although booking as a tourist can be done easily today, it was more difficult in those days, and the Canadian travel agent refused to issue us air tickets for the Toronto-to-Havana flight because we did not have hotel bookings for the entire trip. Ultimately, a hasty fax two minutes before office closing managed to get us on a flight.

Arriving in Cuba on 23 July, we were told that time was too short to arrange for scientific visas, but that friends in high places had worked out a special arrangement for us. We were to surrender our passports in exchange for a handwritten note by an important person. That would supposedly get us anywhere in Cuba. In the meantime, our visas would be processed. What the heck, it was only the Cold War and we were two Americans wandering around remote areas in a communist country without passports. Neither Richard nor I objected.

The goal of our trip was to make exploratory collections throughout southern Oriente. This included the far eastern region of Baracoa and Maisi, areas at the edge of the Earth in terms of Cuban infrastructure, and the dry coastal forest strip of

[1] Department of Biology, Pennsylvania State University, University Park, Pennsylvania 16802, USA.

southern Oriente. However, the major objective was a 5–6 day trek to Pico Turquino, located at the western end of the Sierra Maestra. The Cuban members of the expedition were Alberto R. Estrada, an expert on Cuban amphibians and my first contact in Cuba, Octavio Pérez-Beato, curator of herpetology at the Cuban National Museum of Natural History, Antonio Pérez-Asso, millipede expert and expedition cook, Emilio Alfarro, herpetology assistant and jeep driver, and Riberto Arencibia, lizard specialist and assistant to Pérez-Beato. While in Oriente we had assistance from local biologists, jeeps, and drivers.

Richard and I flew on Cubana Airlines from Havana to Guantánamo, while the rest of the group made the 27-hour, non-stop jeep-journey on land. I am not sure which was the better choice. The Cubana flight, on a 1950s vintage Soviet plane, was non-stop — but took two hours and 15 minutes, which drove home the dimensions of the island. At least flying across Cuba had the advantage of letting us search for potential collecting sites, and one that we saw on this flight, a karst limestone area on the northern slope of the Sierra Maestra, later yielded several new species of reptiles.

Turquino

On 5 August, after two weeks of successful collecting in Oriente, we now faced the last-minute preparations for the long awaited Turquino trek. We would not be the first to collect amphibians and reptiles on Pico Turquino. Of the earlier explorers, Harvard entomologist P. J. Darlington stands out. In June 1936, he collected four new species of frogs on Pico Turquino: the Turquino Stream Eleuth (*Eleutherodactylus turquinensis*), the Turquino White-footed Eleuth (*E. albipes*), the Turquino Red-armed Eleuth (*E. cubanus*, formerly *E. parvus*), and the Oriente Dark-bellied Eleuth (*E. intermedius*). These were formally described by Harvard herpetologists Thomas Barbour and Benjamin Shreve in 1937. Unfortunately, the first three species had not been seen since, despite several ascents of the mountain by herpetologists. One reason for this, we hypothesized, was that the old trail used by Darlington and his primary collecting spot, Cueva del Aura, had long since been abandoned. If the species were localized, other routes to the peak might easily have missed them. Thus, in addition to looking for new species, we had added the objective of rediscovering Darlington's species.

We had two jeeps at our disposal, and our take-off point for the trek was a campground on the south coast, at the southeastern base of Turquino, called La Mula (the mule). The summer vacation season was in full swing. Between the all-night partiers, including a make-shift band playing rumba (Cuban-African music), large rats invading the campsite, and rain at 2:30 AM, none of us got much sleep. We assembled at 5 AM for the short ride, about 12 kilometers, east to the tiny coastal town of Las Cuevas. This village is located on the southwestern flank of Pico Turquino and was the place where most ascents of the mountain began. In Darlington's time, the trail started directly south of Turquino at Potrerillo. Because of the biological interest in Turquino, I decided to document the place names and the locations of the different routes of ascent in a publication with Albert Schwartz in 1991 (see Suggested Readings).

We all assembled with our gear at Las Cuevas, and were greeted by the mulero (guide with mules). After taking one look at all of our equipment, he decided that four mules were not adequate. He had to go look for at least one more mule. This was bad news, because we knew how muleros operated: they frequently let their mules run wild over a large area and often required at least a day to retrieve one. So, our chances of doing the hike that day went down the drain; the uncertainty of summer afternoon weather precluded a later start. Fortunately, we learned that the trail was drivable by our jeeps as far as a shelter called La Esmajagua, elevation 560 m (Figure 2). There, we could at least do some collecting. This shaved a little time off our hike the next day. Unfortunately, the next day was a repeat of the first, with the mulero delayed until early afternoon, postponing our ascent another day.

FIGURE 2. Loading the mules with equipment at La Esmajagua on the southwestern slope of Pico Turquino (© S. Blair Hedges).

Finally, on 8 August, the mulero was waiting with his mules at 5 AM. Although we all hiked on foot and carried backpacks, the five mules still were not enough to carry all of our camping and collecting equipment. The mulero had to make two trips up the mountain, but we had all day, so that was not a problem. The hike to Pico Cuba, which was our base camp for exploring Turquino for the next three days, took only three hours from La Esmajagua. Although the campsite was referred to as "Pico Cuba" by the mulero and others, it was actually a 20-minute hike below Pico Cuba on an open, rounded knob on the southwestern flank of Pico Turquino, elevation 1720 m. A small shelter, a lot of ferns, and a hiker-polluted stream graced the site. We pitched our tents and found a clean source of water down a steep slope.

Frog Mysteries

The first mission was to collect everything around the base camp and get to know the local fauna. The White-fanned Anole (*Anolis homolechis*) was very common during the day, mostly on the ground, and the Turquino Green-mottled Anole (*A. altitudinalis*) was found sleeping at night on vegetation. The frogs, however, presented us with two mysteries that dominated the remainder of our expedition.

The abundant frog on Pico Turquino, calling from low ferns and vegetation at night, appeared to be the Cuban Telegraph Eleuth (*Eleutherodactylus auriculatus*). This is a small species, with a male snout-to-vent length of only 18–20 mm. However, these tiny frogs were not living up to their name by emitting a telegraphic series of clicks, but instead had a single-note "tink" call. At first, we thought that the lower temperatures at this elevation were responsible for slowing down the call, but the temperature at Pico Cuba in early August was not that cool, plus the frogs looked slightly different, more rugged in appearance. Unfortunately, we did not have "proof" that it was a new species, which would entail finding it living together with its closest relative, in "sympatry." All of the *E. auriculatus*–like frogs on Pico Turquino looked the same and made the same sound. This situation simply meant that more detective work was needed.

The second mystery involved a larger species, the Bromeliad Eleuth (*E. varians*). Normally, this is a medium-sized frog (25 mm male snout-to-vent length), tan in color, and capable of producing a loud, 1–3 note hollow metallic "tock" call. As the name implies, these frogs often occur in bromeliads, epiphytic plants that retain small pools of water, but also may be found on leaves, sometimes high in the trees. During our first night at base camp, we found two frogs that resembled the Bromeliad Eleuth sitting on leaves about two meters above the ground. That wasn't surprising, except that we never heard any call that resembled that of *E. varians*. Instead, I kept hearing a repetitious call that I had never heard anywhere else in Cuba. The quality of the notes (hollow "tock's") resembled that of *E. varians*, but they were emitted in a rapid series of 14–22 notes over one or two seconds, much like a machine gun. Based on my description of the call, Alberto assured me that it must have been the Cuban Telegraph Eleuth, which was supposed to occur there and also had a rapid call. However, I had collected many specimens of that species and knew the call very well. The sound coming from high in the trees on Pico Turquino was louder and had a different quality.

This mystery was beginning to gnaw at me. I desperately wanted to find the animal making the repetitious call. With some help from my colleagues, we triangulated on two frogs. As is typical when collecting frogs, a lot of time is spent inching closer and closer to the sound, waiting each time for the frog to forget about the disturbance and start calling again. Sometimes, 20–30 minutes or more can go by between each call, especially as you get closer. Consequently, the entire process becomes an all-night experience. This is what we did that night, with two individual frogs. In both cases, we were unsuccessful, but at least we determined that the sound was coming from high in trees and probably from bromeliads. This was very uncharacteristic habitat for the Cuban Telegraph Eleuth, so we were inclined to rule out that species.

We had only been on the mountain one day and we already had two unsolved frog mysteries. On the second day, Richard, Alberto, Emilio, and I decided to make a small expedition on our own, exploring the highest areas and then camping in a small tent at the top. The latter was necessary because Pico Real-Turquino, the highest point on Pico Turquino (1974 m), was still an exhausting hour-long hike up a very steep trail from the Pico Cuba base camp, and we wanted to reduce the amount of hiking to leave more time for collecting. The rest of the group (Octavio, Tony, and Riberto) would hike up for a day trip, doing general collecting, and return to the base camp that night.

FIGURE 3. Alberto Estrada, Richard Thomas, and Blair Hedges at the highest point in Cuba, Pico Real-Turquino (1974 m), standing in front of the monument to José Marti (9 August 1989). Richard Thomas is holding a galleta (a Cuban cracker), a major source of carbohydrates during this expedition (© S. Blair Hedges).

With enormous backpacks, the three of us set our sights on the peak. Because we were camping, we had to bring food, water, tent, and other basics besides the necessary collecting equipment. The top is a rounded knob with a small clearing and a statue of Cuba's famous 19th-century writer and patriot, José Martí (Figure 3). The statue was a likeness of his head on top of rocks and concrete, about two meters high. Because the cloud forest canopy was 5–10 m high, the view was not so great. We immediately had lunch and set up camp a bit back in the forest. It was now about two in the afternoon and we had only a few hours of daylight left. We decided to split up, with Alberto and Emilio collecting around the peak and Richard and I heading due north of the peak for several kilometers along a narrow, forested ridge called Estribo Turquino.

Hiking along Estribo Turquino, our main activity was cutting bromeliads and examining them for frogs. Some misguided conservationists believe that any collecting activity, especially involving the cutting of plants, is at odds with everything they hold dear. In contrast, systematists and other reality-based conservationists readily see the benefit in removing small, imperceptible numbers of individuals from the population for science. The reason is actually quite simple: the policies that are enacted to protect biodiversity ultimately are based on the data generated by systematists — and photographs just don't work. You need scales, DNA, skin, and bones for primary systematic research.

At first we didn't find any frogs, and Richard walked ahead while I lagged behind cutting more bromeliads. Finally, I found one that contained two strange-looking frogs. They had flattened bodies, as if perfectly adapted to fit between bromeliad leaves. Their coloration was striking. They were yellowish tan with distinct, dark brown heads, and unlike anything I had ever seen (or knew about) in Cuba. In fact, the first thing that came to mind was Jamaica's Yellow Bromeliad Frog (*Hyla marianae*), a species in a different family (Hylidae), but sharing the same habits and thus clearly convergent. I carefully checked between the toes and found no webbing, so these Turquino frogs were eleuths and not hylids. I was certain that it was a new species.

I quickly caught up with Richard and told him the good news. Together, we collected about 25 adults, juveniles, and eggs from one clump of bromeliads. That was enough to make a formal description and also learn something about reproduction in the species. All were bicolored, with dark snouts and light bodies. However, I noticed that the dark snout color disappeared about 30 minutes after we removed the frogs from their bromeliads, indicating a behavioral color change. This made sense, simply because the frogs normally back into the bromeliad leaf axils, exposing only their snouts, and thus a dark snout would hide the frog against the dark axil. On the other hand, a predator would easily see a dark-snouted frog outside of a bromeliad.

After returning to the tent, Alberto correctly pointed out that, aside from the dark snouts, the new frogs were similar in some ways to the common Cuban species that we collected on leaves the first night, *Eleutherodactylus varians*. Thus, the plot thickened. We had what appeared to be typical *E. varians*, a dark-snouted bromeliad frog, and an unidentified repetitious call. With that information, we speculated that the dark-snouted frog was the one making the repetitious call, mainly because the frog was abundant in bromeliads and the call seemed to be localized in those same bromeliads. The two plain-colored *E. varians*–like frogs collected the first night may have been individuals of the bromeliad frog that had lost their dark snouts because they were outside of the bromeliads. At least at the time, the idea seemed reasonable.

We also encountered the *E. auriculatus*–like frog at the peak that night, making the same sound as at the base camp. We never heard the telegraphic call of typical *E. auriculatus*. Fortunately, we found one of Darlington's species, the Turquino White-footed Eleuth, in the leaf litter. This is a stocky, robust species and in that sense resembles other high-elevation leaf-litter eleuths in the West Indies.

It rained most of the night and we got very little sleep. The next morning, we hiked back down to the Pico Cuba base camp and began preparing for a risky hike down the mountain at night.

Nocturnal Descent

We decided that the only way to solve the mysteries surrounding these Turquino frogs would be to hike down to La Esmajagua after dark. Our hope was that we would find the high elevation species (*E. auriculatus*–like and *E. varians*–like) and their low elevation counterparts occurring together, proving that they were different species. Although gene sequences of the frogs from the different populations would be obtained later and compared, it would still leave open the species question. We wanted to solve the riddle then and there. A nocturnal descent was best because we knew the different forms had different calls, and this allowed us to conduct an acoustic survey without having to collect the frogs.

Richard, Alberto, Emilio, and I would make the descent while the rest of the group would come down the next morning. Going down a mountain sounds easy but, in this case, it was quite risky. The trail was steep, rocky, and slippery from rain, and we were carrying loaded backpacks and had slung considerable quantities of electronic gear (tape recorder, microphones, headlamps, and cables) around our necks. Moreover, we were exhausted from the previous day and night of collecting at the peak. Large pillowcases were hanging from our belts, each filled with plastic bags containing animals from different localities. We headed out at 5:40 PM, which gave us some light to begin our descent. Still, we needed two hours to reach Pico Cardero (1285 m). We knew we were in for a long night.

A slight rise in the trail at Pico Cardero gave us the opportunity to put the heavy packs down and do some collecting. We found the Oriente Stream Eleuth (*E. cuneatus*, formerly *E. sierramaestrae*), emitting a chirpy call in secretive locations on the ground, and the Oriente Spiny Eleuth (*E. gundlachi*) and Cuban Long Legged Eleuth (*E. dimidiatus*) on the forest floor. I traced the same repetitious call that we heard up on Turquino to a bromeliad several meters above the ground, and saw the frog making the sound as he was sitting on a leaf of the bromeliad. It was the same species that we found in bromeliads at Estribo Turquino. In addition, the *E. auriculatus*–like frog was making the same call that it gave higher up on the mountain. Thus the change

FIGURE 4. Members of the joint U.S.-Cuban Turquino Expedition at Santiago, Cuba, after their return from the mountain trek; from left to right: Riberto Arencibia, Antonio Pérez-Asso, Alberto Estrada, Emilio Alfaro, Octavio Pérez-Beato, Richard Thomas, and Blair Hedges (© S. Blair Hedges).

in elevation had no effect on the calls of our two mystery frogs, which was an important clue.

At 10 PM, we started down the trail and, after only 5–10 minutes, we got exactly what we hoped for: typical *E. varians* (1–3 notes/call) were heard calling sympatrically with the 14–22 note *E. varians*–like frog. I immediately broke out the tape recorder to document these calls, which confirmed that we had a new species. The two species were sympatric from that point (1200 m) down to about 820 m. Then, the second mystery was solved: we found the zone of sympatry between lowland *E. auriculatus* and the *E. auriculatus*–like frog at about 1100–900 m, confirming the identity of the other new species. This kind of information would not have been possible had we decided to hike in daylight hours. If we had had any champagne on hand — and no more hiking to do — we would have celebrated at that point.

We arrived at La Esmajagua at about 1 AM, tired and sore and miserable from the harrowing seven-hour nocturnal descent, but pleased that our extra efforts and detective work had paid off. It was now 11 August, and we were fortunate enough to collect additional new species elsewhere in Oriente before our return to Havana on the 17th. Back at the Museum of Natural History in Havana, we were greeted with a banner reading "Victorious Return of the Baracoa-Turquino Expedition" (Figure 4). The famous Cuban naturalist Orlando Garrido had made the banner and was anxious to hear about the success of our trip. Garrido also is famous for his unique sense of humor; he wrote the banner in Italian! After meeting with Orlando and administrators of several institutions over the next few days, we finally left Cuba on 20 August.

FIGURE 5. Dark-Faced Bromeliad Eleuth (*Eleutherodactylus melacara*) (top), from Estribo Turquino (1770 m); the Turquino Fern Eleuth (*Eleutherodactylus glamyrus*), from Pico Real-Turquino (1974 m) (© S. Blair Hedges).

Epilogue

Alberto, Richard, and I published the description of the *E. varians*–like frog in 1992, along with two others collected on that trip. We named it *E. melacara*, the Dark Faced Bromeliad Eleuth (*melacara* is Latin for dark face). My collaborator in the study of frog chromosomes, Professor James Bogart of the University of Guelph, found that the new species had 24 chromosomes compared to 18 for typical *E. varians*, further evidence that we had indeed discovered a different species. The *E. auriculatus* like frog required some additional study before Alberto and I described it in 1997, calling it *E. glamyrus*, the Turquino Fern Eleuth (*glamyrus* is Greek for bleary-eyed or watery-eyed, referring to the misty cloudy forest habitat of this species). Our subsequent expeditions to Oriente revealed that *E. melacara* and *E. glamyrus* (Figure 5) are widely distributed throughout the higher elevations of the Sierra Maestra.

We did finally rediscover Darlington's frogs, which had not been seen since the 1930s: *E. cubanus* and *E. turquinensis*. They were collected on later trips to the Sierra Maestra, but never in abundance. Deforestation continues in the Sierra Maestra, largely for planting crops and removal of wood, even within so-called protected areas. Assuming that this continues unabated, the unique flora and fauna of the region will soon disappear — before we even discover what all is there.

Acknowledgments

In addition to the participants of the expedition,

mentioned in the essay, I especially thank Gilberto Silva Taboada, National Museum of Natural History of Cuba, for his organizational skills and efforts in making that trip a reality. The National Museum and the government of Santiago de Cuba Province each provided valuable logistical support and the U.S. National Science Foundation provided financial support.

Suggested Reading

Barbour, T. and B. Shreve. 1937. Novitates Cubanae. *Bull. Mus. Comp. Zool., Harvard Coll.* 80:377–387.

Estrada, A. R. and S. B. Hedges. 1997. A new species of frog from the Sierra Maestra, eastern Cuba (Leptodactylidae, *Eleutherodactylus*). *Jour. Herpetol.* 31:364–368.

Hedges, S. B., A. R. Estrada, and R. Thomas. 1992. Three new species of *Eleutherodactylus* from eastern Cuba, with notes on vocalizations of other species (Anura, Leptodactylidae). *Herpetol. Monogr.* 6:68–83.

Hedges, S. B., L. Gonzalez, and A. R. Estrada. 1995. Rediscovery of the Cuban frogs *Eleutherodactylus cubanus* and *E. turquinensis* (Anura: Leptodactylidae). *Carib. Jour. Sci.* 31:327–332.

Schwartz, A. and S. B. Hedges. 1991. An elevational transect of Lepidoptera on Pico Turquino, Cuba. *Carib. Jour. Sci.* 27:130–138.

Hedges, S. B. 2003. The frog detectives of Pico Turquino, p. 51–57. *In* R. W. Henderson and R. Powell (eds.), Islands and the Sea: Essays on Herpetological Exploration in the West Indies. Society for the Study of Amphibians and Reptiles, Ithaca (New York). Contributions to Herpetology, volume 20.

Discovery and Rediscovery

ORLANDO H. GARRIDO[1]

The lizard genus *Anolis* and the frog genus *Eleutherodactylus* are the most speciose genera of vertebrates in the West Indies — and, as the late Ernest E. Williams stated, for anoline lizards, the time of discovery has not yet ended. He most certainly was correct in his prediction, since practically any expedition to a remote, previously unexplored montane area in Cuba reveals new forms to science. In modern times, the number of species listed for Cuba by Rodolfo Ruibal in 1961 was 23 (including species assigned to the anoline genus *Chamaeleolis*, which often is subsumed within *Anolis*). Fourteen years later, in 1975, Albert Schwartz and Richard Thomas included 40 species in their catalogue. Only nine years later, Miguel L. Jaume and I listed 44. In 1991, Albert Schwartz and Robert W. Henderson cited 46. The most recent checklist, published in 1996 by Robert Powell, Robert W. Henderson, Kraig Adler, and Harold A. Dundee, in a volume dedicated to Albert Schwartz, included a total of 53 species of *Anolis*. Three more species were added in an addendum (Powell and Henderson 1999) and another four have been named since. Because one species was relegated to the synonymy of another, the current number of Cuban anoles totals 59, a phenomenal increase of over 250% in the brief span of 40 years.

The original descriptions of all named taxa provide details of morphological characters, pattern, coloration, size, etc. Many of them contain data concerning natural history, recording observations, for example, of behavior and habitat use. However, because most editors prefer descriptions that are as concise as possible, primarily to save space, many accounts of new species are devoid of many interesting details. In fact, some of the most interesting data on many newly described species are kept in drawers, not published, and eventually lost. Herein, I present such information, interesting because it sheds additional light on poorly known species, but also because it illustrates the joys and vicissitudes of the field work that have attracted so many of us to research in natural history.

The Rediscovery of the "Chipojo Prieto"

The genus *Chamaeleolis*, proposed by Duméril and Bibron in 1837, is a bizarre genus of slow-moving giant twig anoles endemic to Cuba. Prior to 1967, only one species, *C. chamaeleonides* (Figure 1), was recognized. Since then, three more species (*C. agueroi*, *C. barbatus*, and *C. guamuhaya*) have been described, and another (*C. porcus*) has been recognized as a valid taxon and variously placed in the genera *Anolis* or *Chamaeleolis*.

When Barbour and Ramsden (1919) wrote their book on Cuban herpetology, they apparently overlooked Gundlach's (1880) book, in which he recognized *C. porcus* based on an original description by Cope. Gundlach stated that he did not know this species, obviously not realizing that he had collected

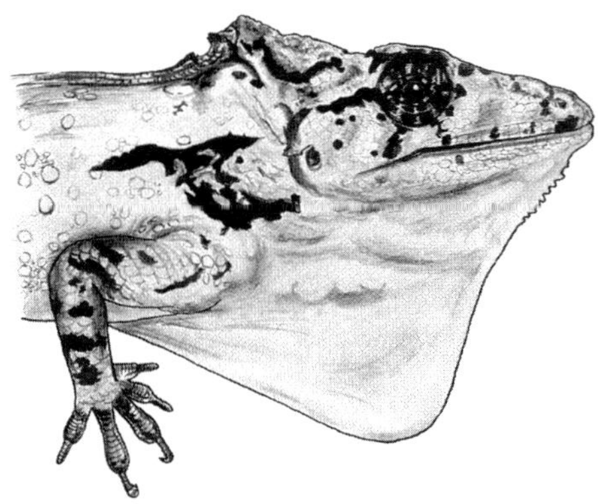

FIGURE 1. *Chamaeleolis* (often assigned to *Anolis*) *chamaeleonides* from the Sierra de Cubitas, Los Paredones, Camagüey Province, Cuba (from Schwartz and Henderson 1985).

[1] Calle 60 No. 1706, entre 17 y 19, Playa, La Habana, CP 11300, Cuba.

FIGURE 2. Oriente Bearded Anole (*Chamaeleolis* [= *Anolis*] *porcus*) from 4 km W La Maquina, Cuba (© S. Blair Hedges).

two juveniles. Apparently, because of the small size of his specimens, he had not noticed any differences in the size and shape of the gular scales. Rendering the oversight by Barbour and Ramsden even more inexplicable is the fact that specimens collected in the area of Guantánamo and Santiago de Cuba, which are attributable to *C. porcus*, had been deposited in Ramsden's collection, which at that time was in Santiago de Cuba.

When the collection was transferred to the recently created Museum of Natural History of Havana in the early 1960s, I had the opportunity to organize and examine it in its entirety. I noticed with great surprise that two species were represented, which was obvious because I was familiar with the western, broadly distributed *C. chamaeleonides*. I wrote to Albert Schwartz, detailing my findings and inviting him to collaborate on a revision of the genus. I subsequently went to Baracoa in search of additional specimens and to get personally acquainted with the eastern species. Al and I eventually (1967–1968) clarified aspects of taxonomy, synonymy, behavior, type localities, and distribution of species within the genus.

While in Baracoa, I visited Anfiloquio (Rubio) Suárez, a farmer-naturalist who knew more about animals than anyone else in the region. He even had a small zoo at his farm, raising hutías (*Capromys*) and all kinds of birds. When I asked him if he had ever seen a Chipojo Prieto, he suddenly paled, stood abruptly, and replied: "Don't even mention them, I get goose bumps just thinking about it. Long ago, I was leaning on a tree when one of those beasts landed upon my shoulder. I almost fainted, but quickly got rid of it." He went on to say that the only place in the area where he knew they were was La Florida: "I certainly won't take you there, but Angel will do it." Angel was one of his employees and turned out to be one of the greatest field guides I ever had. He could climb a tree like a monkey, discover lizards and birds anywhere, and could find hutías in the most obscure hiding places.

Together we traveled to La Florida, a mountainous area not too far from Baracoa. We were lucky enough to secure several species of birds and lizards, including five individuals of *C. porcus* (Figure 2). On the way back, we stopped in a small village where a number of curious children gathered to see our captive lizards, which we were keeping in two wire cages. While they were staring at the cages, a woman came running, abruptly pulling the kids away from the cages, and shouting: "Don't ever get close to those animals! If they bite you, they will inject venom and you won't be able to get rid of them until it thunders." After hearing this, I tried to explain to the group why this was an erroneous concept. After my little speech, the kids quickly assembled again around the cages; but immediately, the woman repeated, this time even more hysterically, her entire rhetoric.

I had no choice but to demonstrate the veracity of my claims. With Angel's help, I retrieved two individuals from their cage, forced them to open their mouths, and hung them from my ear lobes. I admit that I really had to work hard at pretending to feel no pain. In fact, what I felt was akin to two pli-ers tightening on my ears. After a minute or so, An-gel forced the animals to open their mouths and re-leased them (or me?). Then I calmly addressed the astonished audience: "As you can see, nothing happened, I have no wounds, no venom was injected, and I didn't need a clap of thunder to get rid of them." When the lesson was done, we proceeded back to Baracoa. I've not returned to that area to this day.

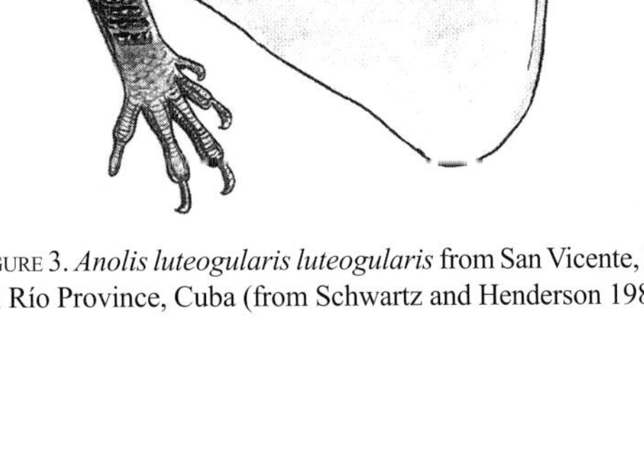

FIGURE 3. *Anolis luteogularis luteogularis* from San Vicente, Pinar del Río Province, Cuba (from Schwartz and Henderson 1985).

FIGURE 4. *Anolis equestris* from west of Gibara, Holguín Province, Cuba (photograph by Richard E. Glor).

Discovery of a "Dwarf Giant"

During the 1960s and 1970s, one of my main goals was to explore the cays of the Cuban Archipielago, and I devoted considerable time and effort making full surveys of the vertebrates found on the many previously unexplored islets. Since these expeditions involved a great deal of collecting, I usually was accompanied in the field by one or two taxidermists. Rarities were always welcomed, but the discovery of giant anoles assured the success of any expedition.

When I visited Cayo Francés (Archipiélago Sabana-Camagüey) for the first time, I was accompanied by the acarologist and taxidermist Jorge de la Cruz. Years before, during previous expeditions, we had found chipojos (giant anoles) on Cayo Conuco (north of Caibarién), Cayo Cantiles (Archipiélago de los Canarreos), and Cayo Real (Cayos de San Felipe). The latter two constituted two very distinct subspecies of *Anolis luteogularis* (Figure 3).

During our first day on Cayo Francés, about 10 AM, we found our first chipojo accidentally. We heard a loud chorus of warblers screaming. As we had heard on other occasions, we thought that a snake or a Pygmy Owl had caught one of them. When we approached, we were surprised to see that the protagonist of the alarm was a chipojo. Based on its size, we assumed it was a juvenile. Later, we found two more individuals of the same size. I said to Jorge: "Tomorrow we must find some adults, because this population looks quite interesting." The next day we devoted our efforts to finding more chipojos. Before noon, we had collected three more individuals, but all were as small as those collected the previous day. While Jorge was squeezing their vents in order to determine their sexes, he detected a swollen abdomen in one of the females. "These are surely eggs," he said to me, "let's find out for sure." He made a small cut in the abdomen and two fully developed eggs were exposed. "This animal is an adult female, not a juvenile. This is a new species, a dwarf chipojo, no doubt about it."

At first I was reluctant to accept his diagnosis. I thought that we were most likely dealing with a distinct geographic race of *Anolis equestris* (Figure 4); but, three days later, during our visit to Cayo Santamaría, we collected two distinctly different taxa of giant anoles — and all my doubts dissipated. One form was obviously allied to *A. equestris*, but smaller and blue in coloration instead of green. It later was described as *A. equestris potior*. The second form was a different population of the same dwarf we had secured on Cayo Francés. Unfortunately, that first specimen was lost with an entire shipment of reptiles that I had sent to Al Schwartz. Fortunately, on a second trip to both keys, we secured additional specimens, and I subsequently named the dwarf chipojo *Anolis pigmaequestris* (Figure 5).

The Discovery of the "Bayoya de Onaney"

One evening, Onaney Muñiz , the renowned Cuban botanist, came to my house to show me a lizard he had pickled in alcohol. After only a quick glance, I easily identified the lizard as a member of the genus *Leiocephalus*. When I pulled it out of the jar, however, to my astonishment, I could not identify the animal as a member of any of the known taxa from Cuba: "Where the hell did you get that animal? This

FIGURE 5. *Anolis pigmaequestris* (from Rodríguez Schettino 1999).

is something entirely different." Onaney explained: "On a previous visit to Loma de Macambo [some 40 km west of the bay of Guantánamo], I had discovered a new palm on the summit. I decided to go back looking for more specimens. At the very top of the hill, the soil is covered by a sort of a smooth limestone (not rough like the typical 'diente perro' found along the coasts). On that rock surface, I saw some lizards running away, and I managed to get one." That specimen, an adult female, eventually became the holotype of *Leiocephalus onaneyi* (Figure 6). Intrigued, I asked him to take me on a special collecting trip to Macambo in order to obtain a series of these lizards. He graciously agreed.

Only the two of us went to Macambo. At the base of the hill, Onaney said: "There is a shorter way to get to the summit, but I presume you will be interested in exploring the whole hill, so we'll take the long way." I should mention that Macambo is located near the southern coast between Guantánamo Bay and Punta de Maisí (the eastern tip of Cuba). The whole region is very dry, xerophytic, and full of all sorts of cacti and thorny shrubs. Like other deserts throughout much of the world, it's very hot during the day and rainfall is scanty (as a matter of fact, this is the driest area in Cuba). The La Salle brothers, León and Alain José, in "La Flora de Cuba," referred to the eastern tip of Cuba as "el infierno de Maisí" ("the inferno of Maisí"). After my visit to Macambo, I've often referred to the hill as "el penthouse de Lucifer" ("Lucifer´s penthouse").

We began our climb from a peasant hut near the road at the base of the hill. The climbing was extremely hazardous and slow, primarily because of the abundance of savage plants, each of which seemed intent on doing us harm. Along the way, I found *Leiocephalus raviceps* only on the lower slopes, *L. macropus* on most of the slopes, but *L. onaneyi* was only at the very top. Not far from the summit, I stumbled on a rock and rolled down about ten meters, slightly hurting my ankle. We decided that Onaney should keep going. I gave him a small wire cage and my gun loaded with mustard seed shot. As he ascended, he advised me to wait right in that spot.

It was very hot indeed, and neither of us had brought a canteen. We had both been bragging about how "tough" we could be when in the field. As I waited, I turned as many rocks as I could searching for tenebrionid beetles. Time passed, and after two and a half hours of waiting in the heat, I decided to descend on my own despite Onaney's advice. When I finally — and safely — reached the farmer's hut, I first drank half of all the water he had in a large container, then I sat down to wait. About half an hour later, Onaney arrived. He looked miserable, soaking wet from sweating, his clothes were torn, and his naturally tanned skin looked even darker than usual. He didn't say a word, got himself a glass, sat on a stool, and proceeded to drink the rest of the water.

It was about 2 PM and we had eaten nothing all day. After resting for another half hour, we took the jeep and headed back to civilization. Along the way, we stopped at a small roadside "diner" that had nothing left but eggs with rice. After the meal, Onaney began to feel dizzy. We managed to get to Santiago in time to take the evening flight back to Havana. Since Onaney lived only a block and a half from my house, we went directly to his place from the airport. His mother was waiting, so I said: "I hope you're feeling better when I see you tomor-

FIGURE 6. *Leiocephalus onaneyi* (from Rodríguez Schettino 1999).

row." The next morning, I went by his house only to find that his mother was very frightened. She told me that, during the night, Onaney was feeling very bad, shivering, and running a high temperature. She had taken him to the hospital, where he stayed for three days being treated for dehydration and sunstroke. Fortunately, he recovered without any further complications. On that terrible morning, however, Onaney, despite fatigue and fainting, had managed to collect the three individuals that enabled me to describe the species.

Months later, I returned to Macambo, this time with Jorge de la Cruz. Our intent was to secure some additional material. This time, we planned to take the shortcut to the top. After climbing not even 30 m, Jorge was exhausted, I had been stung by eight wasps, and our legs were scratched and cut and full of cactus spines. Recalling my previous trip, we wisely gave up, sadly setting aside our goal, and returned home safely — but with empty hands.

The Rediscovery of *Anolis spectrum*

Perfecto Hernández was a peasant and former speleological guide to caves in Pinar del Río. Some years ago, he invited my wife and me, my father-in-law, and my three-year-old son, to spend several vacation days with him and his wife at his home in Valle de Pica Pica, Pío Domingo, Cordillera de los Órganos. Since I had never done any collecting in that area, I thought this would be a good opportunity to explore some of the nearby mogotes (hummocks). We stayed about six days, and I devoted most of my time to collecting insects.

One evening at about sunset, I got a sweep-net and went to collect insects at the foot of a nearby mogote. In one of my sweeps, I noticed that I had inadvertently caught a small, thin lizard. Thinking it was *Anolis alutaceus*, I considered turning it loose but, since I had never collected any lizards in the area, I kept it alive in a plastic bag.

That night after dinner, I began to sort through the material I had collected, and I took a closer look at that lizard. I realized then that it was not all that small and that it had a profile much like that of a crocodile. When I examined it even more closely with my magnifying lens, I noticed with great surprise at least three characters that immediately distinguished this lizard from *A. alutaceus*. It had brown eyes instead of blue, the dorsal scales were much larger and keeled, as were the ventrals, and then there was that diminutive crocodile profile.

FIGURE 7. *Anolis spectrum* (from Rodríguez Schettino 1999).

Since I had never seen a lizard quite like that, I thought that it could be *A. spectrum* (Figure 7), but that species, discovered by Gundlach a century ago, had only been reported by Rodolfo Ruibal from Sierra de Trinidad in south-central Cuba.

The next morning I was eager to secure a good series of this strange lizard, so I returned to the same spot where I had caught the lizard. I was there around ten o'clock, a perfect time to collect lizards because the sun is high in the sky. I thought to myself that this was certainly a great day for collecting. Never before or after have I found so many species of lizards in less than two hours. By noon, I had captured *Chamaeleolis chamaeleonides*, *Anolis vermiculatus*, *A. luteogularis*, *A. bartschi*, *A. porcatus*, *A. angusticeps*, *A. alutaceus*, *A. ophiolepis*, *A. sagrei*, *A. homolechis*, *A. allogus*, and *A. mestrei*, but not a single individual of the presumptive *A. spectrum*; worse than that, I hadn't even seen one.

I went back for lunch, prepared the specimens I had collected, and then returned around 3 o'clock to my collecting area. This was still a very good time for collecting, the sun was still high. My focus this time was entirely on finding *A. spectrum*. I stayed for about three hours until the sun set. I saw several more lizards of the species I had collected that morning, but did not see a single individual of the desired species. I was certainly a bit disappointed, but I was convinced that that species could not be that rare, and was optimistic of finding it the next day.

The following morning, I again surveyed the area at prime time, and came away disappointed once again. By this time I was determined to find the elusive species, so I decided to search much later in the evening than I had the day before. I saw nothing in the first half hour, but around seven o'clock, when the area was heavily shaded, I glimpsed a jumping lizard about five meters ahead of me down the path. Unfortunately, it hid so well among the twigs and dense vines that I couldn't find it despite spending some considerable time searching. I kept walking slowly down the bushy path at the base of the mogote. Two more individuals jumped away from me, but I again failed to find them. It was quite dark and I decided to suspend my search for that evening. Since I had at least seen some animals, I had a feeling that I would succeed next day. Certainly the lizard was not that rare, its activity period appeared to be quite different than those of most anoles, and I had stumbled upon a strategy I had every reason to think would work: look several meters ahead waiting for a lizard to jump, then follow it carefully to the exact landing site.

The next day, I spent the morning and part of the afternoon climbing mogotes, and then headed to the previous day's spot around six o'clock. I walked very slowly, anticipating movement four or five meters ahead. When the first lizard jumped into the brush, I kept my eyes directed at the spot among the twigs and vines where I had seen it land. I approached slowly, kneeled down, and began to search, not moving a muscle, only my eyes. Soon, I spotted a twig-like brownish lizard sitting quietly and staring back at me. Attempting to catch it barehanded among the twigs, vines, and leaves undoubtedly would have been futile. I remember thinking: "I don't care if I squash or damage you, the important thing is to get you." I quickly slapped my hand to the ground, pressing hard, and crushing bunches of twigs and leaves — but I knew I had him. Then I very carefully began to pull away the trapped debris, stick by stick, twig by twig, and leaf by leaf, until I uncovered the lizard.

Having succeeded once, I employed the same method again and again until I had secured an adequate series. I felt very proud indeed; I had wasted lots of time, but I had been patient — and I was successful in the end.

Back in Havana, I immediately compared my newly collected animals with the specimen of *A. spectrum* in Gundlach's old collection. No doubt about it, the lizard was exactly the same as those I had collected. Content, I did not think about this species again for some time.

About a year later, on a collecting trip to the Sierra de Trinidad with the top two Cuban entomologists, Fernando de Zayas and Pastor Alayo, we were working at the "Cafetal de Gaviñas," a site not very distant from Topes de Collantes, where Ruibal had previously reported *A. spectrum*. While trying to catch an *A. isolepis*, I saw some jumping lizards and immediately recalled my previous experiences with *A. spectrum* at Pica Pica. This time, the lizards were much easier to collect; in fact, most tried to escape by running, instead of jumping and remaining motionless, rendering them easier to follow and to grab.

Days later, back in the office, while sorting and cataloguing the newly collected material, I noticed that the lizards from the "Cafetal de Gaviñas" showed a marked sexual dimorphism. Males were indeed quite similar to *A. spectrum*, but the females were more like *A. cyanopleurus*. After examining them properly, I reached the conclusion that those Trinidad lizards identified as *A. spectrum* by Ruibal, were not the same as those I had collected in Pinar del Río or the same as Gundlach's specimen.

I wrote Albert Schwartz and encouraged him to undertake a review of the species, especially because he could readily acquire and examine the very same specimens that Ruibal and Williams had studied at the Museum of Comparative Zoology at Harvard. We eventually collaborated on a revision of the species, confirmed the identity of *A. spectrum*, proposed a new type locality, and described as a new species the populations from the Sierra de Guamuhaya (Trinidad Mountains) as *A. vanidicus*.

The Discovery of *Anolis fugitivus*

While working in the lab studying the behavior of some anoles in captivity, Jorge de la Cruz brought me two female lizards obtained in eastern Cuba. They were collected during an entomological expedition to Monte Iberia by Jorge, Luis de Armas, J. Hostounsky, and J. Krecek. I identified those lizards as grass anoles in the *Anolis cyanopleurus* group. Since the new specimens were easily distinguished from the two known species in that group (*A. cyanopleurus* and *A. cupeyalensis*), they inspired me to organize a special expedition to the eastern tip of the island in order to collect additional specimens, and to investigate these lizards' natural history.

In June 1972, the late entomologist Fernando de Zayas, Luis Roberto Hernández, another entomologist, Fernando Ruiz, a specialist in frogs, and I, all members of the Instituto de Biología, went on a collecting trip to eastern Cuba, from the environs of Monte Iberia near La Melba to the very tip of the island, Punta de Maisí. The trip was indeed very successful; several taxa new to science were discovered, both insects and reptiles — and we collected a series of the new lizard that I eventually named in 1975.

After reaching Nibujón along the northern coast, we went south, establishing a base camp in an abandoned hut at Nuevo Mundo, near a logging camp (Aserradero). The workers gathered at noon for lunch, and we were invited to join them. Zayas described the place we stayed as the prettiest landscape he had ever seen in Cuba. Flitting Collared and Black swifts filled the sky, and were the last creatures we saw before preparing dinner and going to sleep.

Early the next morning I was birding, secure in the knowledge that lizards become active much later in the day. As the morning progressed, I worked toward the spot where Jorge had advised me to look for lizards. It was a wonderful day and the sun was shining brightly. Although I was intently looking for any movement, nothing seemed disturbed by my presence, and everything was depressingly calm. For more than an hour I searched in vain for grass anoles. I found only *A. alutaceus* among some ferns and *A. allogus* in the woods. I had decided to sit for a moment on top of a mound and carefully scan the area when, about two meters away, I detected some movements on a woven mat of saw grass. In this area, saw grasses have three-edged leaves instead of the two found on similar species from other parts of the island. However, like in all saw grasses, the edges of the leaves cut like a small but very sharp blade if you happen to encounter one too closely. These grasses tend to form dense mats, often extending into the lower branches of surrounding trees. As I cautiously crept closer to the mat, I was rewarded with the silhouette of a little lizard I had not previously encountered in the field.

I said to myself: "This is easier than I thought" — but I soon discovered just how wrong I was. I tried to catch the first lizard in the same manner I usually used, but I had not even touched the animal before it disappeared within the tangled saw grass. I never saw it again. I kept looking, kept seeing individuals, but each time they would escape in the same fashion. I was becoming increasingly frustrated.

Fortunately, these lizards were rather common, although they continued to defy my best efforts to capture one. Finally, I decided that the only way to catch one was going to be the hard way. When I next spotted a lizard, I quickly used both hands to flatten and crush the entire mat of saw grass on which the lizard was perched. It worked, but at a terrible price. My hands were quickly shredded, and

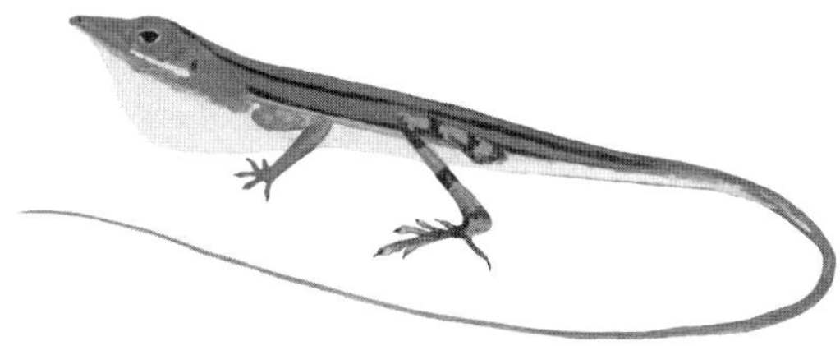

FIGURE 8. *Anolis fugitivus* (from Rodríguez Schettino 1999).

my arms, face, and legs (I usually wear only shorts in the field) suffered as well on those occasions when I practically had to "dive" among the vines. Catching these lizards was a hot, hard, generally miserable, and certainly painful task — but rewarding. By noon, I had secured an adequate series of 16 specimens of both sexes, enough to describe the new species, *A. fugitivus* (Figure 8).

When I joined Jorge for lunch, he looked astonished, staring at me as if I were a ghost. "What the hell happened to you?" Then he burst out laughing. My face, hands, and legs were literally covered with small pink stripes from the many little cuts I had endured. Fortunately, the cook had saved our food because the lunch hour had long since passed. As we were eating, the cook quietly called Jorge aside and asked him very politely what sort of sickness I had, since he had not observed those symptoms before. After Jorge explained the situation, the cook approached me, staring with wide eyes, and said: "Unbelievable. You must be out of your mind!"

Acknowledgments

I thank the directors of the Institute of Biology of the Academy of Natural Sciences, La Habana, for providing permits and equipment for fieldwork, the directors of the Cooperativa Pesquera de Caibarién for providing us with boats, crew, food, and lodging for visiting keys of the northern Archipiélago Sabana-Camagüey, Rubio Suárez for his kind hospitality while in Baracoa, and the botanist Onaney Muñiz for organizing the expedition to Macambo.

Suggested Readings

Barbour, T. and C. T. Ramsden. 1919. The herpetology of Cuba. *Mem. Mus. Comp. Zoöl., Harvard Coll.* 47:71–213.

Díaz, L. M., N. Navarro, and O. H. Garrido. 1998. Nueva especie de *Chamaeleolis* (Sauria: Iguanidae) de la Meseta de Cabo Cruz, Granma, Cuba. *Avicennia* 8/9:27–34.

Garrido, O. H. 1973. Nueva especie de *Leiocephalus* (Lacertilia: Iguanidae) para Cuba. *Poeyana* 116:1–19.

Garrido, O. H. 1975. Distribución y variación del complejo *Anolis cyanopleurus* (Lacertilia: Iguanidae) en Cuba. *Poeyana* 143:1–60.

Garrido, O. H. 1975. Nuevos reptiles del archipiélago cubano. *Poeyana* 141:1–58.

Garrido, O. H. 1982. Descripción de una nueva especie cubana de *Chamaeleolis* (Lacertilia: Iguanidae), con notas sobre su comportamiento. *Poeyana* 236:1–25.

Garrido, O. H. and M. L. Jaume. 1984. Catálogo descriptivo de los anfibios y reptiles de Cuba. *Doñana, Acta Vertebrata* 11:5–128.

Garrido, O. H., O. Pérez Beato, and L. V. Moreno. 1991. Nueva especie de *Chamaeleolis* (Lacertilia: Iguanidae) para Cuba. *Carib. Jour. Sci.* 27:162–168.

Garrido, O. H. and A. Schwartz. 1968. Cuban lizards of the genus *Chamaeleolis*. *Quart. Jour. Florida Acad. Sci.* 30:197–220.

Gundlach, J. 1880. *Contribución a la Herpetología Cubana*. G. Montiel, La Habana.

Peters, G. 1970. Zur Taxonomie und Zoogeographie der kubanischen anolinen Eidechsen (Reptilia, Iguanidae). *Mitt. Zool. Mus. Berlin* 46:197–234.

Powell, R., R. W. Henderson, K. Adler, and H. A. Dundee. 1996. An annotated checklist of West Indian amphibians and reptiles, p. 51–93 + 8 pl. *In* R. Powell and R. W. Henderson (eds.), *Contributions to West Indian Herpetology: A Tribute to Albert Schwartz*. Society for the Study of Amphibians and Reptiles, Contributions to Herpetology, vol. 12. Ithaca, New York.

Ruibal, R. 1964. An annotated checklist and key to the anoline lizards of Cuba. *Bull. Mus. Comp. Zool., Harvard Coll.* 130:475–520.

Schwartz, A. and R. W. Henderson. 1985. *A Guide to the Identification of Amphibians and Reptiles of the West Indies Exclusive of Hispaniola*. Milwaukee Public Museum, Milwaukee, Wisconsin.

Schwartz, A. and R. W. Henderson. 1991. *Amphibians and Reptiles of the West Indies: Descriptions, Distributions, and Natural History*. Univ. Florida Press, Gainesville.

Schwartz, A. and R. Thomas. 1975. A check-list of West Indian amphibians and reptiles. *Carnegie Mus. Nat. Hist. Spec. Publ.* 1:1–216.

Garrido, O. H. 2003. Discovery and Rediscovery, p. 59–66. *In* R. W. Henderson and R. Powell (eds.), Islands and the Sea: Essays on Herpetological Exploration in the West Indies. Society for the Study of Amphibians and Reptiles, Ithaca (New York). Contributions to Herpetology, volume 20.

Conserving the Remarkable Reptiles of Guantanamo Bay

ALLISON C. ALBERTS[1]

When looking for a suitable place to study some of the world's rarest reptiles, a relatively isolated military base might at first seem like a surprising choice. However, the United States Naval Base at Guantanamo Bay, Cuba, like many military installations worldwide, supports large areas of undisturbed natural habitat and is home to an impressive abundance of wildlife. Guantanamo Bay lies in Oriente Province in the southeastern corner of Cuba. It is the oldest overseas United States military base, originally acquired as a coaling station in 1903, but reaffirmed by treaty in 1934 after the American fleet aided Cuba during the Spanish-American war. Because only 11% of the 52-km^2 base is actively used by the military, large areas of near-pristine cactus and thorn scrub, buttonwood and *Phyllostylon* forest, and mangrove tidal thickets remain.

Guantanamo lies in the rain shadow of the looming Sierra Maestra. Consequently, this starkly beautiful region is very dry, receiving only 40–75 cm of rain each year. With air temperatures averaging 28–30°C throughout the year, it's no wonder that as many as 30 terrestrial species of amphibians and reptiles, 21 of which live nowhere outside of Cuba, make Guantanamo their home (Figure 1). Among these, the Cuban Iguana (*Cyclura nubila*) is one of the largest, undoubtedly the most visible, and certainly the most charismatic. No one completes a tour of duty at "Gitmo" without getting to know these inquisitive, prehistoric-looking giants.

FIGURE 1. Typical undisturbed coastal habitat at the United States Naval Base, Guantanamo Bay, Cuba.

Giant Iguanas

Remarkably, these robust lizards (Figure 2) have somehow managed to survive millennia of hurricanes and hunting by native peoples. Unfortunately, outside the fenceline, iguana populations are now declining because their habitat is being eliminated by development or degraded by invasive species. A host of exotics, including mongooses, dogs, feral cats, and black rats, prey heavily on juvenile iguanas, and, in many areas, domestic ungulates are consuming the native vegetation on which iguanas critically depend. Since the early 1990s, my colleagues and I at the San Diego Zoo have had the privilege

[1] Division of Applied Conservation, Zoological Society of San Diego, San Diego, California 92112, USA.

FIGURE 2. Adult male Cuban Iguana, *Cyclura nubila,* feeding on Touch-Me-Not, *Malpighia* spp. at Guantanamo Bay (© John Phillips, Zoological Society of San Diego).

of visiting Guantanamo Bay many times, working to design conservation strategies for Cuban Iguanas and other threatened reptiles.

Getting to Guantanamo is no easy undertaking, but thanks to the United States Navy, we're able to fly in style aboard the base commander's C-12 aircraft, with a pleasant layover in Jamaica that allows us to spend time with friends and colleagues at the tiny yet fascinating Hope Zoo in Kingston. Since the rediscovery of the Jamaican Iguana (thought to be extinct since 1940) in a rugged limestone-floored forest outside Kingston, aptly named the Hellshire Hills, the Hope Zoo has been working to establish a headstarting program for young iguanas. A dedicated facility was built on zoo grounds, which currently houses 100 baby iguanas — all being raised in a safe environment until they are old enough to survive on their own in the wild. To date, 26 iguanas sporting radiotransmitters have been returned to Hellshire, all have established stable home ranges, and at least some released females appear to be nesting successfully. Education is a major part of the mission of the Hope Zoo, and they have recently completed a new learning center on zoo grounds that attracts thousands of school children each year. The brightly painted sign at the zoo's iguana exhibit sends a simple message that says it all: "Nuff Respect Due."

After a short plane flight out of Kingston, Cuba's rugged coastline comes into view, and one can begin to appreciate the strategic and economic importance of the many deep and sheltered harbors afforded by Guantanamo Bay. With the hazy mountains and bustling Guantanamo City in the distance, we board a military ferry to cross the bay, a journey which begins in calm waters favored by West Indian Manatees, but becomes increasingly choppy as the crossing progresses. After about 45 minutes, we arrive on the windward side of the naval base, piled high with gear, and ready to begin our research.

Lizards and Howitzers

We initially began our field studies of Cuban Iguanas in 1993 along a windswept section of rocky coast called Firing Point (Figure 3), with the hope of gaining an understanding of the basic biology of these lizards, as well as developing practical strategies for their recovery. We quickly came to appreciate the origin of the site's name when we first experienced the thunderous testing of nearby howitzers, an event to which the local iguanas appear amazingly and blithely oblivious. Although frustrating trial and error characterized our early attempts to capture these often feisty 1.5-m-long lizards, the process was ultimately made easier by our fortuitous discovery that red grapes, like red flowers, were a favored food item and could easily be used to lure unsuspecting iguanas into hand nets. Over the course of a year, we captured each adult iguana at Firing Point once a month, and recorded its body length, weight, head size, and the diameter of the femoral scent glands lining the thighs, all of which seem to be important for successful defense

FIGURE 3. Study site at Firing Point, Guantanamo Bay, prior to 1995. About 60 adult iguanas inhabited the area shown here.

of territories. After placing the animals in canvas bags to quiet them, we also collected a blood sample from the tail for hormone analysis.

For the iguanas, the daily routine varies little, consisting of a vigorous burst of social activity after morning emergence, followed by relatively peaceful migration to nearby areas of vegetation for midday feeding, and ending with a second bout of social interaction late in the afternoon prior to entering nighttime refuges. Most adult males regularly engage in aggressive interactions with other males. These usually consist of a lumbering chase, but occasionally prolonged pushing, hissing, and biting matches ensue. Almost without exception, the most dominant males have the highest testosterone levels, are larger in body and head size, and possess more active scent glands than lower ranking males. These despots defend small but superior territories that each overlap the ranges of several females. Lower ranking males do not defend territories, instead moving between the territories of dominant males, from which they are continually evicted. Headbobbing, chases, and mouth gaping, the typical weaponry of iguana territoriality, are used almost exclusively by dominant males. These males also spend more time close to females during the breeding season, courting them with gentle nudges and headnods. Although dominant males appear to have better access to mates than subordinate males, they comprised no more than a third of all adult males at Firing Point.

The following year, we conducted an experiment to determine if temporary relocation of dominant males would increase the chances that sexually mature but genetically underrepresented subordinate male iguanas would have the opportunity to mate. This strategy represented a unique approach to lizard conservation, and has the potential to serve as an important management tool for critically small populations. For the duration of the breeding season, we moved the five most dominant males from Firing Point to a large outdoor holding enclosure at the base Army Veterinary Clinic. Within a few days, the five largest previously subordinate males began defending territories that were nearly spatially identical to those vacated by the relocated individuals. In the absence of the dominant males, subordinate males courted females vigorously and their testosterone levels rose dramatically. At the close of the breeding season, the previously dominant males regained their previous territories within two days, although the battles required were among the longest and most intense that we had ever observed. Temporary removal of dominant males may represent a valuable management tool for small or otherwise genetically compromised populations by potentially increasing the number of males contributing to the gene pool.

The World Intrudes

When you work at the same site year after year, you begin to take for granted that it will never change. Such had been the case for Firing Point and its enviably observable iguanas until the fateful Spring of 1995, when over 50,000 Cuban and Haitian migrants arrived unexpectedly at Guantanamo Bay, aboard literally anything that could float. In response to the crisis, extensive areas of forested habitat were graded along the coastline in order to construct temporary tented housing. Although military officials did an admirable job of keeping people and wildlife at a safe distance, Operation Sea Signal nevertheless resulted in a noticeable reduction in the local iguana population. At the height of the refugee crisis, the dry tropical forest surrounding Firing Point was reduced to less than 5% of its former extent, and we could only find a single forlorn adult male iguana inhabiting the area.

Our deep dismay at seeing the destruction of Firing Point's ancient cactus forest was slightly tempered by the realization that we now had the unique opportunity to study if and how rapidly iguanas would recolonize the area. Our aim was to document the extent to which this dry tropical forest ecosystem was sufficiently resilient to recover from severe disturbance, and to study the ecological role that iguanas might play in this process. By mid-1999, we were heartened to find that over 25 adult iguanas had resettled the area, and all appeared to be growing well and in good health. For both males and females, the relationship between body mass and body length had improved to the point where it was comparable to that of healthy undisturbed iguanas, indicating that despite the loss of significant habitat, iguanas were able to forage effectively following disturbance. At the same time, vegetation at the site showed definite signs of regeneration and a number of new seedlings emerged.

Given that iguanas relish the fruits and flowers of many plants, we believed that they might play an important role in the recovery of native vegetation. Past research by us and others has shown that seeds that have passed through the digestive tracts of iguanas tend to sprout sooner and produce seedlings that grow faster than seeds that have not. In addition, repetitive cropping by iguanas, evidenced by an abundance of oddly-shaped flat-topped bushes along Guantanamo's limestone coastline, is likely to stimulate the development of new foliage. By aiding germination, providing nutrients to developing seedlings, promoting plant growth, and dispersing seeds into new areas, iguanas play a key role in keeping forest ecosystems healthy. As so often happens with scientific endeavors, one study leads to a host of new avenues for research. Seeing the recovery process first hand at Firing Point, a site that was so intimately familiar to us, led to a new line of inquiry into the relationship between local iguana populations throughout the base and the vegetation on which they depend. As unglamorous as it may seem, surveying different habitats to determine the density of local iguana populations is an important first step in assessing their ecological relationship with local plant communities.

After identifying eight key habitat types on the base, we returned to Guantanamo in 2000 to carry out a series of morning and afternoon walking surveys using a technique called distance sampling. This involved a team of three slogging through varying densities of thorn scrub (one must stay on an absolute straight-line course to meet the strict assumptions of this survey methodology — meaning that, as much as we wanted to, we could never cheat and go around those nasty-looking cactus patches), for a half-mile or so per transect. Whenever an iguana was sighted, we paused our sweaty march to cheer, note the size and sex of the animal, and measure its perpendicular distance from the transect line. We used a Global Positioning System (GPS), which communicates with a series of navigational satellites orbiting the earth, to record the latitude and longitude of the start and finish of each transect line. The morning breeze, which blessedly and predictably picked up around 1100 each day, was a welcome break from the stifling heat. Additional delights included the occasional delicate pink orchid suspended in midair amid impenetrable thorn scrub, or the bright green, yellow, and azure hues of a sleeping giant anole, or chipojo, in a *Capparis* tree.

Despite hazardously steep terrain, we successfully completed 24 transect lines, and determined that iguanas are approximately four times as abundant along the coastline as inland. Most likely, this is due to the many holes and crevices in the crumbling limestone that characterize the shoreline. These provide a host of hiding places and refuges from the heat, particularly for young iguanas. Not surprisingly, iguanas are three to ten times as abundant in natural areas as in those that have been impacted by people. Perhaps because suitable refuges are fewer and farther between, habitat disturbance in the form of construction and replacement of native vegetation by introduced grasses appears to have twice the negative impact on inland iguana populations as on those along the coast. During one of our transects, we did come across an unusual and unexpected sight — an adult male iguana with the telltale remains of a diminutive yellow juvenile Cuban Grassquit in his mouth, testament to the fact that iguanas are not always vegetarian in their food choices.

Snakes, Turtles, and Banana Rats

With financial support from the United States Department of Defense, we have recently been able to expand our work to include population surveys not only for iguanas, but also for two other ecologically important reptiles on the base, boas and sea turtles. Taking on this additional task meant no rest for the weary — to make the most of our limited time on the base, each full day of iguana surveys was followed by an evening searching for Cuban Boas and monitoring sea turtle nesting beaches.

The most efficient way to search for Cuban Boas (*Epicrates angulifer*) — a truly impressive snake, with rainbow iridescence on its scales and a body length of up to nearly four meters — is to slowly drive roads just after sundown. Nevertheless, one of our most intriguing boa encounters occurred when we were on foot. Researcher Glenn Gerber's practiced eye spotted a Lignumvitae "Tree of Life" which had been hollowed out by lightening — the perfect hiding place for a young male boa. We marked the snake with a microchip just under the skin for permanent identification, and collected a small blood sample for genetic analysis. On future trips, our goal is to use surgically implanted radiotransmitters to track the movements of these elusive and little-known snakes in order to better understand their home range dynamics and habitat requirements. We are especially interested in documenting the effects of an ongoing relocation program, in which boas are removed from urban areas by base personnel (Figure 4) and released into more natural surroundings.

In recent years, populations of large rodents appear to be burgeoning on the base. These animals are known to science as hutias, but to most base residents as banana rats, owing to the distinctive curvature of their scat. Although one can't help but smile at the ubiquitous hutias, which traipse around the base in ungainly family groups that communicate constantly with chirps and whistles, their exploding population is causing significant problems by destroying vegetation, chewing through radiator hoses, and perpetrating other undesirable antics. Cuban Boas probably play a key role in naturally regulating hutia populations, so we are keen to further investigate the ecological relationship between these two species.

After about 2100 hours, when nightly boa catching became significantly less likely, we headed out to investigate a series of beaches on the base that are potentially used by sea turtles for nesting. Because they are relatively undisturbed and protected from hunting, Guantanamo's scenic beaches are of regional importance for sea turtle conservation. Fortunately, it's fairly easy to distinguish which species of turtles are using these beaches simply by examining their tracks. The nesting crawl of all sea turtles generally resembles the path of a miniature bulldozer, with Hawksbill tracks showing an alternating flipper pattern and Green Turtles a more symmetrical, opposite pattern. Because Leatherbacks are one of the largest living reptiles, surpassed in size only by some crocodilians, their tracks are impossible to miss.

FIGURE 4. United States Marine with Cuban Boa, *Epicrates angulifer*, at Cuzco Beach nature area, Guantanamo Bay.

To date, we have mapped ten beaches on the base using GPS and surveyed them for evidence of sea turtle nesting activity. Eight of the ten beaches are used by Hawksbill Turtles, one by Green Turtles, and one by Leatherback Turtles, information that will help guide the Navy in successfully managing these beaches for both people and turtles. Two of the beaches have evidence of more than 100 nesting pits each, and appear to be especially important for Hawksbills. Given that Hawksbill populations are under intense hunting pressure throughout much of the Caribbean, it is crucial that these beaches be carefully managed to minimize human impacts. Together with natural resources personnel on the base, we are committed to evaluating alternative management strategies and raising public awareness about sea turtles and their nesting habits.

Without a doubt, the highlight of our sea turtle work at Guantanamo Bay was a particularly warm and humid night at Pebble Beach in June. This is a small, rocky beach, but one favored by turtles nonetheless. We hadn't dared hope that we would actually encounter a living turtle, but to our delight, as we approached the beach we heard the rasping sound of a female Hawksbill's plastron scraping over the rocks as she lurched onto land. With a special flashlight that emitted only a dim red beam, we were able to scrunch forward on our bellies close enough to observe the nesting process. She selected a site under an aged Sea Grape tree that, judging from soil erosion that revealed old eggshells encrusted many layers deep, has probably been used for this purpose for decades. Oblivious to the eggs of previous nesters, which she dug up and flipped out of the nest cavity with impunity, this particular female took about an hour to dig her nest. With remarkable dexterity, she used her rear flippers to reach down, scoop out flipper-fulls of sand, and carefully deposit them nearby. When the nest cavity was deep enough to suit her, she proceeded to lay over a hundred glistening spherical eggs, and then gently back-filled the nest cavity. Her task com-

FIGURE 5. Iguana crossing signs installed at key locations throughout the naval base have helped reduce road casualties.

plete, she never paused to rest, instead making a direct line to the water and disappearing into the moonlit waves.

Reptile Outreach

Public education is key to the success of any conservation program. Over the years, we have attempted to educate both military servicemen and civilians through open lectures at our field sites, classroom presentations to elementary and secondary students, endangered species pamphlets, and newspaper and radio interviews. In response to a rash of unfortunate incidents in the mid-1990s, we designed a set of "Iguana Crossing" signs (Figure 5) for high traffic areas where collisions had occurred. We have been thrilled to see that over 20 of these signs have been erected at critical locations, and to learn from our local collaborators that iguana road casualties are now a rare occurrence. In fact, the only "problem" with the signs has been their unanticipated popularity as impromptu souvenirs — an issue subsequently addressed by offering scaled-down versions for sale in a local giftshop.

With the help of the base TV station, which advertises regularly for us on the event scroller, we have had excellent turnouts for our semi-annual iguana demonstrations (Figure 6). These presentations are held in the evenings at Iguana House, a relatively remote building near the coast that always seems to have an iguana or two hanging around. Usually about 80 or so people attend, some of them getting their first-ever close up view of an iguana, with a chance to touch one if they wish. While we try to spread the message that iguanas should be neither harassed nor fed (an ever-present problem), we also spend a significant amount of time answering questions. These range from why iguanas bob their heads to how the large population of feral cats on the base is impacting them. Interest and curiosity among base residents about iguanas and other native reptiles continues to be intense, and past interviews and surveys have shown that most people's feelings about them are positive, boding well for the continuing successful co-existence of humans and wildlife on the base.

Public interest in Guantanamo's extraordinary wildlife continues to be high, as evidenced by the more than 75 dedicated volunteers who have helped us with our work over the years. Part of the message we hope to send is that protecting wildlife need not conflict with the military mission. In fact, healthy natural systems are often those best suited to supporting the security and training functions central to military operations. For such a small piece of land, Guantanamo Bay has experienced an unusually tumultuous history. Despite its drastic transformation from the early days of open commerce to its more recent role as a safe haven for Cuban and Haitian refugees, the extraordinary biodiversity of Guantanamo has managed to survive intact. Whatever the future may hold, we can only hope

FIGURE 6. Annual iguana demonstrations are held for military and civilian personnel and their families.

that with awareness and dedicated stewardship, the distinctive reptiles, other native wildlife, and the natural habitats of Guantanamo Bay will continue to flourish.

Suggested Reading

Alberts, A. C. (ed.). 2000. *West Indian Iguanas: Status Survey and Conservation Action Plan*. IUCN—the World Conservation Union, Gland, Switzerland.

Alberts, A. C., T. D. Grant, G. P. Gerber, K. E. Comer, P. J. Tolson, J. M. Lemm, and D. Boyer. 2001. Critical reptile species management on the U.S. Naval Base, Guantanamo Bay, Cuba. Report to the United States Navy for Project No. 62470–00–M–5219. 101 p.

Alberts, A. C., J. M. Lemm, A. M. Perry, L. A. Morici, and J. A. Phillips. 2002. Temporary alteration of local social structure in a threatened population of Cuban Iguanas (*Cyclura nubila*). *Behav. Ecol. Sociobiol.* 51:324–335.

Berovides, V. A. 1980. Notas sobre la ecologia de la iguana (*Cyclura nubila)* en Cayo Rosario. *Cien. Biol.* 5:112–115.

Bjorndal, K. A. (ed.). 1995. *Biology and Conservation of Sea Turtles*. Smithsonian Institution Press, Washington, D.C.

Eckert, K. L., K. A. Bjorndal, F. Alberto Abreu-Grobois, and M. Donnelly. 1999. *Research and Management Techniques for the Conservation of Sea Turtles*. IUCN/SSC Marine Turtle Specialist Group Publication No. 4. Gland, Switzerland.

Estrada, A. R. and R. Ruibal. 1999. A review of Cuban herpetology, p. 31–62. *In* B. I. Crother (ed.), *Caribbean Amphibians and Reptiles*. Academic Press, San Diego, California.

Hartley, L. M., R. E. Glor, A. L. Sproston, R. Powell, and J. S. Parmerlee, Jr. 2000. Germination rates of seeds consumed by two species of Rock Iguanas (*Cyclura* spp.) in the Dominican Republic. *Carib. Jour. Sci.* 36:149–151.

Lando, R. V. and E. E. Williams. 1969. Notes on the herpetology of the U.S. Naval Base at Guantanamo Bay, Cuba. *Stud. Fauna Curaçao Other Carib. Isl.* 31:159–201.

Lemm, J. M. and A. C. Alberts. 2000. Reptiles and amphibians of Guantánamo Bay. *Reptiles* 8:10–25.

Perera, A. 1985a. Datos sobre la dieta de *Cyclura nubila* (Sauria: Iguanidae) en los alrededores de Cayo Largo del Sur, Cuba. *Poeyana* 291:1–12.

Perera, A. 1985b. Datos sobre abundancia y actividad de *Cyclura nubila* (Sauria: Iguanidae) en los alrededores de Cayo Largo del Sur, Cuba. *Poeyana* 288:1–17.

Rodríguez Schettino, L. (ed.). 1999. *The Iguanid Lizards of Cuba*. Univ. Florida Press, Gainesville, Florida.

Sedaghatkish, G. and E. Roca. 1999. *Rapid Ecological Assessment: U.S. Naval Station, Guantanamo Bay, Cuba*. The Nature Conservancy, Washington, D.C.

Silva Lee, A. 1996. *Cuba Natural*. Pangaea, Saint Paul, Minnesota.

Tolson, P. J. and R. W. Henderson. 1993. The *Natural History of West Indian Boas*. R & A Publishing Ltd., Taunton, Somerset, England.

Alberts, A. C. 2003. Conserving the remarkable reptiles of Guantanamo Bay, p. 67–73. *In* R. W. Henderson and R. Powell (eds.), Islands and the Sea: Essays on Herpetological Exploration in the West Indies. Society for the Study of Amphibians and Reptiles, Ithaca (New York). Contributions to Herpetology, volume 20.

Of Woodpeckers and Frogs

Alberto R. Estrada[1]

When I started to work professionally on the herpetology of the Cuban Archipelago in 1982, initiating a major research project in the field of zoology was a real challenge for any Cuban scientific agency or institution. The big state budgets of the early socialist years were no longer available. My initiation as a researcher in the now extinct Institute of Zoology brought on a long period of financial poverty, for me and for the entire scientific infrastructure, that had developed from 1959 to 1980.

My aspirations to field biology were inspired by the scientific articles of Stanley Rand on the ecology of anoline lizards, by the many contributions of Albert Schwartz on nearly all aspects of Antillean herpetology, and by personally meeting Orlando H. Garrido. Unfortunately, my dreams of retracing the steps of Thomas Barbour, P. J. Darlington, and Schwartz were not to be realized — but the random destiny of a scientist under the capricious direction of a totalitarian government can work miracles — among them my involvement in a long ornithological search that led to some incredible herpetology.

An Ornithological Quest

In 1985, I was working with the Empresa Nacional de Protección de la Flora y la Fauna (ENPFF), a department within the Ministry of Agriculture. In that same year, a colleague, Giraldo Alayón García, had been designated to serve as the host for the visit of Lester L. Short, principal curator of Ornithology at the American Museum of Natural History and world-renowned specialist on woodpeckers. Dr. Short was accompanied on this journey by George Reynard, an expert on bird bioacoustics.

Commander Universo Sánchez, an old retired guerrilla and an aficionado of the wild fauna, had invited these American visitors. Orlando H. Garrido, who was to serve as the host, was attending a scientific meeting in Perú. Consequently, Alayón, a well-known arachnologist in the Caribbean area and an observer of birds for many years, found himself enrolled in an expedition financed by the American Museum and for which he had been able to make only some last-second preparations. The objective of the trip was to find the Ivory-billed Woodpecker (*Campephilus principalis*), a species extirpated in the United States and whose status in Cuba had been unknown since the mid 1950s.

At this time, one might question the relationships of a project that involved two American ornithologists searching for a potentially extinct species of bird with a famous Cuban arachnologist and a totally unknown Cuban herpetologist. During one of our routine trips to Topes de Collantes in the Sierra de Trinidad in south-central Cuba, Alayón described his brief six-day jaunt with the American ornithologist in February 1985 and their plan for a more extensive expedition in spring 1986 (Figure

Figure 1. United States / Cuba Expedition, Ojito de Agua, April 1986: (first row, left to right) Alfonso Silva-Lee, photographer; Emma Romeu, journalist; George Reynard, ornithologist from Cornell University; Alberto Garzón, guide; Jennifer Horner-Short, ornithologist from the National Museum of Nairobi, Kenya; (second row) Aracelio Navarro, guide; the author; Giraldo Alayón, biologist from ENPFF; Noelio Pober, mule driver from La Melba; Roberto Gamboa, truck driver from MINAGRI; Lester L. Short, ornithologist and curator emeritus, American Museum of Natural History (photograph by Carlos Peña).

[1] Model Institutions for Excellence Project, Universidad Metropolitana, P.O. Box 21150, San Juan, Puerto Rico 00928, USA.

FIGURE 2. Male Smallwood's Giant Anole (*Anolis smallwoodi*) from El Yarey camp (photograph by H. Huneker).

1). From that conversation came the idea that Alayón and I, with the sponsorship of Garrido, proposed to our boss. Our brainstorming envisioned a project in which a group of Cuban researchers (logically headed by us) would conduct a series of rigorous investigations that would clarify the status of the Ivory-billed Woodpecker since 1956, the year in which the Lamb couple had tracked more than half a dozen pairs in an unpopulated area in the Sierra Cuchillas del Moa and Cuchillas del Toa in eastern Cuba.

Specifically, we sought to determine how much the habitat described by the Lambs had changed. They had ventured into an extensive forest of endemic pines (*Pinus cubensis*) that was surrounded by a mosaic of biomes, including tropical hardwood rainforests, xerophytic chaparral that Cubans called charrascales, and wild mountains, from which the great rivers of Cuba originate. Thus, the imminent extinction of a most impressive bird created an ideal opportunity to explore one of the most unreachable and least-explored regions of Cuba — and a dream come true for a herpetologist and an arachnologist.

FIGURE 3. Male Tiger Anole (*Anolis rubribarbus*) from the western slope of Pico Toldo (photograph by H. Huneker).

The First Cuban Expedition

To our surprise, our project was approved, and we began the feverish preparations for an expedition, the importance of which exceeded any with which we had ever been involved. The ENPFF, over which another prominent retired guerrilla commander, Guillermo García, presided, sent me in September 1985 to the province of Holguín to attend to various urgent affairs, among them the presentation of a letter to the second-in-command of the Communist Party in that region (he would be in charge of logistical support). A series of meetings with provincial governmental officers, the Carlos de la Torre Museum of Natural History in Holguín (MHNH), and the provincial delegation of the Minister of Agriculture (MINAGRI) resulted in the provision of necessities for our dream expedition. The fundamental support came from MINAGRI: food, transportation, fuel, hammocks, machetes, batteries, the coordination with local guides, mules to carry the materials to the mountain, and permits for collecting specimens. The MHNH supplied technical personnel: Alejandro Torres, bird curator and photographer, and Eduardo Solana, a veteran taxidermist. In addition, the museum helped us obtain all the materials for collecting and preserving specimens of all kinds.

Communication in the mountains of eastern Cuba was almost non-existent. The inhabitants had little access to telegraph, telephones, or radios. To get information to persons in remote areas, the local radio station would transmit messages. The principal support provided by the government and the party was access to the radio; Alayón recorded a message with the call of the Ivory-billed Woodpecker and instructions for anyone who had seen such a bird to come to our camp or to the local government or party authorities. A description of the Ivory-billed Woodpecker completed the message: a bird a bit larger than a Cuban Crow (*Corvus nasicus*), with black and white plumage, a big ivory beak, and a crest of black or red feathers on its head.

On 12 October we set out in a truck of Russian manufacture toward La Colorada de Farallones, a small community of some 50 houses on the bank of the Moa River. The landscape consisted of steep limestone cliffs, and the intense red soil stained the unpainted walls and floors of houses. The locals were warm and hospitable. Our work began immediately in the form of long walks looking for people who knew of the woodpecker. At the same

FIGURE 4. Nuevo Mundo Damp from the Calentura Arriba Observation Point, February 1993, during the expedition with Martjan Lammertink.

time, my wildest herpetological dreams were becoming a reality. I saw anoles I had never seen before, *Anolis argillaceus* (Dark Bark Anole), *A. smallwoodi* (Smallwood's Giant Anole; Figure 2), and *A. rubribarbus* (Tiger Anole; Figure 3), and the nights were filled with the unfamiliar voices of *Eleutherodactylus ronaldi* (Cuban Khaki Eleuth) and *Bufo taladai* (Cuban Spotted Toad), none of which I had ever heard or seen in their natural environment.

Our efforts to obtain information on the Ivory-billed Woodpecker were fruitful. The northern portion of the area explored by the Lambs in 1956 was totally deforested; the centuries-old pines had given way to crops and pastures. Another important part was flooded behind the dam that collected the waters of the Moa and Calentura rivers (Figure 4). However, a 71-year-old farmer, José Guerrero Sánchez (Figure 5), remembered the Ivory-billed Woodpeckers, and, in his opinion (shared by others we interviewed), the species could only remain on an abandoned farm, Cayo Probado.

Cayo Probado (Figure 6) was south of the area studied by the Lambs and near another site (Ojito de Agua) that would become inextricably linked with my most rewarding experiences as a field biologist. On 16 October, we — Alayón, Solana, Peña, Torres, and I — set out for Cayo Probado. We climbed via a long, steep road quickly baptized by the group as "The hill of three rests," because it had to be climbed in three stages. We found a natural pine forest, but we also discovered a brigade of woodsmen busily felling the few remaining old trees — but we also found, for the first time, something we had only seen in photos: the marks left by Ivory-billed Woodpeckers in their search for food (insect-infested dead trees from which the bark had been stripped in a most peculiar manner). Though excited by the woodpecker signs, I didn't neglect my herpetology; I found a yellowish-orange *Tropidophis melanurus* (Giant Trope) and I collected the *Eleutherodactylus* that would later be described as a new species. Unfortunately, on the morning of the 18th, a storm forced us to break camp and return to our base.

When we got back to Farallones, we had news from La Melba, a mining area 30 km to the southeast on the shores of the Jaguaní River and on the other side of a high plateau. The message made reference to a man named Alberto Garzón Martínez, who said that he had seen an Ivory-billed Woodpecker in June 1985 in the vicinity of Ojito de Agua. We quickly set off for La Melba, but the storm never let up and we feared that the road to La Melba would be blocked to vehicle transportation. Soggy and tired, we arrived at dusk but were unable to find Alberto Garzón. I walked through the village listening to the choirs of *Eleutherodactylus auriculatus* (Cuban Telegraph Eleuth) and *E. ronaldi*, and I was able to see the Arroyo Bueno unload vast volumes of turbulent water into the flooded, but relatively quiet Río Jaguaní. The local authorities provided cover in an agricultural workers' shelter, but we decided to suspend our activities and return to the provincial capital the next day.

The expedition ended on an unfinished note; we were unable to interview Garzón, but we knew

FIGURE 5. José Guerrero Sánchez, an old farmer, shared memories of Ivory-billed Woodpeckers, La Melba, spring 1991 (photograph by Martjan Lammertink).

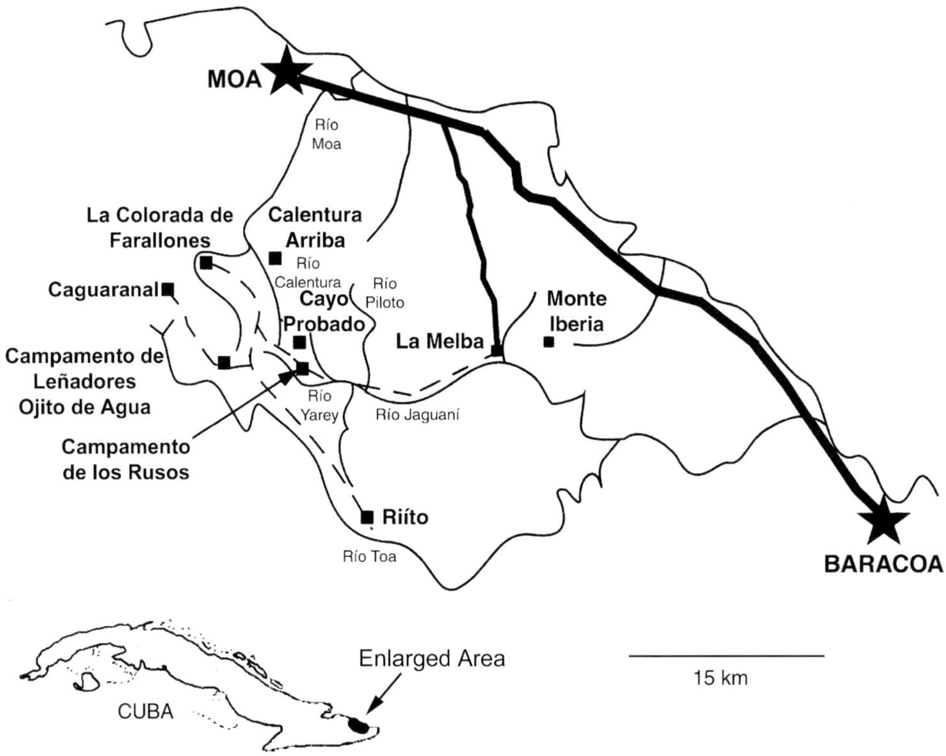

FIGURE 6. Area in which the search for the Ivory-billed Woodpecker was centered.

where to find him, and we had identified the habitat where we had to look for woodpeckers. On the plus side, we had collected dozens of samples of different amphibians, reptiles, and birds, had found evidence of the presence of endangered endemic mammals, the Almiquíes (*Solenodon cubanus*) and hutias (*Capromys* sp.), and a considerable amount of invertebrate material awaited its destiny in the MHNH or our research collections in Havana.

The First Encounter

Excited by the results of the October experience, Alayón and I (Figure 7) returned to Holguín in March 1986 accompanied by Eduardo Solana and Carlos Peña of the MHNH and with the MINAGRI taxi driver, Roberto Gamboa. We arrived at La Melba in a Russian jeep on 12 March. Our interview with Garzón was very fruitful. He recounted a hunt for *machos jíbaros* (feral pigs) in June 1985 during which he saw a single male Ivory-billed Woodpecker in the vicinity of the campamento de leñadores (woodsmen's camp) in Ojito de Agua. We showed Garzón color drawings of all of the woodpeckers known from Cuba, except the Ivory-

FIGURE 7. The author near Farallones de Moa, September 1986 (photograph by Francisco Conejero).

billed. He studied them very carefully and said: "I am sorry, comrades, but the bird that I saw does not look like any of these." Then we showed him a drawing of the Pileated Woodpecker from the United States (*Dryocopus pileatus*), and he said: "This one is very similar!" However, when he saw a drawing of a pair of Ivory-billed Woodpeckers, the miner exclaimed: "Man! Why didn't you show me this before? This is the animal you want. The one that I saw at Ojito is like this one; it had the red crest."

In this manner, Alberto Garzón and his father-in-law, Aracelio Navarro, became members of our group — with the authorization from their bosses at the Merceditas mine in La Melba. On 12 March, we traveled from La Melba to Sagua de Tánamo, west of Moa, and from there south to Guantánamo, the capital of the easternmost province in Cuba. From Guantánamo, we continued northeast, passing through Yateras to Riíto, an agro-forest community some 15 km to the south of Ojito de Agua. Torrential rains had rendered the red-dirt forest roads impassable, so we decided to spend the night at Riíto before continuing on to the campamento de leñadores.

Our experience was similar to that on the first trip in October — natural pine forest, alternating with thorny charrascales, and much evidence of recent deforestation. Recumbent pine trunks, 8–10 m long and 30–50 cm in diameter, paralleled the main road awaiting their fate. Tracks of enormous trucks and tractors were everywhere.

At noon, we arrived at the campamento de leñadores at Ojito de Agua, a rustic pine cabin with a tar roof. A dark and humid room was a barracks for woodsmen, and a small separate bedroom for the cook and her partner. The rest area, kitchen, and dining room had a roof, but no walls. The woodsmen welcomed us with abundant coffee made with water from the stream. After talking with the old woodsmen from the area, who undoubtedly knew the bird, part of our group went to the Río Yarey in the southeastern part of Ojito, and Solana, Alayón, Garzón, and I continued northwest to the area where Garzón had seen the male in June 1985. We walked silently, exchanging only low murmurs with our guide. After trekking up and down a slope, we heard a "pent-pent-pent-pent," undoubtedly the call of an Ivory-billed Woodpecker. It came from the depths of a ravine some 100 m downhill from our position. We hid and waited for what seemed an endless period, then we slowly, carefully, and silently continued our reconnaissance. We were looking for large dead or dying trees where a woodpecker might feed, nest, or rest.

Although anticipation was high, no woodpeckers materialized. As we headed back to camp, I was about 50 m ahead of the group, crossing the clearing where we had heard the bird, and climbing a steep slope. By now it was 4:40 PM, and the afternoon sun was low in the sky. All of a sudden, an enormous black and white bird passed some 30 meters in front of me, flying low and almost scraping the road first to my left, then to my right before disappearing behind a slope to our west. The entire episode lasted only 10 seconds, but I knew beyond any doubt that I had just seen an Ivory-billed Woodpecker. Further, I was sure it was a female, for I had not seen any red in its plumage. I was breathless because of the steep slope and unable to speak for some seconds, but then I called to the rest of the group still on the slope below me. We separated and waited impatiently for about 20 minutes, and then we all heard a distant "pent-pent." This call and the one we had heard previously, along with the sighting of a female, led us to believe that a male was nearby — but we saw no more woodpeckers that evening.

For two days, we continued to explore the area where I had seen the female — without much luck. We did find some marks in the dead trees that were still standing that might have been made by feeding Ivory-billed Woodpeckers, and I continued my collection of anoles and *Eleutherodactylus* frogs (Figure 8). Still, we decided to move on, and set out early for the source of the Yarey River, in an area known as campamento de los Rusos, near Cayo Probado. The morning was cloudy, I was walking ahead with Alberto Garzón and Aracelio.

FIGURE 8. The author taking notes on amphibians and reptiles collected around the campamento de los Rusos, March 1987 (photograph by Antonio Pérez Asso).

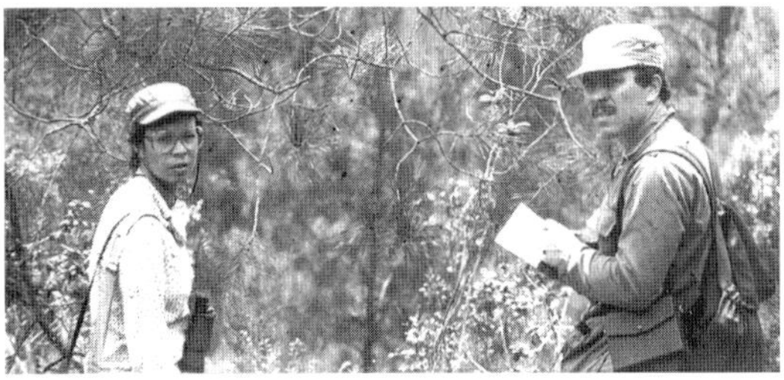

FIGURE 9. Aymeé Posada and her husband, Giraldo Alayón, at the Lester Short Observation Point, 16 March 1987, after the last sighting of an Ivory-billed Woodpecker at campamento de los Rusos.

Alayón was about 100 m behind us with Solana and Elio Bleth, a guide who had been sent from Riíto to join us in our quest. A little after nine, a ruckus caused by a couple of crows attracted the attention of Alayón (Figure 9), who focused his binoculars on a dry pine some 200 m in the ravine to their right. What he saw was two crows harassing a female Ivory-billed Woodpecker, which was moving around one of the uppermost branches. Alayón observed this spectacle for several seconds until the woodpecker disappeared into the depths of the ravine.

Excitement abounded, for we had invested many hours in a vigil that lasted days and covered a variety of sites, none of them easily accessible. Although we neither saw nor heard any more woodpeckers, the adrenaline had kept us going. Then reality hit, we were exhausted and hungry, so we decided to call it a trip. We sent Carlos and the driver ahead to Riíto to get our jeep, the comforts of which rivaled a modern luxury car right then. As we drove to Holguin that evening, we began to compose the press release for that same night.

FIGURE 10. United States / Cuba Expedition, April 1986, campamento de los Rusos, near Ojito de Agua: Jennifer Horner-Short and Lester L. Short at the Lester Short Observation Point.

Confirmation

Only a little more than two weeks after our initial sightings of Ivory-billed Woodpeckers in the vicinity of Ojito de Agua, Dr. Lester Short and his wife, Dr. Jennifer Horner (Figure 10), arrived with George Reynard, all of them loaded with recorders, microphones, and parabolic reflectors. Alayón and I served as guides to campamento de leñadores. Also accompanying us were Gamboa from the MINAGRI, Peña and Solana from the MHNH, and our guides, Garzón, Navarro, and Bleth (Figures 1 and 11). This expedition lasted until 18 April 1986 and made the following sightings of Ivory-billed Woodpeckers: (1) 6 April, 12:45 PM, Jennifer Horner, unidentified sex in Yarey Canyon; (2) 7 April, 9:45 AM, Alberto Garzón and Elio Bleth, a pair in a dry pine in Yarey Canyon; (3) 10 April, 7:00 AM, George Reynard at some 300 m from his observation post near campamento de los Rusos, a female flying over Yarey Canyon; (4) 12 April, 7:20 AM, Lester Short in the vicinity of campamento de los Rusos, a female; (5) 14 April, 10:45 AM, George Reynard at more than 200 m from his observation post, a female flying over Yarey Canyon; (6) 16 April, 7:20 AM, Lester Short, at 6 m along one of the streams at the source of the Yarey River, a male that flew over his head and disappeared high in the forest upstream; (7) 16 April, 7:22 AM, Jennifer Horner, in the same ravine that had been walked by Short, a male flying over the pines that flanked the stream. Additional data collected by this expedition included recordings of the Bee Hummingbird (*Mellisuga helenae*) and many other calls of various birds and amphibians.

Although the expedition was exhausting, our American colleagues had confirmed our sightings — but now they faced the unpleasant reality that these birds, possibly two and maybe as many as four, were unlikely to last long. They were trapped in the last vestige of suitable habitat, around which the forest had been reduced to practically nothing. In his report to the Museum of Natural History in Havana, Dr. Short spoke to the difficulty of saving the species, but recommended that its presence should become the focus of a campaign to protect the remaining pine forest at Ojito de Agua and the many plants and animals that called it home.

Repercussions

The newspaper coverage of our sightings in March and April 1986 generated considerable controversy. Although most specialists, reporters, and even the

FIGURE 11. United States / Cuba Expedition, April 1986, campamento de los Rusos, near Ojito de Agua: (left to right) Aracelio Navarro, Elio Bleth, and Alberto Garzón during a break.

general public rejoiced at finding what was feared to be an extinct species, Félix Guerra, a writer for *Bohemia* (a magazine published in Havana), cast doubts on our discovery, describing it as "a fortunate coincidence that occurred a few weeks before Dr. Short's visit." Generally, however, both the Cuban and American press reported the events as told by the participants.

Immediate consequences of these expeditions included a cessation of all deforestation in Ojito de Agua and the modification of a plan to construct a road that was to enter the Farallones and Riíto areas. The Cuban Academy of Sciences (ACC, for its Spanish acronym) became involved in the project, although it had been indifferent to our efforts to that point. The Academy took over the project and quickly designated the region of Cuchillas de Moa and Toa as a Biosphere Reserve. Some other governmental agencies (e.g., Comisión Nacional de Medio-Ambiente) and various ministers responded to interest expressed by a number of international organizations (e.g., the International Council for the Preservation of Birds, or ICPB) by taking an increasingly active role in subsequent activities. The ICPB sent Christoph Imboden, its president at the time, Lester Short, and other well-known experts to Havana. Faustino Pérez, another retired guerrilla commander, received them, presented them with keys to the city, and organized a last-minute tour of Ojito de Agua with Alayón serving as a host.

In the meantime, promises by the ACC to do everything possible to protect the Ojito de Agua region elicited protests in the forestry section of the MINAGRI, which had been promoting the timber industry in the area (Figure 12). Also, the security agencies of the local and regional cells of the Communist Party became heavily involved in all scientific or cultural projects that involved an interaction of Cuban and foreign professionals (and even more so if the foreigners were from the United States). Collectively, the publicity and the events that followed resulted in some fundamental changes in subsequent research on the Ivory-billed Woodpecker and the Ojito de Agua region.

The Legacy of a Disappearing Species

As biologists, we have long been aware that a number of vertebrate species from Cuba are extinct or nearly so. Not only had individuals of some species not been seen in some time, critical habitats that once were common had been dramatically reduced or entirely eliminated, even in the most remote areas. Species on this sad list include the Cuban Macaw (*Ara tricolor*), the Zapata Rail (*Cyanolimnas cerverai*), the Hook-billed Kite (*Chondrohierax uncinatus*), the Ratt Hutia (*Mesocapromys auritus*), Pigmy Hutia (*Capromys nanus*), and San Felipe's Hutia (*Mesocapromys sanfelipensis*), as well as the Cuban Solenodon (*Solenodon cubanus*). Most disappeared without a whimper, and an uncaring public didn't even blink. In contrast, I was able to say farewell to the Ivory-billed Woodpecker, to feel its presence in the air, to hear its enigmatic call, to record the marks left after it fed, and to see with my own eyes, at least for a few seconds, one of the last individuals of the species, a final dying spark of a star that is fading forever. It was the most unforgettable experience of my life.

Unfounded rumors abounded, but a number of subsequent expeditions to the region (some continuing even today) failed to find additional birds. Our scientific report had clearly stated the sad reality that the Ivory-billed Woodpecker was extinct in Cuba. Nevertheless, of at least 10 expeditions in which I participated, six were specifically devoted to finding evidence of more individuals in the region around the Ojito de Agua and in other seemingly favorable habitats, including relatively distant areas such as the Sierra Maestra.

FIGURE 12. A Russian truck loaded with pine logs cut by Castellanos' woodsmen brigade, Ojito de Agua, March 1986.

FIGURE 13. Eggs and hatchling Dark Bark Anole (*Anolis argillaceus*) from Ojito de Agua (photograph by H. Huneker).

Although failures in regard to woodpeckers, nine expeditions to the Ojito de Agua area, additional forays to Pino del Agua and Pico Bayamesa, in the Sierra Maestra, provided me with the inestimable opportunity to search for amphibians and reptiles in the few remaining tracts of natural habitat left in Cuba. After a collection of anoles from Cayo Probado and campamento de los Rusos, that I described with Orlando H. Garrido, my first really exciting herpetological discovery was *Anolis inexpectatus* (Pine Bush Anole). During a later expedition with Lester Short, I discovered the communal egg-laying sites of *A. argillaceus* (Dark Bark Anole; Figure 13), took a series of photographs, and followed the development of the eggs through hatching. I collected a small boa of the genus *Tropidophis*, which was almost certainly a new species, in the vicinity of Cayo Probado. This specimen, and another collected at the Báez River in Baracoa by a joint Cuban-U.S. expedition, served as types for the description of *Tropidophis fuscus* (Cuban Dusky Trope). In addition, the vast collections and accumulated observations from the early expeditions led to the publication of a list of the amphibians and reptiles from the Cuchillas de Moa and Toa in collaboration with our colleagues from those expeditions.

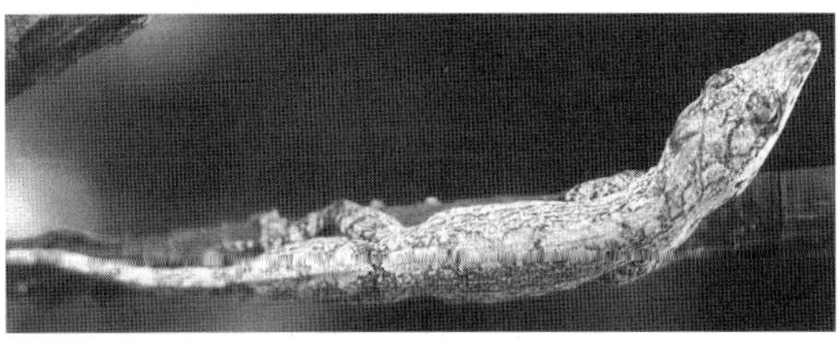

FIGURE 14. Male Alayón's Twig Anole (*Anolis alayoni*) from El Yarey camp (photograph by H. Huneker).

I mentioned previously that I had collected some samples of a new species of *Eleutherodactylus* during the first expedition in 1985. I found more specimens during subsequent expeditions in 1986 and 1987. Similarly, in different areas around Ojito de Agua and La Melba, I had acquired a series of twig anoles that were very similar to *Anolis angusticeps* (Cuban Twig Anole). These and additional specimens collected during the Cuban–United States Herpetological Expedition to the province of Guantánamo in 1990 led to the descriptions of *E. toa* (Oriente Pallid Eleuth), and *A. alayoni* (Alayon's Twig Anole; Figure 14). Later, in the vicinity of La Melba and the Jaguaní Reserve to the east of Ojito de Agua, I made a series of observations of egg-laying by *Diploglossus nigropunctatus* (Cuban Pale-necked Galliwasp) and an equally extensive series of specimens and nests of *Eleutherodactylus guantanamera* (Guantanamera Eleuth), which then was considered to be a population of *E. v. varians*, the Western Bromeliad Eleuth). Those data led to descriptive notes on the egg laying sites, embryonic development, hatchlings, and male parental care in the latter species.

The Extinction of Our Project

My critical opinions of the conservation policies of Cuban governmental agencies led to my being designated as a "non-appropriate person" in 1987. From an ideological point of view, the government couldn't tolerate my central role in the well-publicized search for the woodpecker, my participation in expeditions with or without foreigners, and my coauthorship of the resulting reports. Subsequently, by explicit orders from the President Minister of the ACC, Rosa E. Simeón, I was excluded from participation in an expedition organized by the National Geographic Society in February 1988. Although I participated in other expeditions to eastern Cuba between September 1986 and March 1989, the original project conceived by Giraldo Alayón and me had been taken away, and I specifically was excluded from any active role. My arbitrary retirement from the project and the inherent contradictions that emanated from these purely political decisions were attributable to the highest level of the National Enterprise for the Protection of the Flora and Fauna and its president, Guillermo García — who continued to exclude me from various projects. I was forced to quit my position in June 1989, and was without a job for many months.

FIGURE 15. Martjan Lammertink in the cloud forest on Monte Iberia, April 1993.

A New Focus and More New Species

How was my career rejuvenated (at least temporarily)? Martjan Lammertink (Figure 15), then a 19-year-old birdwatcher from the Netherlands, had explored the region around Ojito de Agua in the spring of 1991 with only the help of a guide from Farallones de Moa and copies of my maps and notes. He had contacted me in Havana while he made arrangements for his visa. At the end of the trip, having received all kinds of evasive answers to questions posed to other participants in the old project, he asked me to accompany him on a new expedition in 1993. At that time, I was working at the MINAGRI Instituto de Investigaciones Forestales in the Zapata Swamp. I told him that I would seriously consider his offer, but I would have to do it without the sponsorship of any official agency.

Obviously, I agreed. Thus, my last expedition to Ojito de Agua was in early 1993. The goal of this expedition was to fill in some informational gaps that had not been addressed by the earlier expeditions in which I had participated and by others organized by various Cuban and American groups. Two potentially favorable woodpecker habitats had not been explored. An area with enormous trees, including some old pines, and a considerable number of standing dead trees was known from near the source of the Jaguaní River on the high plateau of Toldo Peak directly east of Cayo Probado. Another timbered area was known to remain in the western part of Ojito in the vicinity of the source of the streams that fed the Toa River.

We went directly to those sites, where Lammertink, now a biology student at the University of Amsterdam, Huub Huneker, a Dutch landscape gardener, Carlos Peña from the MHNH, a guide from Farallones de Moa, and I set up some observation posts at Cayo Probado, campamento de los Rusos, and Ojito de Agua. In addition, we explored the regions around Cupeyal and Caguaranal at the source of the Toa and the steep canyons of Jaguaní (Figure 16) to the west of the Toldo and Piloto rivers on the slopes of the southern high plateau. We found no evidence of Ivory-billed Woodpeckers, neither active nor old, nor did we find many trees of appropriate age and size, much less the promised centenarian stands of natural pine. From recordings, Lammertink had learned to imitate to perfection the call of the Ivory-billed Woodpecker using a trumpet

FIGURE 16. Canyon of the Río Jaguaní at the base of Pico Toldo.

mouthpiece adapted to a small toy cornet (Figure 17). He could emit a call audible at hundreds of meters, but we never got a response.

FIGURE 17. Martjan Lammertink imitating the call of an Ivory-billed Woodpecker on Monte Iberia, Las Tetas de Julia are in the background, April 1993.

FIGURE 18. Lunch in the rain at campamento de leñadores, February 1993: (left to right) Martjan Lammertink; Carlos Peña, MHNH; the author; Huub Huneker; Noelio Pober, mule driver from La Melba (photograph by H. Huneker).

FIGURE 19. The Río Jaguaní at Mina El Yarey.

FIGURE 20. The Ojito de Agua pine forest from one of several observation points during the expedition in February/March 1993.

A few years earlier, an enormous piece of pine forest to the north of campamento de leñadores (Figure 18) at Ojito de Agua had burned. A large number of dead pines were left standing by MINAGRI enterprises, based on recommendations in the reports of various Cuban and American expeditions that had visited the area. This area was located between the sources of the Jaguaní and Toa rivers (Figure 19), around the same springs as those that fed the streams at Ojito de Agua (Figure 20) and Culebra de Hacha. The latter entered Farallones Cave and emerged from the cavern as the Moa River. The pines did not bear signs of the Ivory-billed Woodpecker, but big grubs of wood-eating beetles and other invertebrates were abundant there. The size and nature of the area's standing trunks were not suitable for nests or shelter, but the habitat might have been a feeding area. Our observation posts and calls produced nothing.

I became more convinced than ever that the individuals seen fleetingly in 1986 were the last of their kind. Fortunately, much of the herpetofauna in the region remained a mystery (Figure 21). During the long nights in camp, I recorded the call of a sibling species to *Eleutherodactylus auriculatus*. Although morphologically very similar, these frogs had such different calls that, on several occasions, I had crawled out of my sleeping bag in an effort to identify the mysterious voice. I was able to collect an adequate series, describe their live coloration, and record a number of individuals, including calling males vocalizing in concert with *E. auriculatus*. As an appropriate tribute to the bird, the search for which had led to the discovery of the frog, we (Blair Hedges and I) named it *Eleutherodactylus principalis* (Oriente Greenish-Yellow Eleuth).

From Monte Iberia to Pico Bayamesa

After having said farewell to our partners of six long weeks and a well-deserved rest in Havana, Martjan and I went back to the MHNH and from there, with the help of its director, Alfredo Rams, and his *Lada* (a Russian car based on a 1970s-era Italian Fiat), we went to La Melba. The objective of this trip was to explore the high plateau northeast of that village and which is known as Monte or Alto de Iberia. Some relictual hardwood forest with scattered stands of old pine were reputed to exist there. Martjan and I camped near Arroyo Sucio (Figure 22), a small tributary of the Jaguaní River. Our results were similar to those obtained on other

FIGURE 21. Cuban Side-blotched Curlytail (*Leiocephalus macropus asbolomus*; top), Giant Trope (*Tropidophis melanurus*; middle), and Black and White Racer (*Antillophis andreae*; bottom) from El Yarey camp (photographs by H. Huneker).

recent trips to the region. Some years before, a competition among woodsmen had destroyed the taller trees in the primary forest, leaving only fallen trunks, 40–50 cm in diameter and 4–5 m long, scattered along some clearings and trails (Figure 23). These were now covered with a dense mantle of trailing ferns that rendered walking in the area next to impossible. Under a constant drizzle, we covered almost all of the southeastern area of Alto de Iberia using the main trails that cross near two conical, rocky promontories called Las Tetas de Julia (Julia's Teats, Figure 17).

There, along the banks of Arroyo Sucio, under the mantle of trailing ferns, we found some small frogs that produced a light hiss that reminded me of the voice of *Eleutherodactylus limbatus* (Yellow Striped Pigmy Eleuth). With considerable effort, I managed to catch two of these diminutive frogs under the ferns, one of them with an egg. The daytime choruses were similar to those of *E. limbatus*, and their general appearance made me think that they were the same species that I had collected in La Melba and Farallones de Moa — but at night, I happened to see an active individual crawling on a dry leaf. It moved its feet slowly, like a gecko, and my attention was riveted on the intense purple color of its body and the brilliant metallic copper of the lines that ran from its snout to its vent. These colors did not coincide with the brown and yellow-orange tones of *E. limbatus*! A more detailed examination of coloring and morphology left no doubt that it was a new species, subsequently named *Eleutherodactylus iberia* (Iberia Eleuth). It equaled the record of the smallest tetrapod in the world (Figure 24). A

FIGURE 23. The author sitting on the trunk of a log felled during a competition of woodcutters in the 1980s on Monte Iberia, April 1993 (photograph by Martjan Lammertink).

FIGURE 22. Martjan Lammertink in the camp at Arroyo Sucio on the western slope of Monte Iberia, April 1993.

FIGURE 24. Iberia Eleuth (*Eleutherodactylus iberia*) with an egg (photograph by Martjan Lammertink).

FIGURE 25. A pair of small Tetas de Julia Eleuths (*Eleutherodactylus tetajulia*) in the stem of a hollow fern, April 1993 (photograph by Martjan Lammertink).

photograph of this small frog on a Cuban coin was published many times, with venues including *National Geographic Magazine*, *The New York Times*, and the world-wide web.

While resting on our walk near Las Tetas de Julia, Martjan called my attention to the vocalization of an insect or frog that he had not heard before. It was a light "tui-tui-tui-tui" that emanated from the leaf litter of the cloudforest. After hours of intense searching under the interminable drizzle, I found a pair of small frogs and an egg clutch inside the hollow stem of a fern. These frogs had a voice and appearance that were reminiscent of *Eleutherodactylus intermedius* (Oriente Dark-bellied Eleuth) from the Sierra Maestra, but it was undoubtedly a new species. In my field notes, I called it the Tetas de Julia Eleuth. A year later, Blair Hedges was looking at my notes and commented on the funny name — and, later, we formally described it as *Eleutherodactylus tetajulia* (Figure 25).

At Monte Iberia, I also was finally able to record *Eleutherodactylus toa* (Oriente Pallid Eleuth), which I had heard for the first time during the last night of our previous expedition at the confluence of the Yarey and Jaguaní rivers. Unfortunately, I had been unable to record it then because my recording equipment had been damaged in a torrential storm that almost flattened our camp.

We left La Melba for Baracoa, and from there we went to Santiago de Cuba, where we spent the night. The next day, we traveled on to Guisa, from where we planned to climb the northern slope of the Sierra Maestra in order to reach the foothills of Pico Bayamesa, and eventually get to the region called Pinar del Millón, on the northwestern slope of the peak. Although no report from Gundlach's time (mid-19th century) mentioned Ivory-billed Woodpeckers in any part of the Sierra Maestra, the existence of composite pine and hardwood forest in an isolated and sparsely populated region had the potential for a pleasant surprise.

After overcoming some logistical difficulties, we climbed Pino del Agua Arriba, on the northcentral slope of the longest and highest mountain range in Cuba. Our goal was a former training camp for Communist Party youth, who, in the early 1960s, had alternated ideological indoctrination with reforestation work. One species of native pine had been of particular interest; *Pinus maestrensi*, the largest of all native species, which could reach 2 m in diameter and 20 m of height. In the 1990s, the camp was only a memory, and a few cement floors of houses and barracks and some scrap iron from abandoned vehicles or agricultural machinery remained as silent reminders of the youth camps.

We camped for eight days on a road that led to an abandoned small town that once had been Nuevo Mundo. Results were negative. The forest was all that had been promised, but we found no evidence of Ivory-billed Woodpeckers. We continued up the slope of Pico Bayamesa (1,730 m), but we quickly became aware that it would be impossible for us to proceed without an appropriate guide. We had intended to reach Pinar del Millón, on the southwestern slope (Figure 26), but the evidence that we already had did not justify continuing the search.

That short trip to the region contributed enormously to my herpetological endeavors. I recorded calls of *Eleutherodactylus auriculatus* and *E. glamyrus* (Turquino Fern Eleuth). I collected a female *E. melacara* (Dark-faced Bromeliad Eleuth),

FIGURE 26. Pinar del Millón from the top of Pico Bayamesa (1,730 m), Sierra Maestra, May 1993 (photograph by Martjan Lammertink).

which had not been collected before. I also found this species and recorded the call almost 30 km to the east of Pico Turquino, which dramatically extended the species' known distribution. I also recorded the calls of *E. ricordii* (Oriente Yellow Spotted Eleuth) and *E. cuneatus* (Oriente Stream Eleuth), and I collected important series of specimens that led to descriptions of several new species.

Epilogue

I would return to Sierra Maestra a year later, accompanied by Richard Thomas, Blair Hedges, and some Cuban colleagues. This time, we were not searching for the Ivory-billed Woodpecker, a quest that had revealed so many of the region's secrets to me. Instead, the goal was purely herpetological, and I was able to focus exclusively on my first love — not knowing at the time that it would be my last expedition in Cuba before going into political exile.

Acknowledgments

In addition to the participants in the various expeditions, I thank my wife Celia A. Puerta, for her strong support throughout my career. Also, I express a very special "thank you" to Dr. Gladys Bonilla, who translated the original Spanish manuscript, and to Robert Powell, Martjan Lammertink, and Manuel Acevedo, who made a number of suggestions that improved the final version of the manuscript.

Suggested Reading

Alayón, G., A. R. Estrada, and A. Torres. 1987. Lista de las aves observadas en la Reservade la Biosfera "Cuchillas del Toa," provincias de Holguín y Guantánamo, Cuba. *Garciana* 6:1–13.

Estrada, A. R. 1987. *Anolis argillaceus* (Sauria: Iguanidae) un nuevo caso de puestas comunales en anolinos cubanos. *Poeyana* 353:1–9.

Estrada, A. R. 1990. Las puestas de *Eleutherodactylus varians* (Gundlach et Peters). *Biología* 4:163–167.

Estrada, A. R. and G. Alayón. 1986. Reporte de expedición. Búsqueda del Carpintero Real. *El Volante Migratorio, Perú* 6:15.

Estrada, A. R. and G. Alayón. 1986. La existencia del Carpintero Real o Pico de Marfil en Cuba es realidad no un sueño. *El Volante Migratorio, Perú* 7:25–27.

Estrada, A. R., G. Alayón, A. Pérez-Asso, C. Peña, and E. Solana. 1987. Lista preliminar de anfibios y reptiles de las Cuchillas de Moa y Toa. *Garciana* 8:3–4.

Estrada, A. R. and S. B. Hedges. 1991. Nueva especie de *Eleutherodactylus* (Anura: Leptodactylidae) de la Región Oriental de Cuba. *Carib. Jour. Sci.* 27:139–145.

Estrada, A. R. and S. B. Hedges. 1995. A new species of *Anolis* (Sauria: Iguanidae) from eastern Cuba. *Carib. Jour. Sci.* 31: 65–72.

Estrada, A. R. and S. B. Hedges. 1996. A new frog of the genus *Eleutherodactylus* from eastern Cuba (Anura: Leptodactylidae). *Herpetologica* 52:435–439.

Estrada, A. R. and S. B. Hedges. 1996. At the lower size limit in tetrapods: a new diminutive frog of the genus *Eleutherodactylus* from Cuba (Anura, Leptodactylidae). *Copeia* 1996: 852–859.

Estrada, A. R. and S. B. Hedges. 1997. Nueva especie de *Eleutherodactylus* (Anura: Leptodactylidae) del macizo Sagua-Baracoa, Cuba. *Carib. Jour. Sci.* 33:222–226.

Estrada, A. R. and S. B. Hedges. 1997. A new species of frog from the Sierra Maestra, Cuba (Leptodactylidae, *Eleutherodactylus*). *Jour. Herpetol.* 31:364–368.

Garrido, O. H. and A. R. Estrada. 1989. Nueva especie del complejo *Anolis alutaceus* (Lacertilia: Iguanidae) para Cuba. *Biología* 3:57–66.

Gundlach, J. 1876. *Contribución a la Ornitología Cubana*. Imprenta La Antilla, La Habana.

Hedges, S. B. and O. H. Garrido. 1992. A new species of *Tropidophis* from Cuba (Serpentes, Tropidophidae). *Copeia* 1992: 820–825.

Lammertink, M. and A. R. Estrada. 1995. Status of the Ivory-billed Woodpecker *Campephillus principalis* in Cuba. *Bird Conserv. Internatl.* 5:53–59.

Estrada, A. R. 2003. Of woodpeckers and frogs, p. 75–87. *In* R. W. Henderson and R. Powell (eds.), Islands and the Sea: Essays on Herpetological Exploration in the West Indies. Society for the Study of Amphibians and Reptiles, Ithaca (New York). Contributions to Herpetology, volume 20.

Cuban Treefrogs, Curiosity, and the West Indies

Walter E. Meshaka, Jr.[1]

In the early 1990s, I began to study the life history of the CubanTreefrog in Florida (Figure 1). I was interested in answering only one very simple question: why is the Cuban Treefrog such a successful colonizing species? One of the things that I had to do to answer this question was to compare the way this species lived in its native geographic range with how it made a living in its introduced range. Not only was I excited about examining museum specimens from Cuba, the Cayman Islands, and the Bahamas, I could hardly wait for the opportunity to see these animals in their native habitats.

A New Land Not Unlike Its Old Home

So, off I went; first, to Grand Cayman Island in July 1992, then to Nassau, Grand Bahamas in November 1992, and finally to Cuba in June 1994. Each trip lasted from several days to a week. Throughout the Cuban Treefrog's world, searching for them often meant searching with and around people. Because I grew up in Miami, I had an understanding of at least some aspects of West Indian culture, having been immersed in them for much of my life.

Of course, the climate is subtropical, and even the hammocks are floristically similar to those of the West Indies. In many ways, the Cuban Treefrog came to a new land not unlike its old home, and I subsequently discovered that its native lands were not so unfamiliar to me. The cultures and surroundings were amenable. Folks were not unlike my neighbors in Miami; if anything, they were a bit friendlier and often more curious about my unusual endeavors. I enjoyed learning how these native West Indians lived and even what they thought of the Cuban Treefrog, or "la rana platanera," as Cubans call it.

Grand Cayman: Frogs, Roaches, and Cigars

Exposed sharp limestone, hammocks, abrupt rainstorms followed by sunshine, and the grating squawk of my old friend, Grand Cayman Island (GCI) in midsummer was for me like the Florida Keys. The culture of GCI was in part a Miami mix of Jamaicans, Caymanians, and unique to the Caymans, a host of proud Scots. Buildings were Key West Old, with stone homes, tin roofs, lighted here and yonder, and each with a cistern. The latter serve as oases for Cuban Treefrogs in an otherwise harsh environment, with little moisture and little cover. Again, like Florida, Grand Cayman was a place of stark contrasts. Other buildings were Miami Beach New, swanky, thick, and all dolled up.

Figure 1. Cuban Treefrogs, *Osteopilus septentrionalis*, in Florida. Populations of this species are native to Cuba, the Bahamas, the Turks and Caicos Islands, and the Cayman Islands, and have been introduced into Florida, Puerto Rico, St. Maarten, St. Barthélemy, and Anguilla.

[1] Section of Zoology and Botany, State Museum of Pennsylvania, 300 North Street, Harrisburg, Pennsylvania 17120, USA.

I was accompanied on this trip by my colleague, Brian Butterfield. We stayed at the home of my childhood buddy, Greg McTaggart — proud member of a Caymanian pioneer family. I immediately made a beeline for the not-so-modern parts of town, where I quickly made friends with gentle folks. Word spread quickly that someone was there to catch frogs, which were held in esteem only one step above the roaches it ate. By the second night, akin to extremely minor celebrity status, I was being shown about by one homeowner after another, each wanting me to take all the Cuban Treefrogs I could. They knew what the frog liked: lighted sections of buildings, especially near woods, cisterns, and awnings. Such fun to smoke cigars and walk about the area at all hours, especially at night, with no cause to fear, catching Cuban Treefrogs, and hearing songs of one of my music heroes, George Jones, playing in the background (Greg had told me that old country music was popular there). A stiff drink, sleep, and back at it the next day with a mid-day siesta (a thoroughly civilized habit) — sheer bliss.

These activities really did attract attention, not only because people were happy to see the frogs carted off, but also because they were afraid of them and were genuinely amazed that we were not only fearless, but were catching them for reasons other than eradication. I was struck by the fact that everyone who had anything to tell me about the frogs always added their belief that the frogs can urinate in your eyes and would cause blindness. I had little success in explaining that, although the skin was highly toxic, urea alone was expelled when the frog was frightened. I was to find that the disdain for this frog and belief in its blinding qualities were widespread in the West Indies. Folks were nonetheless curious about it and about other representatives of the local fauna. Curiously, although disdain of a familiar pest was the norm among West Indians in Miami, non-Latin Floridian reactions to treefrogs ranged from terror, as if it were a snake, to mild curiosity about an animal either utterly unknown or unknown to them in the past.

Frogs were pretty big on GCI and greenish. I could hardly wait to see live ones in the Bahamas.

Grand Bahama: A Curious Scourge

Four months later, I was in the Bahamas staying at the Hotel Casuarina. I was treated very kindly, and taken about far and wide each night, stopping to collect at potential hotspots. Sandy and Hugh Buckner, Pericles Malis, and the chief of police could not have been kinder.

By day, I ripped through palm boots and other hiding places. I saw my first live Racer, *Alsophis vudii*, during one of these daytime forays. It was periscoped and hunting in the grass among some palm trees, much in the same way as North American racers. Bark Anoles (*Anolis distichus*), Cuban Brown Anoles (*A. sagrei*), and Curlytails (*Leiocephalus carinatus*) were about. This was November, getting drier and a little cooler, and one night I heard a chorus of Leopard Frogs (*Rana sphenocephala*). These all were common themes to my hometown and GCI. The plants, native and cultivated, were familiar, as were many elements of the herpetofauna. The pace was not so hectic, it was hot, and the food was delicious. Lastly, of course, the Cuban Treefrog was considered something of a curious scourge for the same reasons mentioned earlier.

Cuba: Abundant Contrasts

Cuba, the prodigious cradle of West Indian herpetofaunal diversity and the homeland of my favorite frog, was my last stop in June 1994. Long-awaited, I flew out of Miami to attend a zoology conference in Habana. In fact, I met Blair Hedges and Richard Thomas at the airport waiting on the same flight. In Miami, a fond association for me has always been Royal Poincianas in full bloom through the month of June and spindly tall Royal Palm Trees. This is precisely what I saw as we touched down in Habana, and both trees were prevalent across town. I was happy to be there and conversed easily with folks.

Right off, I could see that my feelings for Habana would be of mixed emotions when I left. On the one hand, people were friendly, gentle, and because communism is not so old there (as compared to its history in eastern Europe), I encountered no shortage of entrepreneurial spirit. Children were obedient and educated. With a cafecito each morning while awaiting my ride, I watched from the porch of the El Presidente Hotel as school children in clean uniforms sat politely in rows while their teachers read to them outside the school across the street. Buildings were beautiful old Castilian Spanish, vegetation was lush, climate was humid-warm (to my liking), and Cuban food is second only to my own Lebanese cuisine. Although most everything is scrumptious, my favorite is lechon assado.

The not-so-nice part of Cuba was the way in which the whole country is frozen in time — as if time had been standing still since 1960. Lots and lots of 50-year-old cars, many dead on the sides of the road because a carburetor broke 12 years ago. This was a metaphor for economic life in general. Objects one would consider fancy were chipped or worn, old without upkeep or replacement. The city was economically depressed. Considering that a month's salary of most workers bought a chicken, and hardly anyone could afford to eat meat, the economic lot for most Cubans was depressing. How awful, when even a small investment by United States standards would jumpstart a stalled economy, with hard-working people just waiting for the opportunity. Still, people with whom I spoke blamed no one, but everyone wished that conditions would improve. This backdrop, I'm sure, explains the Miami Cubans who live a cultural version of the "depression baby," growing up to save every rubber band and jar.

In general, however, contrasts abound. Religion is somewhat restricted in Cuba, whereas Miami exiles are strongly faithful. Daily fare for most Cubans is beans and rice; in contrast, even the smallest Cuban restaurants in Miami serve abundant meals of meats and fish and offer rich desserts. Political expression is harshly curtailed in Cuba, whereas exiles participate actively in the political process.

Anyway, there I was, knocking around in a city new but, at the same time, very familiar to me. I met all sorts of people, many with kin in southern Florida. My first few days were in a hotel on the outskirts of town built years ago for the Pan American games. For three nights, I shared cigars with new acquaintances — and grabbed every "rana platanera" I could find. The bravest any local ever got was to approach one close enough to point and call me. I also saw tarantulas at night and, during the day, the Cuban Brown and Green anoles (*Anolis porcatus*) were ubiquitous. Both species are established in southern Florida, as is the very un-anole-like Knight Anole (*A. equestris*), an enemy of the Cuban Treefrog. I saw no Knight Anoles, nor did I encounter *Chamaeleolis*, an ancient anole ecologically analogous to the Old World chameleons.

To make transportation to the meeting at the Palacios de la Convencióes easier, I was moved to El Presidente Hotel in the heart of Habana, just blocks from the famous Malecon (the seaside thoroughfare). Of course, I preferred rural areas for the frog, but the city was an experience as well. Other than one sour area I decided to avoid, Habana was safe and much to my liking. So, I went off to the meetings each day. During breaks, I'd be off to catch treefrogs. They often rested in exposed situations because the daily rain showers would bring frogs out of their retreats. Luis Moreno, an elderly Cuban scientist and gentleman, and I jumped into the fountains like kids. Barefoot and with pants rolled up, we caught adults and tadpoles, and even found calling males right after a fierce but short storm. These storms are common during the Cuban summers, and are known locally as Manga de Viento.

My Kind of Place

I sure liked it there. I kept having great dreams of taking the railway across the spine of Cuba looking for frogs in towns I know only from museum labels and maps. And why? Very simply, because, within the region of the wet-dry season subtropical system of southern Florida and the Caribbean, resides a series of common threads relating to the fauna, flora, and life histories of these species and likewise common threads of human customs, food, and language. I would return gladly, over and over again, to work in that region. Taken together, the West Indies are home to a most curious and diverse herpetofauna, living in a physical and cultural setting so very much to my liking.

Suggested Reading

Meshaka, W. E., Jr. 2001. *The Cuban Treefrog in Florida.* Univ. Press of Florida, Gainesville, Florida.

Silva Lee, A. 1996. *Cuba Natural.* Pangaea, Saint Paul, Minnesota.

Meshaka, W. E., Jr. 2003. Cuban treefrogs, curiosity, and the West Indies, p. 89–91. *In* R. W. Henderson and R. Powell (eds.), Islands and the Sea: Essays on Herpetological Exploration in the West Indies. Society for the Study of Amphibians and Reptiles, Ithaca (New York). Contributions to Herpetology, volume 20.

My Love Affair with Cuban Reptiles

Lourdes Rodríguez Schettino[1]

When I was a little child, my parents took me to interesting places throughout Cuba. Although at that time I was not interested in amphibians and reptiles, I enjoyed the beautiful landscapes and collected small stones and flowers from each place we visited. This modest collection may have been indicative of my burgeoning affection for all things natural. Later, during my undergraduate course of study, my favorite academic subject was zoology. Upon completion of my degree, I joined the Faculty of Biology at the University of Havana. There, I had several good teachers who guided me into the field of vertebrate zoology, and, when I finished my predoctoral studies, I began working in the Institute of Zoology of the Academy of Sciences of Cuba as a researcher in herpetology.

I have been interested specifically in herpetology since my professor, Orlando H. Garrido, exposed me to the many Cuban species and noted that much has yet to be discovered in this field. Sadly, many Cubans do not like or are afraid of herps. The aversion to the frogs is widespread, people saying that they are ugly, cold, and slimy. Some farmers pass on the story that giant anoles bite humans and cause fevers. Others explain in all seriousness that the Cuban Boa will swallow the legs of a person sleeping in the shade of a tree. The most astonishing story, however, is that the Cuban Boa will suck at the breast of a nursing mother while sticking the tip of its tail in the mouth of the infant, effectively deceiving both mother and child.

Inoffensive and Even Beautiful

The truth is that Cuban herps are inoffensive creatures, without venom or toxins of any kind, and generally inclined to escape at the first indication of human presence. They neither attack nor pursue anybody. Moreover, they are beautiful and interesting in their habits and behavior. For instance, the very tiny frogs of the genus *Eleutherodactylus* are at the lower size limit for vertebrates. *Eleutherodactylus iberia*, a species from the mountains of northeastern Cuba, may be the second smallest of all amphibians, with male snout-vent length ranging from 9.6–10.0 mm. Also, contrary to the pattern seen in most frogs, all of the Cuban species in this genus (48) bypass the free-living tadpole stage and instead undergo direct development within the egg. Every naturalist should witness the spectacle when such diminutive frogs emerge from the egg.

Most impressive are the lizards of the genus *Anolis*. Many species inhabit Cuba, and most are exceedingly abundant, living mainly on the trunks of trees, but also on cliffs, rocks, and in caves. Many are beautifully patterned, and males and some females exhibit amazing, brilliantly colored dewlaps that identify each species. It's a wonderful sight when a male fully extends his dewlap, warning other males to stay away or, conversely, advertising his availability for female attention.

Two noteworthy species of *Anolis* have a transverse fold of skin in the gular region instead of a

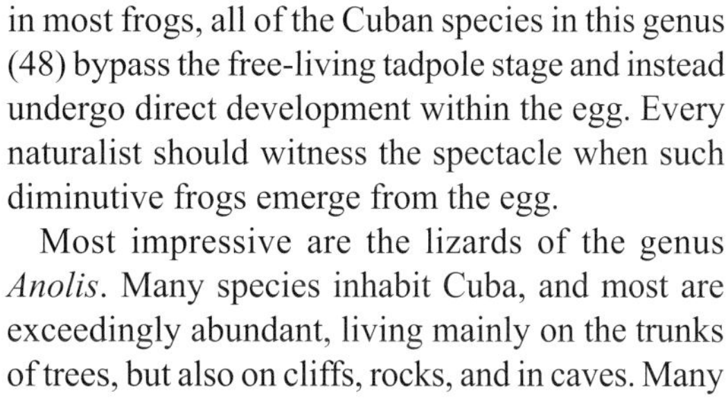

Figure 1. *Anolis bartschi*: male from Cueva del Indio, Pinar del Río Province; arguably the prettiest anole in Cuba.

[1] Instituto de Ecología y Sistemática, CITMA, Calle de Varona Km 3.5, Boyeros, AP 8029, La Habana, CP 10800, Cuba.

FIGURE 2. Panoramic view of the Viñales Valley, Pinar del Río Province, showing the habitat of *Anolis bartschi* in the "mogotes" of Pinar del Río Province.

dewlap. One of these, *A. vermiculatus*, also is quite peculiar in its habitat selection. Living in the gallery forests of the Cordillera de Guaniguanico is not so unusual, but a strict association with rivers and a readiness to dive into the water at the first signal of danger is unique among Cuban anoles. Individuals can stay underwater for up to 50 minutes. The other anole without a dewlap is *A. bartschi* (Figure 1), arguably the prettiest anole in Cuba. It also lives in a rather unusual habitat, the walls and holes of the "mogotes" (hummocks) in the Sierra de los Organos (Figure 2). Both species are vocal and squeak when they are handled or threatened.

Phenomenal Habitat Diversity

The high number of Cuban herps (57 amphibians and 135 reptiles are currently recognized) is largely attributable to the great variety of available habitats that range from well-preserved cloud, montane rain, and evergreen forests to swamps and mangrove forests, from the most heterogeneous to the most homogeneous types of vegetation, including those places modified by humans. Politically, Cuba is divided into 14 provinces and the Isla de la Juventud off the southern coast of La Habana Province. If one considers the geographic distribution and endemism of reptiles, 12 biotic zones (Figure 3) are

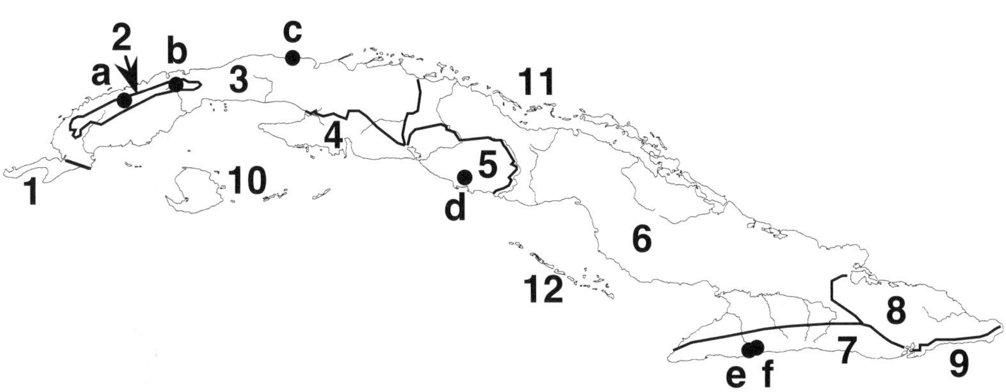

FIGURE 3. Map of Cuba showing biotic zones (indicated by numbers) and the main localities mentioned in the text (indicated by lower case letters): (1) Península de Guanhacabibes; (2) Cordillera de Guaniguanico; (3) western lowlands; (4) Península de Zapata; (5) Macizo de Gamuhaya; (6) central eastern lowlands; (7) Sierra Maestra; (8) Macizo de Sagua-Baracoa; (9) southeastern coastal plain; (10) Archipiélago de los Canarreos; (11) Archipiélago de Sabana-Camagüey; (12) Archipiélago de las 12 Leguas; (a) Cueva del Indio, San Vicente, Viñales Valley, Pinar del Río Province; (b) Soroa, Pinar del Río Province; (c) Playa Jibacoa, La Habana Province; (d) Topes de Collantes, Sancti Spíritus Province; (e) La Mula, Santiago de Cuba Province; (f) Pico Turquino, Santiago de Cuba Province.

evident, with those involving mountain ranges supporting higher number of species and higher levels of endemism than those located in lowlands. Some of this is attributable to the habitat heterogeneity associated with topographic features, but the mountains also have more well-preserved habitats and mosaics of vegetation patterns that support a lot of microhabitats where many species can find shelter and food. Some subset of almost all of these biotic zones is protected within at least one of the six Biosphere Reserves that have been recognized by the UNESCO and the IUCN for Cuba.

Splendid Landscapes — and Herps

Perhaps the most beautiful landscapes are found in some coastal zones in the southeastern region of the eastern provinces, where the slopes of the highest mountains, the Sierra Maestra, extend almost to the seashore, and in the mouths of rivers flowing from the mountains. The latter are characterized by peaceful beaches with stones worn smooth by the combined effects of the sea and rivers. Although tempting, swimming is not allowed at these sites because the deep waters close to shore are home to dangerous riptides and currents. Still, they're nice and peaceful, great places to rest, listening to the placid murmur of the sea and enjoying the pleasant views. One of the most beautiful of these is La Mula, at the mouth of the Río Turquino, where the fresh water emanating from the highest elevation in Cuba doesn't mix with the sea water until well off the coast.

However, places where mountains meet the sea are not the only scenic spots. Many others are just as wonderful in their own way. For example, the rivers high in the mountains feature pure, crystal-clear, and very cold water cascading over rocks and stones. Also impressive are the "mogotes" that border the Viñales Valley. These are vertically walled, perforated limestone hills that support a unique type of vegetation and an equally unique and well-adapted fauna. Rendering these and other similarly marvelous places even more wonderful is the fact that all are inhabited by herps, many of which are endemic to a region or even to a specific locality.

These splendid landscapes — and their herpetofaunas — have attracted many Cuban and foreign herpetologists over the years. Like me, they've been attracted by the animals but, in the process of studying our reptiles, they have visited many localities, from coastal regions to the highest mountains, from the Península de Guanahacabibes at the western tip of the island to Maisí near the eastern extreme, from the Península de Hicacos in the north to the southern region at Cabo Cruz, and to many of the offshore islets and cays. Thus, herpetology has allowed them (and me) to become better acquainted with my country and, in the process, to meet and establish friendly relationships with many good people.

Unforeseen Consequences of Herpetology

Lizards of the genus *Anolis* are my favorites among the many Cuban herps. One of them, *A. lucius* (Figure 4), found on rocks, cliffs, the entrances of caves, and tree trunks, always in close relation to limestone, is widely distributed from eastern (La Habana Province) to western (Granma Province) regions of the island. Its most extraordinary feature consists of three blue, black-bordered, semi-transparent scales in its lower eyelids that work as "sunglasses" when the eyes are closed at the moment the lizards emerge from their shaded shelters into the bright sunlight. At Playa Jibacoa, on the northeastern coast of La Habana Province, some of my students and I studied this lizard during monthly visits over an 11-year period.

We learned that *A. lucius* feeds mainly on ants, breeds from March to October, lays its eggs in communal sites firmly attached to irregularities in the cave walls and to each other, and squeaks indignantly when captured or escaping into its retreats. Studying this species was fun, both because of its unique characteristics and because the study site is rather nice.

The study site is a cluster of small caves in the middle of a cliff about 12 m above sea level. To get to the caves one climbs a stairway, from the top of which one has a terrific view over the canopy of the coastal vegetation to the beautiful beach below.

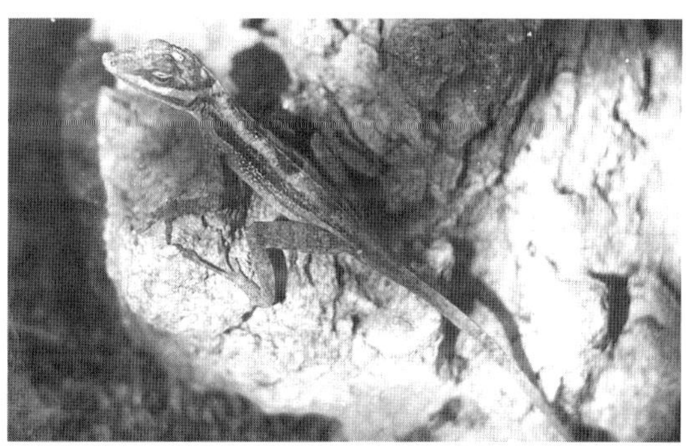

FIGURE 4. *Anolis lucius*: female from Playa Jibacoa, La Habana Province.

Anolis lucius is easy to see on the rocks and walls of the caves and the collecting is quite pleasant — but sometimes with unforeseen results. During the first years of fieldwork, two of my students fell in love, got married, and, after a little time, a daughter was born — an infrequently reported and potentially distracting consequence of herpetology!

Sea Grapes and *Guazuma* Trees

Anolis argenteolus is in many ways similar to *A. lucius*. They resemble each other in their habits and in some characteristics of their habitat use and ecology. *Anolis argenteolus* lives in the east-central portion of Cuba from the eastern region of Camagüey Province to the eastern tip of the island. It also has "sunglasses," but they are composed of two (instead of three) semi-transparent scales. We conducted a long-term study of *A. argenteolus* at La Mula (Figure 5), that most marvelous place on the southeastern Cuban coast. When I visited the site for the first time in April 1980, I found an extensive, dense sea grape forest consisting of old, thick trees with large branches occupied by several species of lizards. Although that site became a goal in its own right over the years, that first visit was part of a more extensive expedition into the Sierra del Turquino.

We were conducting a survey of vertebrates, the first step of which was a 12-day stay at Alto de La Emajagua, a small, flat plateau at an elevation of about 800 m along the southern trail to Pico Real (or Turquino), which reaches an elevation of 1972 m. Our camp was situated alongside a well-forested glen surrounding the springs that feed the Río Turquino. We had two tents, a large one for the six men and a small one for me, the only woman in the group. As pretty as the site was, these were unpleasant days for me because the wet conditions aggravated my asthma. Nevertheless, it was a very successful trip that yielded many herps, including a new species of lizard. In retrospect, I've largely forgotten my misery and tend instead to fondly remember the lizards.

FIGURE 5. Habitat of *Anolis argenteolus* in the sea grape forest at La Mula, Santiago de Cuba Province.

Subsequently, Orlando Garrido and two others of our party took off on a two-day ascent of Pico Cuba (1872 m) while three of us remained to continue surveying the area around the camp. We were about 200 m from the camp when what appeared to me to be a non-typical male *A. angusticeps* on a twig of a *Guazuma tomentosa* tree. Raúl Cabrera captured it and another similar male in a nearby tree. I kept these two lizards alive until Garrido returned. I told him that I had two lizards that "are and are not *A. angusticeps*." He asked: "How can that be? Let me see them." When he saw the lizards, he immediately exclaimed: "This is a new species! Let's find some more specimens." A careful search revealed eight more individuals on the twigs of a big *Guazuma* tree right behind my tent. These lizards became the type series for *A. guazuma,* described by Garrido in 1983. This pretty little anole spends most of its time flattened on the peripheral twigs of trees, but quickly falls to the ground at any sign of danger.

After a few days we moved our camp to the sea grape forest at La Mula. Both the weather and the fieldwork were good. We found many individuals of *A. argenteolus* living on the sea grape trees at heights ranging from 1 to 2 m, whereas *A. jubar* was abundant nearer the ground (at heights <1 m).

***Anolis argenteolus* at La Mula**

I next visited La Mula with Stan Rand and Dale Marcellini in July 1985. Some of the trees had been cut in 1982 to establish a rustic camping site among the sea grapes. *Anolis argenteolus* was still the dominant species, but *A. jubar* was noticeably less abundant than it had been during our first trip. We were unable to explain why one species appeared to have been affected more than the other by the changes in habitat. Perhaps, because *A. argenteolus* is exceedingly cryptic and difficult to see, it was less likely to be disturbed by the increased number of human visitors to the site. In contrast, *A. jubar* is dark brown to black, contrasts dramatically with the bark of the trees and may have become a common target of human activities.

From September 1985 to November 1986, I visited La Mula bimonthly to study *A. argenteolus*. I found that these lizards feed mainly on ants, breed

FIGURE 6. *Anolis ahli*: male from Topes de Collantes, Sancti Spriritus Province (photograph by Kevin de Queiroz).

all year round, lay eggs in communal sites firmly attached to crevices in trees and to each other, and squeak when handled or alarmed. Although the camping site remains and has been expanded to include small cabins, the locality is still very nice. During my most recent visit in June 1999, *A. argenteolus* and other species were abundant, but *A. jubar* had almost entirely disappeared from the area.

A Ghost?

Topes de Collantes is another special place. It lies about 800 m above sea level in the Sierra del Escambray. Pine forest is the predominant type of vegetation. A large, solid building, sitting on top of the mountain, serves as a hospital for people suffering from stress or from respiratory and rheumatic ailments. In addition, three very nice hotels provide opportunities for resting or for enjoying horseback rides through the beautiful surroundings, the most spectacular features of which are the splendid falls of the Río Caburní.

In May 1979, I conducted an herpetological census at this site with Alberto Valdés. We collected *Anolis ahli* (Figure 6), a species endemic to the Sierra Guamuhaya, finding most specimens in a deep forested glen behind one of the hotels. We were pleased because one of the goals of the trip was to enrich the holdings of endemic species in the herpetological collection at the Institute of Zoology (now Institute of Ecology and Systematics). Later, when all the specimens were in the collection, our excellent curator, Luis Moreno, noticed some traits in one of the lizards that did not match exactly with *A. ahli*. He informed Garrido, who immediately described it as a new species, *A. delafuentei*.

This was the second time that I was present for the discovery of a new species, but I hope that it will not be the last. Curiously, in May 1981, Moreno returned to the same place looking for more lizards of this species and only found *A. ahli*. Also, I was there in April 1994 and again in June 1995 with Kevin de Queiroz; we carefully searched the glen by day and night, and we did not find any *A. delafuentei*. Steven Poe and I were there once more in October 2001, and again failed to find *A. delafuentei*. What has happened? Is this a "ghost" or a true species? Is it merely rare or has it become extinct? Future research will tell.

Anolis bartschi

As noted, *Anolis bartschi* may well be the most beautiful of all Cuban anoles, and its habitat is simi-

FIGURE 7. *Anolis bartschi*: male from Cueva del Indio, Pinar del Río Province.

FIGURE 8. Habitat of *Anolis vermiculatus* in a gallery forest at Soroa, Pinar del Río Province.

larly wonderful. These lizards live on the walls of the "mogotes" in the Sierra de los Organos. A particularly special "mogote" in San Vicente, Viñales Valley, is bisected by a river, creating a big and interesting cave, the Cueva del Indio. *Anolis bartschi* is very abundant on the walls of this hill (Figure 7).

At that locality in October 1981, Stan Rand and I observed a fight between a tethered and a resident male. Both males raised their limbs, swelled their throats, opened their mouths, and attacked before the "intruder" tried to escape (Stan considerately pulled the tethered male to safety before any substantial damage was done). Later, as a result of bimonthly visits from May 1986 to April 1987, we learned that *A. bartschi* feeds on insects, mollusks, flowers, and fruits, breeds from February through September, and lays its eggs in communal sites in cracks in the walls of the mogote.

Twelve Sympatric Anoles!

However, in spite of the many wonderful sites where I've engaged in herpetological fieldwork, my absolute favorite is the gallery forest at Soroa, in the Sierra del Rosario (Figure 8). Many species occur in the relatively small and accessible area. Of particular interest is *Anolis vermiculatus*, the Cuban Caiman Lizard, whose name comes from its resemblance to a small crocodile and from its previously mentioned willingness to escape danger by jumping into the flowing water of the river.

During the months of March, May, and September 1984, June, July, and December 1986, and February 1987, my student, Mercedes Martínez, and I studied some aspects of the species' ecology, collecting data along the shores of the Río Manantiales, which encircles the hotel at Soroa and falls in a pretty cascade 18 m high. We found *A. vermiculatus* within 5–6 m of the water, on trunks, branches, and roots of trees and on rocks. It feeds on big insects, spiders, fresh water shrimp, mollusks, small fishes, amphibians and reptiles, and flowers and fruits. Reproductive activity extends from April to August, but the egg-laying sites remain a mystery.

As special as *A. vermiculatus* is, it is not the only reason for conducting fieldwork at Soroa. In a plot of approximately 5,000 m^2 that extends from

FIGURE 9. *Anolis vermiculatus* on a rock over the water of the Río Manantiales, Soroa, Pinar del Río Province.

FIGURE 10. *Anolis mestrei*: a male from Soroa, Pinar de Río Province, fully extending his dewlap (photograph by Kevin de Queiroz).

behind the tourist cabins to the waterfall, we have found 12 species of anoles (e.g., Figure 10) in the gallery forest! Diverse microhabitats allow each species to occupy its own "home" without interference from others. My first visit to the site with herpetological goals was in February 1976, but I've returned many times, most recently in October 2001. Every time I introduce this amazing assemblage of species to somebody new, they've been amazed to see the clearly differentiated microhabitat in which each species lives. I have been pleased to reveal this treasure to a series of researchers like Stan Rand (1981, 1983), Ernest Williams and Dale Marcellini (1983), M. Lizana (1986), Kevin de Queiroz (1994, 1995, 1997), Jonathan Losos, Paul Hertz, and Leo Fleishman (1997), and Steve Poe (1999, 2001).

Each Trip Only Gets Better

I am extraordinarily fortunate to have been able to conduct fieldwork throughout Cuba. Many of the trips have been difficult, and preparations and logistics often are almost overwhelming, but, looking back, recollections of the problems encountered fade in the face of the good memories of interesting lizards, special places, and nice people. Each trip has surpassed the previous, not only in terms of scientific data but for the remembrances of adventures in this paradise that is Cuba. I am already looking forward to my next trip.

Suggested Reading

Buide, M. S. 1985. *Reptiles de Cuba*. Editorial Gente Nueva, La Habana.

Estrada, A. R. and S. B. Hedges. 1996. At the lower size limit in tetrapods: a new diminutive frog from Cuba (Leptodactylidae: *Eleutherodactylus*). *Copeia* 1996:852–859.

Rivalta González, V. 2000. Checklist and bibliography of Cuban amphibians (Anura). *Smithsonian Herpetol. Info. Serv.* 124:1–20.

Rodríguez Schettino, L. 1985. Distribución altitudinal de los iguánidos en la Sierra del Turquino, Cuba. *Cien. Biol.* 14:59–66.

Rodríguez Schettino, L. 1993. Areas faunísticas de Cuba según la distribución ecogeográfica actual y el endemismo de los reptiles. *Poeyana* 436:1–17.

Rodríguez Schettino, L. (ed.). 1999. *The Iguanid Lizards of Cuba*. Univ. Press of Florida, Gainesville.

Rodríguez Schettino, L., D. L. Marcellini, and J. Novo. 1987. Algunos aspectos ecológicos sobre *Anolis vermiculatus* (Sauria: Iguanidae) en Soroa, Pinar del Rio, Cuba. *Poeyana* 343:1–9.

Rodríguez Schettino, L. and M. Martínez Reyes. 1992. Hábitos alimentarios de *Anolis bartschi* en San Vicente, Pinar del Río, Cuba. *Cien. Biol.* 25:30–40.

Rodríguez Schettino, L. and M. Martínez Reyes. 1994. Características tróficas de una población de *Anolis lucius* (Iguania: Polychridae) en la costa septentrional de Cuba. *Avicennia* 1:67–77.

Rodríguez Schettino, L. and M. Martínez Reyes. 1996. Algunos aspectos de la ecología trófica de *Anolis argenteolus* (Sauria: Polychridae) en una población de la costa suroriental de Cuba. *Biotropica* 28:252–257.

Schwartz, A. and R. W. Henderson. 1991. *Amphibians and Reptiles of the West Indies: Descriptions, Distributions, and Natural History*. Univ. Florida Press, Gainesville.

Rodríguez Schettino, L. 2003. My love affair with Cuban reptiles, p. 93–99. *In* R. W. Henderson and R. Powell (eds.), Islands and the Sea: Essays on Herpetological Exploration in the West Indies. Society for the Study of Amphibians and Reptiles, Ithaca (New York). Contributions to Herpetology, volume 20.

The Pleasure of the Adventure: Searching for Reptiles in Eastern Cuba

ANSEL FONG G.[1]

Describing in a few pages my feelings about the West Indies — and, more specifically, my homeland of Cuba — is a most difficult task. Similarly, I have a hard time describing exactly what attracts me to my work as a scientist. It is impossible to explain why I walk to work every day with the same unbridled joy and curiosity as a pupil walking to his first day at school — and I find it equally impossible to say why I return time after time to my fieldwork — even to distant and nearly inaccessible sites. Obviously, I derive considerable pleasure from my work, which is linked inextricably to this island, this fauna, and this life. The best examples to illustrate "my experience in this tropical paradise" have occurred during fieldtrips to El Toldo (Figure 1), a high plateau in eastern Cuba. These trips have involved great adventures, scientific discoveries, beauty, and friendships.

A Hard Trip and a Magnificent Morning

"We begin monitoring El Toldo on September 25," said Dr. Nicasio Viña, director of the center where I work, on an afternoon in 1996. Little did I realize then that these words would mark the beginning of one of the most unforgettable experiences I have had in my career as a herpetologist.

[1] BIOECO, Museo de Historia Natural "Tomas Romay," Enramadas # 601, Santiago de Cuba 90100, Cuba.

FIGURE 1. El Toldo resembles a mountain that has had its top sliced off (photograph by N. Viña).

FIGURE 2. Our caravan of vehicles crawling toward El Toldo (photograph by Luis O. Melián, courtesy of N. Viña).

El Toldo is really a mountain massif, called a high plateau by geographers, resembling a mountain that has had its top sliced off. It sits in the northern part of Cuba's Oriente region, about 25 km southeast of the city of Moa in Holguín Province. The city is known as the "red city," due to the characteristic color of the soils in the surrounding area. From Moa, the mountain seems permanently covered by an immense cloud.

At midday our caravan of vehicles (Figure 2) started to climb El Toldo. We had a GAZ 66 Russian truck, a WAZ jeep, a fuel tanker, and later a Japanese Komatzu bulldozer. At first, the road looked easy, extending like a long, narrow, red band threaded across the landscape. But, with time, the road became steeper and muddier, slowing our progress to a crawl. Eventually, the jeep had to be towed by the tanker and it, in turn, had to be pulled by the bulldozer. At times, the scene reminded me of the famous Kon Tiki dragging its lifeboat through a rough sea. After a long five hours, we finally arrived at our camp.

The next morning helped erase memories of the difficult journey. A magnificent sunrise — the sun providing enough warmth without burning — created a shimmering kaleidoscope of color as light refracted through the dewdrops hanging off every blade and branch. A light wind brought with it a fresh woodland smell of pine and earth. Birds added to the magical moment with a cheerful mixture of trills, whistles, and cries.

FIGURE 3. The Wright's Trope (*Tropidophis wrighti*), a very peaceful and beautiful Cuban snake (photograph by A. Fong).

Our first day at El Toldo was full of new discoveries. At 7 AM, we left for our first exploratory journey, and we couldn't take a step without seeing or hearing something new. We first noted the birds, and here we were blessed with some of the best singers in Cuba. King of the mountains is the Cuban Solitaire (*Myadestes elisabeth*), with its long melancholic call that sounds like the cry of a thwarted lover. Another great voice is the Cuban Bullfinch (*Melopyrrha nigra*), with a melodic voice that emerges from the cacophony of background noise. It fluttered around us, alongside the Cuban Hummingbirds (*Mellisuga helenae*) and Oriente Warblers (*Teretistris fornsi*).

We continued deeper into the forest and combinations of colors flashed in front of our eyes: the blue, white, and red of the majestic Cuban Trogon (*Priotelus temnurus*); the green, white, and red of the dumpy little Cuban Tody (*Todus multicolor*); and the grey, green, and yellow of the Oriente Warbler. Colorful Cuban Parrots (*Amazona leucocephala*) flew by, while the Stripe-headed Tanager (*Spindalis zena*) showed off its coat of orange, olive-green, yellow, black, and white.

A Most Diverse Fauna

That first day was not unique. Each of the 27 days of the expedition brought us into contact with the unique fauna of the region. Turning over a rock, we'd discover a pretty scorpion, or a little tarantula would poke its face out of a hiding place. Rooting through a bromeliad, we would come across a whole family of rare little ants, carrying food or their own babies, still in the larval or egg stage. A stick insect might appear on your shoulder. If you tried to catch it, it would stop, extend its fore and hindlegs, and pretend it was a dead pine twig, which it resembles exactly in texture and color.

However, I am a herpetologist, and so it was the reptilian and amphibian discoveries that kept me enthralled on these journeys. In one of my first excursions, I found three wonders of the Cuban herpetofauna. The first was a Dusky Trope (*Tropidophis fuscus*), a small, dark snake that measures only

FIGURE 4. The El Toldo Anole (*Anolis toldo*), one of the rarest (and most wanted) lizards in Cuba (photograph by N. Viña).

some 25 cm in total length. Until that moment, only two specimens of this species had ever been seen. The one I found was a female that soon thereafter produced three babies. These, too, were dark, but with two big white spots on the nape — as if they were wearing a headscarf — and had yolk-colored tail-tips. Incredible!

The second rare find was a Wright's Trope (*Tropidophis wrighti*; Figure 3). This small, off-white colored snake is covered with big black spots that remind one of a clown's outfit. The strange pattern and the yellow tail-tip have led to the misguided belief that this little snake is dangerous. Cuba, however, supports no venomous species, and most snakes — particularly this one — are quite peaceful. Wright's Trope is capable only of curling into a ball to protect itself when disturbed.

The third marvel was a Pineland Bush Anole (*Anolis inexpectatus*), a small lizard that is almost impossible to find unless you employ just the right search image. Its brown color, slender habitus, and long tail help this anole disguise itself as a twig. Only its brilliant blue eyes are obvious, in a fashion that would be envied by any girl aspiring to become a model. These little lizards also are capable of disappearing from sight in two or three jumps when threatened.

For exclusive discoveries, the prize goes to a lizard so rare that it didn't even have a common name. We decided to name it the "El Toldo Anole" (*Anolis toldo*; Figure 4), after the place where we found it. That happened on 19 June 1997, during our third trip to El Toldo when we had climbed to the highest point of the peak. After spending most of the morning and part of the afternoon there, the nine of us (two ornithologists, an arachnologist, three geographers, two botanists, and me) were returning to camp, walking single file, tired, but still full of enthusiasm. Luis de Armas, the arachnologist, was walking at his typically casual pace, all the time scanning his surroundings as if trying to undress the forest. Suddenly, he stopped, looked up, and extended his hand into a shrub alongside the trail. When I reached him, he was holding a little green devil that was squeaking, biting, and frantically struggling to escape.

Back at camp, I was able to examine it in greater detail. It was a beautiful green animal bearing dark green spots on its tail, legs, and head. The body spots were larger and aligned in vertical stripes. Like few Cuban lizards, it was emitting short high-pitched squeaks. It turned out to be a new species, and I had the pleasure, along with Orlando H. Garrido, the dean of Cuban herpetologists, to be one of its describers. At present, only this single specimen is known, enlarging the list of very rare (and most wanted) reptiles in Cuba. As far as we know, the species lives only in these mountains, hiding among the twigs of trees and shrubs in the virgin forest.

Facing Nature

Often, the problematic situations, rather than the nice experiences, bring the greatest rewards. This certainly has been the case with my work as a herpetologist, during which I am constantly coming face to face with physical obstacles and logistical challenges I have to conquer with will power, perseverance, and skill — but, if biology consisted only of going to work and sitting at a computer, I would be better off becoming a fireman.

Our trips to El Toldo were perfect examples of the difficult situations faced in the field. Every journey brought its difficulties and a need to find the quickest solution with a minimum of personal and scientific risk. One of our most challenging — and rewarding — experiences occurred when we were caught in the middle of Hurricane George during our fifth and last expedition in 1998. Every day we had been listening to the weather reports and tracking its projected path toward our position, and, finally, when we were certain George would pass close to El Toldo, we decided to stay. We knew we were taking risks being so exposed on the mountain, but we also knew that it was a unique opportunity to conduct research on the behavior of the fauna as the hurricane passed through.

We made the necessary preparations to our camp, set up our equipment, and waited for the storm. Rain started falling in the early hours of 22 September. During the next 36 hours, it would beat down incessantly, requiring us to empty the rain gauge several times as we measured 324 mm of rain. As dawn broke on the morning of the 23rd, we realized that we would not be facing merely heavy rain. The winds picked up speed and, as the situation was getting difficult, we decided to pack our gear, leaving only the most indispensable items. Everything was protected with multiple layers of waterproof material and placed in water-resistant boxes.

When the sky became dark, everyone went to bed to try and get some much-needed rest, foreseeing far worse conditions ahead. By midnight, our predictions were confirmed when a gust of wind, which seemed to have come from the depths of the earth, hit our hut, shaking it like the rattling bones of an old skeleton getting up to walk and, more critically, leaving the front wall at a dangerous angle. We rapidly gathered our things and assumed a combat position. From that moment, the gusts came one after another — we'll never know their actual velocity, but we estimated some in excess of 100 km/h. Looking over the damage to the building, we made the decision to move to a smaller hut in the camp, one that offered less resistance to the winds. For two hours, this refuge protected us from the rain and winds. The walls rattled, the ceiling trembled, but the little house withstood the worst that Hurricane George threw at it.

We were all soaked to the bone and the combination of darkness and feelings of insecurity was like something out of a horror film. Then, suddenly at 2:30 AM, the wind disappeared, the rain stopped, and we were left in an eerie calm — almost like we had suddenly been moved to another part of the globe. The eye of the storm was above us.

A patrol (Figure 5) was organized rapidly and, in half an hour, we had assessed the damage and evaluated the situation. The river — no more than 10 m from the camp — was swollen. The hut where the electric generator had been before we moved it to a safer position was completely destroyed. The front wall of the main hut was leaning dangerously, and some nearby pine trees had been felled by the wind — but the frogs were still croaking, seemingly oblivious to what had just happened.

However, this was not the end. At about 4 AM, the winds and rain began again, the latter coming even stronger than before. Fortunately, the trailing half of George was not as wide, and, by dawn, the storm had almost completely disappeared. Only a light drizzle remained in the morning, as if George was intent on extracting the last drops of rain from the clouds. By noon, the sun was warming the landscape and wiping out the memories of the night before. The birdsongs returned, lizards reappeared, and the *Ameiva auberi* resumed their tireless quests for food. By afternoon, only the fallen trees provided evidence that a hurricane had passed through.

FIGURE 5. Luis Melián led the patrol to evaluate the situation (photograph by N. Viña).

After our return to civilization, we saw the evidence of George's might in pictures from throughout the Caribbean. Many islands, principally Puerto Rico, had borne huge amounts of destruction, whereas Cuba had managed to escape with relatively little damage. Heavy rain (more a benefit than a harm) was the only real effect (Figure 6). Personally, the experience was unforgettable and, for the team, it provided invaluable insights into the effects of one of the most powerful forces in nature.

People and Personalities

The current Cuban population is the product of hundreds of years of ethnic and cultural mixing. The

FIGURE 6. Heavy rain was the only real effect of Hurricane George (photograph by N. Viña).

beliefs, customs, and idiosyncrasies of Europeans, Africans, and Chinese immigrants have combined to form the rich blend found in present-day Cuba. This melting pot of characters is one of the things I love most about my country, and my work brings me into contact with a wide spectrum of personalities. About ten people took part in the trips to El Toldo, and the group was so eclectic (and, at times, eccentric), it would make an excellent cast-list in a García Márquez novel.

Our cook was nicknamed "El Diablo" ("The Devil"). He was a short, thin man about 60 years old. Although very talkative and boisterous, he also was very superstitious and afraid of the unknown. His fear of the dark and of solitude was so great that he often would become terrified after dark, believing that a tiger-cat prowled the forest. In fact, this mythical tiger-cat was the creation of some of my colleagues who convinced the cook that such an animal was exploring the camp at night, leaving tooth and claw marks in the outside walls.

During our second journey, Luis de Armas, our scorpion specialist, was doing surveys to evaluate the knowledge and use of scorpions among the native population. "El Diablo" wanted to be involved, so Luis began asking him some questions.

"Do you know something about scorpions?" asked Luis.

"No, I don't," replied "El Diablo."

"And what about alacranes?" tried Luis, using the other common name for scorpions in Cuba.

"Oh yes, I know everything about alacranes," "El Diablo" responded.

He thought he knew something about everything and was always eager to share his opinions. If you were talking about timber, he would begin talking like a woodcutter. If the conversation drifted to ants, he would tell you which were good ones and which were bad. The only time he couldn't find something to say was when two colleagues held a conversation one night about cladism and its application to zoological groups in Cuba. "El Diablo" sat and listened for a few moments, and then stood and walked away, acting as if he had just witnessed two Martians discussing the merits of Jupiter over Saturn.

Another character was a local Moan called Pedrín. He was a tall man with deeply tanned skin and the physical build of a man who has spent his life doing physical work. He was a rural man who had lived among the flora and fauna of the area and, for him, the best sounds were those of birds, the best smells those of flowering plants, and the best sights those of the green Cuban woodlands.

Pedrín had been a woodcutter since childhood, and the forest held no secrets for him. He didn't fear anything — well, except for spiders and snakes. He feared these two animals so intensely that he would run around the camp screaming if someone tried to approach him with a harmless spider. Several times he refused to come into my room because I had a little Cuban Racer (*Alsophis cantherigerus*) curled up in my hands.

Finally, I will tell you about Luis O. Melián, an ornithologist by profession, but a general handyman by trade. His avocation is helping solve other people's problems. Melián, as everyone knows him, is tall and thin like a Cuban palm with a frond of tangled hair that he never combs. He is over 50 years old, but has the energy of a 25-year-old. He has never missed a single field trip. As an ornithologist, he has yet to encounter a Cuban bird in the field that he couldn't immediately identify.

Always ready for action, he invariably dresses the same way each day: boots, blue jeans, and a long-sleeved shirt. From his neck hang two pairs of glasses — sunglasses and binoculars — and his shoulder sports a clipboard with field data sheets, colored pens, and a compass. A hunting knife is attached to his belt and his back bears a small rucksack. Ah, the Melián rucksack. It is like a magician's hat from which he plucks out an array of hidden items: a toothpaste tube filled with honey, a cassette recorder, a thermos filled with coffee, or entomological forceps. You can have any kind of problem — and Melián will almost always find a solution somewhere in his bag. He also is a man of many talents; as good a mechanic as he is a cook, as talented a taxidermist as he is a painter. One minute, he'd be baking a cake for someone's birthday, the next he'd be fixing an electric generator after everyone else had given it up as a lost cause. Melián is, without a doubt, a modern genius.

With this cast of characters, I have visited El Toldo five times, with each trip lasting 28 days. One-hundred-and-forty days represents a small amount of time in the 40 years that someone might devote to their work — but these 140 days have produced a long lifetime of experiences and stories.

Acknowledgments

I am grateful to Robert W. Henderson for the opportunity to participate in this volume and for encouraging me on many occasions. I am indebted to participants in the El Toldo biodiversity monitoring project for sharing all those experiences — and I am especially grateful to the persons mentioned above for letting me use their stories. I thank Luis Fong for reviewing an early draft and Richard Neill for making detailed idiomatic and editorial suggestions that immeasurably improved the text.

Suggested Reading

Armas, L. F. 1996. La meseta de las cabezadas del río Piloto, Moa: un núcleo de biodiversidad poco conocido. *Cocuyo* 5:5–6.

Armas, L. F. 1999. Observaciones sobre una migración de mariposas (Lepidoptera: Pieridae) en el Toldo, Moa, provincia de Holguín. *Cocuyo* 9:16–17.

Armas, L. F., A. Fong, and F. Rodríguez. 1999. Depredación del alacrán *Centruroides anchorellus* (Scorpiones: Buthidae) por la lagartija *Anolis porcatus* (Iguania: Polychridae). *Cocuyo* 9:14.

Fong, A. and O. H. Garrido. 2000. Nueva especie de *Anolis* (Sauria: Iguanidae) de la región norte de Cuba oriental. *Rev. Biol. Trop*. 48:665–670.

Silva Lee, A. 1999. *Cuba Natural*. Pangaea, St. Paul, Minnesota.

Viña D., N., N. Viña B., A. Fong, L. O. Melián, and F. Rodríguez. 1998. Características fundamentales de la biodiversidad de vertebrados de la altiplanicie de El Toldo, Moa. *Canoa* 4:67–80.

Fong G., A. 2003. The pleasure of the adventure: Searching for reptiles in eastern Cuba, p. 101–106. *In* R. W. Henderson and R. Powell (eds.), Islands and the Sea: Essays on Herpetological Exploration in the West Indies. Society for the Study of Amphibians and Reptiles, Ithaca (New York). Contributions to Herpetology, volume 20.

Jamaica

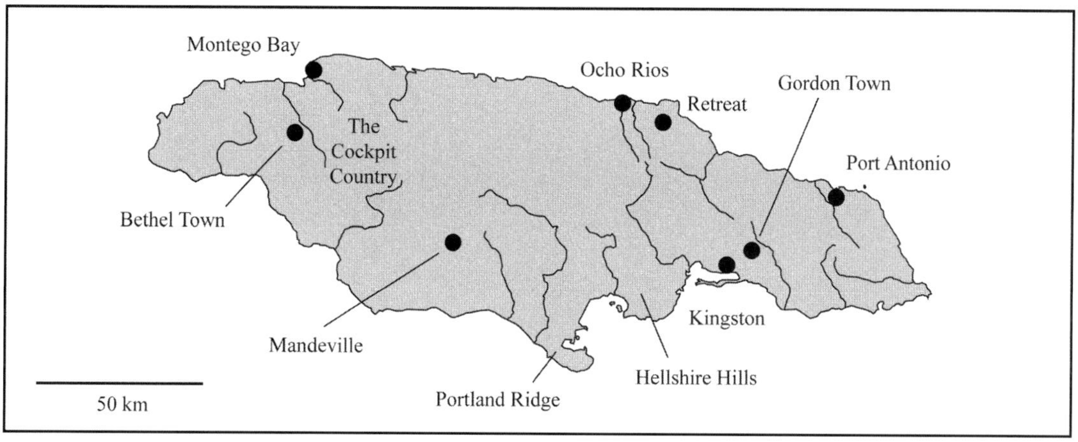

Introduction
Bahamas
Cuba
Jamaica
Hispaniola
Puerto Rico
Lesser Antilles
Indices

Man of the Yard

Brian I. Crother[1]

Collecting expeditions are never uneventful (I'm sure some sort of rule exists somewhere). Part of the reason is surely attributable to the inherent uncertainty in the proposition of collecting. You have goals to acquire this and that, but the specimens and various environmental factors behave independently of your goals — and then there are the people, from officials to locals, who can affect in myriad ways the realization of an expedition. Once in a while, a conjunction of uncertainties occurs and truly trying or surprising events result. Well, to make a longer story short, one brief trip provided one of those special moments when the synergism of uncertainties coalesced into a life-altering experience.

The summer of 1986 saw my last collecting trip to Jamaica. In just over two years, I had visited the island three times: once alone to get my bearings, a second time with the late Joe Slowinski (who spent the better part of three days trying to clean hundreds of seed ticks off his body with ethanol), and a third time with David Hillis and Jay Savage. Jay was my major professor and Dave was a new assistant professor and member of my doctoral committee at the University of Miami in Florida. I guess they figured that I had the place well enough staked out that they could now join in and have some fun. For some reason, rumors abounded to the effect that I rather enjoyed myself on these trips. These rumors just might have originated when someone overheard me wax eloquent on the tasty Red Stripe beer, smooth golden Appleton rum, and tropical beaches. In fact, Jay tells a story (and, for the sake of my reputation, I must argue against its veracity) about a visit to the beach in Montego Bay. When we walked up to one of the many palm-leaf covered huts to order drinks, according to Jay, the bartender extended his hand to me and said, "Brian! It's been a long time! Where have you been?" The story always produced a hearty laugh at my expense. Whatever the reason, they had joined me for a short hit-and-run one-week trip to the most enigmatic isle of the Greater Antilles, Jamaica. Our stated goal was to collect *Arrhyton polylepis*, a very secretive little snake about which little was known.

On my first visit, in the spring of 1985, I had carried with me a very proper letter of introduction that I was to deliver to the Yacht Club at Montego Bay. Eventually this led to a meeting with Mrs. Philip Harvey, a multi-generation lady landowner of British descent who lived in the bush (as she described it) at Woodstock in Westmoreland Parish. She granted me a room for several days, and I quickly developed a fondness for scheduled tea times and cocktail hours. The kerosene-driven electricity simply added to the elegance of the bush. Being a lad from the harbor district of Los Angeles, this was quite surreal. While I worked out of and on her ranch in the gently rolling hills south of Montego Bay, I met her friends, James and Pam Williams and their daughter Annabel. We all got along wonderfully, and Annabel even went into the field with me, dismantling forgotten rock piles and collecting herps (as well as ticks). The Williamses were also descendants of British landowners (15th generation is what I recall) and ranched mainly pork. On this visit we were to be their guests, and Annabel was to be our able guide.

Pam and James were both fairly short with graying sandy blonde hair. They had the taut frames of folks who never retreated from a task. They had proper British accents laced with Jamaican vernacular, which somehow made the seemingly permanent tall white rubber boots appropriate dress. Annabel looked and acted much like their genetic combination. She was in her early twenties and had

[1] Department of Biology, Southeastern Louisiana University, Hammond, Louisiana 70402, USA.

recently graduated from a proper British university. Although her interests were in biology (she pursued a career in aquaculture), she was not a herpetologist — but had no fear of snakes and such. She was fiercely independent, unabashedly confident, and possessed a Bunyanesque work ethic — in short, a most pleasurable sort of companion. Their property was a vast expanse of tall grass fields, steep but forested limestone hills, and ancient rock walls and mounds. Being a hog ranch, there were of course a few muddy hog pens. Maybe the charm of Jamaica (or the Red Stripe) confounded my senses, but I can't recall any pungent odors diffusing from the pens. Like Mrs. Philip Harvey, the Williams family lived in Westmoreland Parish, but near Bethel Town. The main house was called Kew Park and it sat on a hill above the surrounding land. We were housed in the cottage, which was a full-sized house in its own right.

After dinner and pleasantries, we visited a couple of the forested, steep-sided, limestone hills and turned up a few frogs and geckos. The terrain was fairly nasty to negotiate, but we managed. I enjoyed being out, but the beast we sought was not to be found in those parts of Jamaica. So, in the morning we headed north and east to the hills above Port Antonio. Along the way, we collected here and there, and accumulated a number of things. Enough that we had to stop at a Williams family friend's house to process our specimens. They were not home, so we set up on the porch, pulled out the Appleton and Red Stripe, and proceeded to dissect and pickle. This went on for a bottle of rum and a few beers, so when we finished we quickly settled in. The owners arrived and were the kind of hosts that loved to overdo everything. It was wonderful — at least until the gentleman of the house pulled out his bottle of vodka from the freezer and asked us to join him in a few snorts. I don't know about you, but mixing beer and rum is one thing. Adding shots of vodka is more than I can handle. We politely managed to join him for one, but then quickly begged off with the excuse of an early morning. He was clearly disappointed.

The next morning Annabel carted us off to some palm frond-covered cabanas on the beach in a beautiful cove just east of Port Antonio (who says field work requires one to rough it?). This was our field station for the next three days. Accompanied by our rum and beer, we herped day and night and thoroughly enjoyed ourselves. We did not collect any *Arrhyton polylepis*, but we did pen a rhyme that became our motto for the trip and beyond: "Red Stripes in the morning, herps take warning; Red Stripes at night, herpers delight."

On our last day, we headed into the hills searching for the kind of habitat that makes me drool: an old, overgrown field with rock mounds nearly invisible beneath entrenched grasses. As we drove slowly on the steep, winding road, we passed a local who was carrying a mighty load of wood. He was a relatively tall fellow with a wiry frame and a somewhat disheveled look — as if he had just finished the physical task of chopping the wood. We stopped and hollered for the fellow to jump in the back for a lift. He happily obliged. After a few miles of climbing, our passenger knocked on the window and asked to be dropped off. He jumped out, load slung over his back, and headed to a rickety wood and wire gate that opened onto a narrow trail that dropped into a small valley. As we drove away, we looked back at the land and beheld our grail, the rock mounds. Annabel backed up the truck and we asked the good sir if that was his property. Indeed it was, he said. We asked about looking for snakes and he retorted that no snakes were to be found on his land. We asked if we could look anyway, and I suppose he said "sure" mainly because he was grateful for the ride. In hindsight, I think he would have said yes anyway.

Well, we took to the stone piles like stray dogs to a brisket. We were quickly rewarded with the sight of *Celestus* (Figure 1) scurrying and diving deeper into the rocks. The proprietor warned us not to catch those lizards because they were galliwasps and had stingers on their tails. He was mighty astonished to see us start grabbing these beasts and not being stung. Our enthusiasm must have been infectious because our friend, who would not even get near the rocks at first, began to turn over rocks with us. He still would only point out beasts to be collected and would not get near one. We collected *Typhlops* and our goal, *Arrhyton polylepis*, and even a *Peripatus*. The hunting was good. And then, as we tired, our friend turned over another rock and swiftly grabbed a large galliwasp. With legs spread and feet balanced on worn rocks, the man held the saurian in a firm grasp, raised it in the air, and let out a bellowing roar. We stepped back stunned, and then realized collectively what had just happened — and we roared and clapped with him and for him. Hundreds of years and many generations of folklore and

FIGURE 1. *Celestus crusculus*, a Jamaican "Galliwasp" (photograph courtesy of Margaret M. Stewart).

learning were broken down at that moment. His ancestors who had built those rockpiles centuries ago were probably the same ones that first uttered the name galliwasp and gave it meaning. In a few hours, both the rockpiles and the stigma of the galliwasp were torn down — a cultural barrier shattered and a new man born.

We finished up and gave our new friend some gifts for his help and time and property. As we shook hands and exchanged names, he offered his: Norbert Scale, Man of the Yard. We looked at each other after we heard his last name and all silently agreed on the truth of serendipity and synchronicity. Of perhaps greater significance was what else we learned. We smugly speak to one another about our passion for our work and for the animals we study. In this smugness, we take for granted why we work on reptiles and amphibians, on why we enjoy collecting. On that day, not one of us matched the intensity of Norbert Scale. If we had forgotten, Norbert reminded us, and if we never knew it, he revealed to us the true meaning of passion.

A celebration was in order, so we hit Port Antonio on a mission to buy champagne. The best we could do was reconstituted champagne (Yikes! I didn't even know such a thing existed until then.), but we bought a few bottles anyway. We had a hearty meal, and drank and laughed and told stories to deep into the night. Dave and Annabel disappeared for a quick dip in the ocean, and Jay and I sat and reflected on the incredible day. We were to leave in the morning and although we had finished our work, we knew there would be Red Stripes at dawn.

Acknowledgments

Mary White and Mitch Leslie read this tale and provided excellent comments. I definitely welcomed (and needed) their positive reaction to the initial product!

Suggested Reading

Crother, B. I. 1999. Phylogenetic relationships among West Indian xenodontine snakes (Serpentes: Colubridae) with comments on the phylogeny of some mainland xenodontines. *Contemporary Herpetology* 1999 (2):http://dataserver.calacademy.org/herpetology/herpdocs/index.html.

Crother, B. I., and C. Guyer. 1996 Caribbean historical biogeography: was the dispersal-vicariance debate eliminated by an extraterrestrial bolide? *Herpetologica* 52:440–465.

Crother, B. I. 2003. Man of the yard, p. 109–111. *In* R. W. Henderson and R. Powell (eds.), Islands and the Sea: Essays on Herpetological Exploration in the West Indies. Society for the Study of Amphibians and Reptiles, Ithaca (New York). Contributions to Herpetology, volume 20.

Of Mongooses and Lizards

Byron Wilson[1]

Because the West Indies boasts such a diverse and endemic-rich herpetofauna, the region has acted, not surprisingly, as a magnet in attracting herpetologists. Researchers have addressed a wide range of issues — speciation, biogeography, ecology, evolution, behavior, physiology, and, more recently, conservation. Indeed, the West Indies has proven to be a fertile ground for nurturing both herpetological curiosity and herpetologists — and it has been both easy and efficacious to justify a professional focus on the reptiles and amphibians of this region.

However, I suspect that more than just a few herpetologists were first attracted to the West Indies for other reasons — reasons not at all herpetological, and reasons assuredly not work-related. I certainly fall into this latter category. Aside from the undeniable lure of a tropical paradise, I was drawn to the charm and charisma of West Indian people, and to the ease and content of West Indian life. In fact, I was drawn so closely that I had a Caribbean child before even contemplating any serious commitment to working or conducting research in the region.

Jamaica: Enticing Land of Contrasts

My first visit to the Caribbean was in 1988, to visit relatives in Jamaica. I immediately fell in love with essentially everything Jamaican: the land, the food, the culture, and, of course, the people. Jamaicans are unmistakably (and often unavoidably) engaging, and shyness is simply not characteristic of the culture. Jamaicans can and will claim adamantly that no other West Indian island has attained comparable fame and reputation. For such a small nation (roughly 2.6 million inhabitants), Jamaica has certainly produced a disproportionate number of players on the world stage — and, while Jamaicans can rightly call from their ranks a number of influential thinkers and even thinking politicians of historical note, Jamaica has attained such a high international profile primarily through excellence in the arenas of sport, tourism, and culture.

For example, on a per-capita basis, no other nation has come close to producing the number of world champion track and field athletes, and the trend shows no sign of waning (much to the delight of student-athlete recruiters from United States universities). Of course, Jamaica is famous for its tourism. Beginning in the 1930s, and popularized in those early years by Errol Flynn and his debaucherous Hollywood crowd, Jamaica has become nearly synonymous with an ideal tropical vacation.

Undeniably, the single entity most responsible for Jamaica's name recognition is Bob Marley, the brilliant singer and songwriter, who almost single-handedly delivered reggae music to the global theater. Marley broadcast an enduring message of redemption and hope to oppressed underclasses throughout the world. No West Indian has come close to attaining comparable global recognition.

However vibrant and enticing, Jamaican culture does have its downside. It is a decidedly violent country that can make the dubious claim of being among the murder capitals of the world — and, while serious crime is a persistent threat, the threat is exaggerated by non-random incidents (e.g., drug and gang activities), as well as by the international media. In Jamaica, most murders are drug, gang, or domestic in origin. Keep your nose (and privates) clean, and you reside in an entirely different risk category.

The Land

Physically, Jamaica is an island of lush and spectacular beauty. Although its natural environment has taken a severe beating at the hands of man (as have

[1] Department of Life Sciences, University of the West Indies, Mona, Kingston 7, Jamaica & Jamaican Iguana Research and Conservation Group.

all sizable islands in the Caribbean), Jamaica still harbors some natural areas of exquisite splendor. In a handful of isolated but sizable pockets, one can find pristine Caribbean habitats of unsurpassed quality and remarkable purity. In particular, some rugged inland areas have escaped the wrath of man altogether, and thus retain their original, unspoiled value. Unfortunately, these areas continue to shrink under increased human pressures, and whether they will survive far into the new century is unsettled and unsettling.

An Invitation to Help Herps

Following its "rediscovery" in 1990, the Jamaican Iguana (*Cyclura collei*; Figure 1) became the focus of intensive local and international conservation efforts. In 1996, I was invited by Dr. Peter Vogel (Head) and other members of the Jamaican Iguana Research and Conservation Group (JIRCG) to participate in this important program. With funding organized by the Fort Worth Zoo, I began fieldwork in the Hellshire Hills in January 1997.

As for many insular reptiles occurring in otherwise intact habitats, the greatest threat to the Jamaican Iguana's persistence comes from introduced mammalian predators such as the Indian Mongoose. Implicated in the extinction or near-extinction of up to six endemic Jamaican vertebrates, the mongoose has become as infamous for its assaults on island biotas as it was made famous by Kipling for its perceived beneficial role as a killer of venomous snakes. I, despite having dreamed of owning a "pet" mongoose (my very own "Rikki-Tikki-Tavi"), was to become the mongoose's worst nightmare. My mission was that of a conservation mercenary: a confessed killer with a commendable agenda.

FIGURE 1. Nesting female Jamaican Iguanas (*Cyclura collei*) at one of the two known nesting areas in the Hellshire Hills (photographs by Glenn P. Gerber).

Our objective was to discern whether the notorious mongoose could be eliminated from a core conservation zone, and, if so, examine what impact such a removal campaign would have on the naturally occurring fauna, particularly ground-inhabiting reptiles. Our ultimate goal was to determine the efficacy of using mongoose control as a tactic for preserving biodiversity more generally.

Our research task appeared straightforward: (1) conduct appropriate surveys to assess the abundance of various mongoose prey species (reptiles, birds, and large arthropods), (2) remove mongooses from selected experimental plots, and (3) continue survey work to quantify the populational responses of targeted species to the removal of the mongoose. Despite this conceptual simplicity, implementing our research plan in the remote interior of the Hellshire Hills proved to be extraordinarily challenging and extremely entertaining.

The Hellshire Hills

Known more for the beaches and fried fish along its developed southeastern corner, the interior Hellshire Peninsula is essentially uninhabited and virtually unknown. Fringed by about 12 km of empty coastline, the Hellshire Hills form a rugged landscape of honeycombed limestone (karst) that contain, arguably, the third largest tract of undisturbed dry forest remaining in the insular Caribbean. Roughly half the peninsula's approximately 120 km^2 of forest remains in either pristine or nearly pristine condition — home to herps, herpers, and, unfortunately, illegal tree cutters. The northern portion of the peninsula has been severely degraded due to tree cutting for charcoal production, and the wave of destruction is moving steadily southward, into the remaining primary forest that is home to the iguana.

From a Temperate to Tropical Desert

My previous research efforts had focused on desert lizard communities in the American West, so it was appropriate that I ended up in the driest habitat in

FIGURE 2. View from research station ("South Camp") in the interior Hellshire Hills.

this "land of wood and water" — complete with cacti to serve as reminders of arid experiences past. However, research in a Jamaican dry forest is nothing like research in an American desert. American deserts are vast areas containing many potential research sites that often are easily accessible by road or with a 4-wheel drive vehicle. Huge distances are easily negotiated by car, and suitable study areas can often be reached with little or no walking. Critically, one's vehicle can generally be used as a base for remote camping, so life can be lived rather comfortably (e.g., with ice and fresh food).

In contrast, undisturbed Jamaican habitats such as the Hellshire Hills can generally be accessed only by foot. It is an unfortunate reality throughout the tropics that vehicular access leads to rapid destruction of the natural environment. So it is in Jamaica, where access to undisturbed habitats usually entails considerable trekking over invariably rugged terrain: were it any easier, the habitat would have long ago been decimated.

Given the time constraints attending study site access, camping in the bush is the only way to conduct serious research in Jamaica's remaining wild areas. Otherwise, most of the day is consumed by the trek in and out of the study area, leaving little time and very little energy to devote to actual work. Early in the Jamaican Iguana project, a bush hut was erected in the middle of the peninsula to permit extended research stints in the interior forest. Use of this remote bush camp has afforded not only ample time for research activities, but also seemingly timeless periods to revel in a pressure-free, natural existence.

For me, the logistical constraint of living a remote bush life also provided the opportunity to realize a childhood dream: living the life of a Robinson Crusoe "wannabe," complete with a palm-thatched hut, commanding view of a tropical sea, and near-solitary rein over an uncontaminated tropical landscape. Since the project was in need of a more southern base station, in 1997, I constructed a secondary bush hut ("South Camp") on a ridge overlooking the sea and an impressive expanse of virgin forest canopy (Figure 2). While I am quick to point to this simple but ideally located bush hut as my greatest achievement in the field, I must credit Mr. Edwin Duffus (the "re-discoverer" of the Jamaican Iguana) for making it possible. Mr. Duffus showed me the ultimate site for a view-lover's bush hut, taught me how to use (and sharpen) a machete, and served as a technical expert on other key aspects of bush life. Based at this scenic dwelling (Figure 3), I was able to spend periods of up to three weeks without seeing another human — finally putting closure on a perennial dream of sustained natural solitude.

Some Additional Logistical Considerations

Aside from the several hours of walking to access the research site, no water sources exist within the interior forest — nor do sources of any other necessary supplies. Hence, everything for work and sustenance must be carried in by researchers or donkeys. To complicate matters, our research plan demanded that I spend long periods of

FIGURE 3. The author enjoying the view at "South Camp" in the interior Hellshire Hills (photograph by Glenn P. Gerber).

FIGURE 4. View from "South Camp," the silhouette is the remains of an involuntary participant in our exotic mammal removal campaign: a young wild pig chased down and killed by the author using a machete (photograph by Joseph A. Wasilewski).

uninterrupted time monitoring traps in the forest (up to 40 days continuously); hence, aside from transporting research materials and traps, I had to stockpile enough food, rum, etc. to last the duration of long field stints. With no refrigeration, fresh food became a source of daily concern, if not near-total preoccupation.

Searching for food along the coastline proved amusing, if not always nutritionally productive. Indeed, the primal nature of foraging on an empty stomach supplied some fine moments of legitimate feral joy, but just as often proved frustratingly ineffective. Land crabs could often be obtained in the nearby mangrove swamps, and fish and conch from the sea provided occasional relief from the monotony of canned, dried, and salted fare. But ultimately I proved to be a better beggar than forager. My most impressive haul — and the culinary highlight of my time in Hellshire — was a wild pig that Edwin Duffus usurped from a pack of wild dogs. Mr. Duffus generously gave me half the pig, salted and stored in a 20-liter bucket. Despite one minor maggot infestation, the stash proved to be a psychological and protein-rich bonanza: "Lord of the Flies" recreated in civilized fashion. I feasted on that pig for 14 consecutive days (cf. Figure 4).

FIGURE 5. Indian Mongoose (*Herpestes javanicus*) after a close encounter with an iguana conservationist.

Even more so than food, water quickly emerged as the primary limiting factor in central Hellshire. Productive research boiled down to the imperative of water transport, water storage, and extreme water conservation. One can easily drink over a gallon of water each day, with other needs (cooking and washing), a minimum of 2–3 gallons are required per person per day. Our original system involved hiring a charcoal burner to carry in 10 gallons at a time on a donkey. Unfortunately, both donkeys and charcoal burners can be fickle, and work had to be abandoned on several occasions when this supply line broke down. Being thirsty in central Hellshire, with a two-hour walk to the nearest water source, does not constitute a dream Caribbean experience, nor does the 2-hour-plus trek back into the hills carrying five gallons of water. Ultimately, we developed collection systems to harness rainwater from the roofs of our bush huts — thus obviating the need for donkeys, or for researchers functioning as asses.

Beyond Food and Water: the War on Exotics

As elsewhere, the Hellshire Hills mongoose has proven to be rather trap-prone (Figure 5). Using standard live traps, we have now recorded over 40,000 trap days (1 trap day = 1 trap open for 1 day) and have removed over 300 mongooses. As evidenced by a reduction in capture rate (roughly an order of magnitude), we are confident that our removal plot is 100% free of resident mongooses. At present (2001) we are capturing an average of 1–2 mongooses per week — primarily dispersing or widely foraging males. Encouragingly, only twice in five years have I observed free-ranging mongooses within the removal area, whereas observations of mongooses have remained common in the control area. Thus, mongooses are trapped quickly once they contact our trap line, and the maintenance of a mongoose-free zone is feasible.

In addition, and using whatever means necessary, we have also removed around 30 feral cats, several hundred pest rodents, a small number of dogs and pigs, and a growing number of Marine Toads (*Bufo marinus*). In short, we are waging a battle against all invasive animal species for which prohibitive legislation does not exist (i.e., all species except man). Such battles are now widely regarded as crucial conservation measures that will be necessary if

FIGURE 6. Juvenile Jamaican Iguana (*Cyclura collei*) caught in a mongoose trap — the first evidence that exotic predator removal may be having a positive influence on the survival of young iguanas (photograph by Glenn P. Gerber).

we are to save many currently threatened species from extinction. But, unlike small islands where complete eradication of pest species may be practical, our battle in Jamaica amounts to a management exercise with no end in sight. Also, while we strive to wage a humane war, the reality is that the battle is far from attractive. Removing exotic species is a nasty business, and will doubtless long remain the most critical, but unpalatable and often ignored conservation measure.

Does Exotic Predator Control Work?

A major component of our predator control effort has focused on documenting the potential demographic responses of mongoose prey species to the elimination of this potent, non-native predator. Our primary focus is on the terrestrial reptile community, but we have also been enumerating the relative abundance of most large arthropod taxa and one ground bird (the Caribbean Ground Dove). Before initiation of mongoose trapping in May 1997, we established several pitfall trap grids, and generated baseline measures of abundance for the terrestrial fauna over a 48-day trapping period (February–April). This original (1997) trapping regime has been repeated in every subsequent year (1998–2001), with traps being opened and closed on the same calendar days of each year. We are now looking forward to a sixth consecutive season in 2002.

Results to date (as of 2001) are encouraging. The number of juvenile iguana sightings and incidental captures (in mammal traps; Figure 6) have both increased since the initiation of predator control measures, though small sample sizes and the largely anecdotal nature of the observational data preclude drawing any firm conclusions. Similarly, we have documented some promising trends among other reptile species on mongoose-free versus mongoose-infested plots, but await additional long-term data before claiming any definitive successes.

"Waiting" for demonstrable success has proven tiring. Given that we have found over ten lizards in the stomach of a single mongoose, we anticipated being knee-deep in herps before you could say "whack that mongoose." A drought during the 1998–2000 period probably discouraged rapid demographic responses to predator removal, but other factors are likely operational. For example, improved survival and larger body size in some lizard species may have resulted in increased saurian predation on other saurian species — especially predation on younger (smaller) individuals. Hence, we may document significant shifts in population size structure, but not in overall population density. On the other hand, demographic recovery may simply be slower than anticipated, especially for rare species. So, while our original predictions of post-mongoose herp blooms now appear naïve, we remain optimistic that several targeted species will ultimately show dramatic increases in abundance as a result of our predator removal program.

Incidental Discoveries: A Caribbean Bonus

The ever-present potential for new herpetological discoveries further renders the West Indies a stimulating place to conduct field work. So, in spite of so much attention by herpetologists, even relatively well-known islands such as Jamaica continue to yield pleasant surprises.

Within two months of beginning work in early 1997, I was rewarded with a most remarkable and aesthetically appealing find — a specimen of the Blue-tailed Galliwasp (*Celestus duquesneyi*; Figure 7) — an otherwise ordinary anguid lizard, but with tail bands of such a vivid sky-blue that they appear almost artificial. Known only from several specimens collected in the 1930s at Portland Ridge (about 30 km west of Hellshire), the species was unknown from Hellshire, and its overall status had remained a mystery for the previous half century.

Chapman Grant, who described the original Portland Ridge material, noted that area's ecological similarity and geographical proximity to Hellshire, and predicted that *C. duquesneyi* might also occur in Hellshire. Also aware of this connection, and privy to at least one credible report of a "medium-sized" *Celestus* in Hellshire, Peter Vogel was keen to see pitfall traps erected in the hope of capturing this beautiful and enigmatic lizard. Being alone in the forest at the time of the first pitfall capture, I was credited with this herping equivalent of a "garbage goal."

A second notable capture came in 1998, when I stumbled onto a Hellshire representative of the Jamaican "Thunder Snake" (*Tropidophis* sp.). This taxon was unknown from the Hellshire Hills, and the specimen differed somewhat from the other recognized Jamaican subspecies (S.B. Hedges, pers. comm.). The genus in Jamaica is now undergoing a revision that will result in the designation of at least three endemic species, and the Hellshire form may conceivably constitute a fourth species restricted to the Hellshire Hills (S. B. Hedges, pers. comm.).

I collected a second Hellshire *Tropidophis* in 1999, roughly three meters from the site of the 1998 specimen. Not coincidentally, both of these captures were essentially in my hut, in the kitchen area, and both captures were entirely incidental to matters of chemical sustenance (coffee and food). So these discoveries illustrate a truism found in so many enterprises: 99% is just showing up and being there — the other 1% is often luck. In general, my herpetological experiences in Jamaica have been very much about being there. Fortunately, when it comes to pristine Caribbean habitats, just being there can be a full and rewarding experience.

FIGURE 7. Blue-tailed Galliwasps (*Celestus duquesneyi*) from the Hellshire Hills, St. Catherine, Jamaica (bottom photograph courtesy of Peter Vogel).

Acknowledgments

First, I thank Suzette Wolfe Wilson, who brought me to Jamaica. I also thank the three individuals who facilitated my participation in the Jamaican Iguana project: Peter Vogel, for defying his better judgment and allowing a crazy Yank to run amok in the Hellshire forest, Rick Hudson of the Fort Worth Zoo, for generous financial support, and Allison Alberts, for encouraging me to get involved in West Indian iguana work and for encouraging others to help make it happen. I thank Edwin Duffus and Delano Lewis for various forms of assistance, and for some much appreciated deliveries of fresh food. I also thank the "Dread," "Brown Man," "Stamma," and "Mas Sonny" for keeping the supply route viable and friendly. Finally, I thank our local collaborating institutions, especially the Hope Zoo and the National Environment and Planning Agency (NEPA), for helping the field effort in various ways.

Suggested Reading

Alberts, A. C. (ed.). 2000. *West Indian Iguanas: Status Survey and Conservation Action Plan*. IUCN – The World Conservation Union, Gland, Switzerland.

Conservation Breeding Specialist Group (CBSG). 1993. Population and Habitat Viability Analysis for the Jamaican Iguana, Kingston, Jamaica. IUCN/SSC Conservation Breeding Specialist Group, Apple Valley, Minnesota.

Grant, C. 1940. The herpetology of Jamaica II. The reptiles. *Bull. Inst. Jamaica, Sci. Ser.* 1:61–148.

Schwartz, A. and W. M. Carey. 1977. Systematics and evolution in the West Indian iguanid genus Cyclura. *Stud. Fauna Curaçao Carib. Isl.* 53:15–97.

Schwartz, A. and R. W. Henderson. 1991. *Amphibians and Reptiles of the West Indies: Descriptions, Distributions, and Natural History*. Univ. Florida Press, Gainesville.

Tolson, P. J. 2000. Control of introduced species, p. 86–89. *In* A. C. Alberts (ed.), *West Indian Iguanas: Status Survey and Conservation Action Plan*. IUCN – The World Conservation Union, Gland, Switzerland.

Vogel, P. 1994. Evidence of reproduction in a remnant population of the endangered Jamaican Iguana, *Cyclura collei* (Lacertilia, Iguanidae). *Carib. Jour. Sci.* 30:234–241.

Vogel, P. 2000. Jamaican Iguana: *Cyclura collei*, p. 19–22. *In* A. C. Alberts (ed.), *West Indian Iguanas: Status Survey and Conservation Action Plan*. IUCN – The World Conservation Union, Gland, Switzerland.

Vogel, P., R. Nelson, and R. Kerr. 1996. Conservation strategy for the Jamaican Iguana, *Cyclura collei*, p. 395–406. *In* R. Powell and R. W. Henderson (eds.), *Contributions to West Indian Herpetology: A Tribute to Albert Schwartz*. Society for the Study of Amphibians and Reptiles, Contributions to Herpetology, vol. 12, Ithaca, New York.

Wilson, B. S. and P. Vogel. 2000. A survey of the herpetofauna of the Hellshire Hills, Jamaica, including the rediscovery of the Blue-tailed Galliwasp (*Celestus duquesneyi* Grant). *Carib. Jour. Sci.* 36:244–249.

Wilson, B. S., A. C. Alberts, K. Graham, R. Hudson, R. Kerr, D. Lewis, N. Lung, R. Nelson, N. Thompson, and P. Vogel. In press. Survival and reproduction of repatriated Jamaican Iguanas: headstarting as a viable conservation strategy. *In* A. C. Alberts, R. L. Carter, W. K. Hayes, and E. P. Martins (eds.), *Biology and Conservation of Iguanas*. Univ. California Press, Berkeley.

Woodley, J. D. 1980. Survival of the Jamaican Iguana, *Cyclura collei*. *Jour. Herpetol.* 14:45–49.

World Conservation Union, The (IUCN). 1996. *IUCN Red List of Threatened Animals*. IUCN – The World Conservation Union, Gland, Switzerland.

Wilson, B. 2003. Of mongooses and lizards, p. 113–119. *In* R. W. Henderson and R. Powell (eds.), Islands and the Sea: Essays on Herpetological Exploration in the West Indies. Society for the Study of Amphibians and Reptiles, Ithaca (New York). Contributions to Herpetology, volume 20.

Recollections of Jamaica

MARGARET M. STEWART[1]

My first trip to Jamaica was during the summer of 1966. At that time, the island was quiet in terms of socio-political strife, having just obtained independence in 1962, and we sensed no hostility to tourists. Actually, I was not a tourist but an invited instructor in a summer tropical biology course at Ocho Rios, on the north shore.

Frogs Jumping in Every Direction

That was the trip that introduced me to the incredible little frogs of the genus *Eleutherodactylus*. I had been asked to provide field trips and laboratories on the frogs of Jamaica. Near the field station was a coconut grove, with coconut husks piled around the base of each tree. Knowing that the frogs I heard calling at night needed nest and retreat sites, I began turning those husks until I found frogs (Figure 1). I got near the bottom of the pile and then, in amazement, watched frogs jumping in every direction — and not just frogs, but a small snake, a few lizards, and the most incredible array of invertebrates I have ever seen in one spot. There were crickets, cockroaches, ants, termites, beetles, bugs, pillbugs, millipedes, daddy longlegs, colorful long flatworms, spiders of many sorts, land crabs, small scorpions, mites, ticks, huge centipedes, and other invertebrates that I had never seen before.

With so many frogs readily available, I wondered why I had spent so many years trying to catch frogs in the northeastern United States — and I quickly came to the realization that the inflated densities of island anurans in the Caribbean could work to my advantage. Here indeed would be my next field study site. Little did I know that I would be working in the Caribbean studying those *Eleutherodactylus* frogs for the next twenty-odd years.

Much Work to be Done

When I reviewed the literature on the frogs of Jamaica, I saw an absence of ecological work and decided that would be my niche. Although Garth Underwood and Ernest Williams had done a lot of work on *Anolis* lizards, the "eleuths" obviously needed attention. With 15 species and the densities I found, that should be possible. E. R. Dunn's 1926 work on the amphibians of Jamaica and Gardner Lynn's 1940 publication served as starters. Coleman Goin's papers on pattern polymorphism in Jamaican eleuths was the closest I could come to my kind of work. Schwartz and Fowler's "The Anura of Jamaica," which appeared later, was priceless. I am not a systematist; that was up to someone else.

However, I was interested in how such high densities of frogs of a few species allocated resources. On continental land masses, the more typical pattern is to have high species diversity but low species abundance.

Benefits and Costs

Another decided advantage of working on a Caribbean island is the absence of venomous snakes. Having grown up in copperhead country and having recently spent a year in Africa, not having to worry about snakebites was a real benefit. My hus-

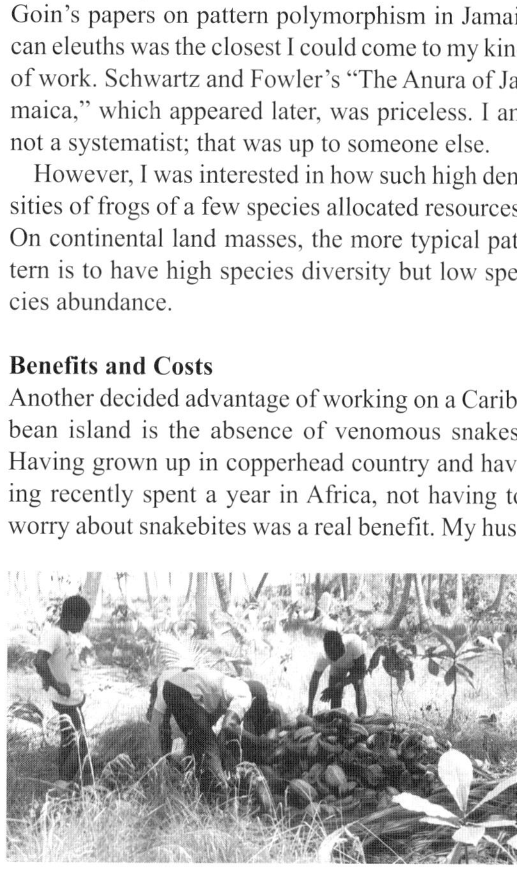

FIGURE 1. Turning a husk pile by the White River near Ocho Rios, April 1973.

[1] Department of Biological Sciences, State University of New York at Albany, Albany, New York 12222, USA.

band, George Martin, and I had sabbaticals coming up soon, so I got busy finding funds and planning an extended research trip to Jamaica, which we did in March–June 1973 and again in December 1975–January 1976.

By the second trip, socio-political problems were making field work in Jamaica a bit dicey. Soon after, my research projects shifted to Puerto Rico, which is a whole different chapter in Caribbean adventures.

FIGURE 2. Bamboo arcade near Worthy Park Estate, April 1973.

Topography = Diversity

Besides the abundance of frogs for study, certainly a major reason to fall in love with Jamaica is the beauty of the landscape. The highly variable topography, from the karst landscape in the west to the Grande Ridge of the Blue Mountains in the east, up to 7,402 feet, provides some of the most beautiful views I have seen anywhere — and that highly variable landscape with so many different habitat types and elevations yields extremely high biological diversity for such a small area. The island is only about 52 x 144 miles (4,232 mi^2, slightly smaller than Connecticut) and, for its size, houses the greatest biological diversity in the Caribbean. Ferns, birds, insects, and bromeliads (called wild pines) in addition to reptiles and amphibians show considerable diversity and high levels of endemism. Jamaica has over 237 known species of orchids and 500 species of ferns. Coral reefs have been some of the finest in the Caribbean (seriously degraded now for several reasons). The Cockpit Country (karst topography), with series of undulating round knolls and depressions, called haystacks, were new to me. When seen from the air, the area is special. On the ground, those small roads with over-arching bamboo arcades (Figure 2) beckon one to explore further.

Finding the Moisture-loving Frogs

I planned to "scout out" the island and its frogs during my 1973 trip. May was supposed to be a wet month, but the rains never came, and we were confronted with dry conditions everywhere. I needed to learn the frogs and their calls and to find suitable work sites where I could live inexpensively and find the species that were amenable for study. On subsequent trips, I then could settle in and concentrate on the ecology and behavior of local species of *Eleutherodactylus*, studies that no one had ever done. Little did I know how difficult reaching that goal would be. Having established field stations where one can work comfortably is priceless. With the exception of Mason River, near Kellits, the stations that Jamaica did have were on the coast, designed for marine studies, not for studying moisture-loving frogs.

Ram-goat Dashalong

I found the Jamaicans to be especially interesting and creative people. One of the first things I noticed about them was their sense of rhythm. I watched the fellows moving luggage at the airport. They made almost a dance of it. Although I did not hear them singing, it seemed as if they turned the baggage into drums and beat the rhythm by slapping their thighs and then the bags. They had turned the tedium of the job into a party. I noticed the same sort of thing with a fellow who was giving me information at the Naseberry Inn. He was sitting on a box and slapping a rhythm on the box as he talked. Then I realized that it was a bass box, and that he probably used it in a band. Later, in early May 1973, when we were working at Mason River, it was cane harvest time. A cane truck could take up most of the road, but the drivers were good at warning oncoming vehicles several curves ahead with their horns — and blowing their horns then turned into a game as they proceeded down the road. I wondered if each driver had his own rhythm and that drivers were communicating via those rhythms. Bus drivers had different rhythms when announcing their impending arrival with their horns. This use of rhythm is reflected in the steel drum bands that have developed throughout the West Indies. The sophistication of musicians in and from Jamaica in developing a musical style, reggae, that is recognized throughout the world, is evidence of strong creative talents. My tastes go more to calypso, but I give the Jamaicans credit for their creative musical talents.

Another evidence of delightful creative talents in Jamaicans comes in their place names. I never learned the patois, but it was definitely a foreign language for us even though English is the basic component. When looking at a map, one is struck by the free and creative use of language. Where else would you find such descriptive names as Wait-a-bit, Put Together, Salt Gut, Quick Step, Look Behind, Me no sen You no come, Time and Patience, Rest and be Thankful, Prickly Pole, Sleepy Tree Gully, Three Fingers at Four Feet, Second Breakfast Spring, and Dick's Last Shirt?

Local names of common plants also reflect creativity in describing the surroundings. My favorite is Ram-goat Dashalong (a runner weed that often takes over trail edges and roadsides). Then there is Piss-a-bed, and I never found why it was so named. Prickly Pole is a spiny palm, *Bactris jamaicana*, of which one has to be aware so as not to grab it. I was amazed to find that almost every native Jamaican could tell you the name of the trees, shrubs, orchids, and other wild flowers in their area — and they all knew the "Whistling Toad," one of the more common eleuths, and the "Bullfrog," *Bufo marinus*.

"Al Lost"

Although Mason River Field Station was in the hot karst country, at 2250 ft., it had its delights. We were surprised to find an interesting sphagnum bog with plants we recognized from home, such as Sundew, *Drosera*, and a stand of cattails. Bracken ferns were tall, branching, and almost climbing. Snoring Frogs (*Osteopilus* [formerly *Calyptahyla*] *crucialis*) were around, so we could hear them and record their calls. An eleuth we had not heard before, *Eleutherodactylus junori*, called in sequence around the garden. It produced a rippling effect as the call of one frog triggered the call of another, then another. Not until I got to Puerto Rico, did I hear this pattern of calling again, with *E. wightmanae*. The station lacked running water, yet we were supposed to use the flush toilets to avoid contamination of the area. That meant carrying water up from a spring at the bottom of the hill. We kidded a lot about having running water only when Al Breisch, my field assistant, ran with it. The last day there, we decided to have a water race between Al and the caretaker's young daughter in which each participant carried an oil can of water on their heads. The winner would be the one to return first with the least amount of water spilled on the way (Figure 3). Al lost.

FIGURE 3. The water race between Al Breisch and the caretaker's daughter; Mason River Field Station, in Cockpit Country, May 1973.

Ants in His Pants

I am not a beach person, but, of course, beaches are what bring most tourists to the island. Instead, I head for the hills and those lush tropical forests that are homes to so many incredible natural communities — and I had to see what the frogs were like on those high peaks. We learned about two places where we could stay in the Blue Mountains. One, at 4,000 ft., was the Pine Grove Inn, at Guava Ridge. The other was an old coffee plantation, Whitfield Hall, at about 4,200 ft., where a good trail led to Blue Mountain Peak. We would be able to see the frog species that lived at high elevations as well as the cloud forests along the high ridges.

After making arrangements via telegraph, we took off from the hot Mason River Field Station with great anticipation and excitement. We were delighted at the thought of cooler climates. We drove up to Guava Ridge via the beautiful road through Gordon Town. We stopped at every stream crossing where there was vegetation and a possibility of seeing frogs. The Maypoles, *Agave sobolifera*, were blooming, so we stopped to photograph those stunning plants.

Somehow, George had gotten into an ant nest and, with a few still in his trousers, we had to stop. Pulloffs were rare along that narrow winding road. We finally found a place to stop safely. Since traffic was nonexistent, he took off his pants and was taking care of the ants when, much to our surprise, a car pulled up behind us. A lady got out of the car and approached us with a friendly, "Are you the Martins?" How could anyone possibly know who we were? And at such an embarrassing moment! We admitted that we were indeed the Martins. The lady, Mrs. Abrams, was to be our hostess at the Pine Grove Inn, and, as we learned later, she was going up to open the inn just for us.

Cooler Climes, Bats, and Cubby

A hotel staff of three was there just for the three of us. We felt well cared for in a cottage of the charming inn that had seen better days. Mrs. Abrams had brought with her the best cake I had eaten since my childhood. The cake and tea helped us recover from the long trip. Later, we were amused by the bartender-waiter who came to take our drink orders. Being a little tired of rum, George ordered scotch. The waiter asked whether he wanted soda or water with his scotch. George requested soda. The waiter answered: "We don't have any soda." We enjoyed our week at Pine Grove. As well as good food and great views, we found many *Eleutherodactylus gossei* and *E. cundalli* (Figure 4) and were able to make observations on their activities.

FIGURE 4. *Eleutherodactylus gossei* (top), pink morph, Pine Grove Inn, May 1973, and *E. cundalli* (bottom), female guarding newly laid eggs in a coconut husk retreat, January 1976.

FIGURE 5. George Martin and Al Breisch enjoy the view from Blue Mt. Peak at sunrise, May 1973.

To get to Whitfield Hall, one had to ford the Yallahs River, and we had to take the necessary provisions with us. The road above Hagley Gap was so bad that I proclaimed that we would stay as long as we could and then leave for good. I was not driving up that road, if one could call it that, again. It was often almost vertical, narrow, winding, with trenches and huge holes, and no shoulders. To make it worse, the local children had a game of hanging on to the back of a vehicle until they could hold on no longer. Our little Volkswagen Beetle just wasn't up to all that.

We stayed at Whitfield Hall almost two weeks, long enough to clean out our own food supply and anything edible in the tiny store at Penlyne Castle (mind you, we found no sign of a castle anywhere). This was not looked upon with favor, for the store had other clients whom it supplied, and owners had to wait until market day in Kingston to replace their stores. We had found a farmer with a field of onions who sold us a big bunch, so, for a while, we lived off fresh onions and cheese. He too was not pleased, for he had to have enough onions to make his trip to the market (walking) worth the effort (even though we paid him top price for his onions). Fortunately, marvelous cheeseberries (the introduced *Rubus ellipticus*) grew in masses along the roads and trails and we devoured many for dietary supplements. Strangely, the locals told us these fruits were not edible. We could buy grapefruit locally, but, alas, they were so sour we could barely eat them. We had saved them to take on our nocturnal trek to the 7,402 ft. peak and didn't realize their lack of sugar until we were hungry and shivering from the cold at sunrise on the top of the mountain.

The trip to the top of Blue Mountain was well worth the effort. The views from the top were spectacular (Figure 5), and, coming back, we passed through the cloud forest in which Tree Fern glades (*Cyathea* spp.; Figure 6) were covered with Old Man's Beard (*Usnea*), lichens, and other epiphytes. It was enchanting. These elfin forests, with their stunted moss-covered trees, were beautiful and eerie, as cloud forests are, even if we were lucky enough to have a clear day. We were hoping to find the high elevation frogs, *E. alticola*, and Al even found a nest with its guarding parent, under stones (Figure 7). Unfortunately, the parent escaped, so we could not sex it. In other species that we found attending nests, the females guarded the eggs, unlike the species in Puerto Rico in which males serve as attendants. An altitudinal zonation of species in the Blue Mountains has resulted in *E. alticola*, *E. nubicola* (Figure 8), and *E. andrewsi* replacing the broader-ranging *E. cundalli*, *E. gossei*, and *E. pantoni*. In spite of dry conditions, we found all these species, photographed them, and recorded their calls.

Whitfield Hall was a strange place. The owner was a retired engineer. The existing structure, built in 1802, was probably the manager's quarters.

FIGURE 7. George Martin and the author looking for *Eleutherodactylus alticola* at 6,000 ft. in fern forest on Blue Mountain, May 1973 (photograph by A. Breisch).

FIGURE 6. Tree Fern glade at 6,000 ft. on Blue Mountain, May 1973 (photograph by A. Breisch).

Cynthia, the cook, seemed to be in charge. The first night, we found small droppings deposited on our pillow cases. We wondered if it might be insect frass or gecko droppings, but as the sun set we learned the answer when hundreds of bats emerged from under the corrugated metal roof. A pet burro, Cubby, was spoiled by everyone. We set up our frog processing work on the steps to the front room, where there was adequate light. Cubby wanted to be part of the act, and we had trouble keeping him out of our notebooks and jars. He was obviously

FIGURE 8. *Eleutherodactylus nubicola*, in a ravine by Whitfield Hall, May 1973.

used to coming right on into the room, and he didn't like it when we urged him to go elsewhere. Cynthia dried coffee beans on big sheets of metal roofing on the ground in the back yard. Cubby often walked right through the coffee. After it dried, Cynthia roasted the coffee in an iron skillet over a fire in the yard. The roasting was not always even. Little did the world know under what conditions the famous and expensive Blue Mountain coffee was prepared. When we left, Cynthia reluctantly sold us a pound to take with us.

Ticks, Ants, and the Rat

Even though we had no need to worry about venomous snakes, other nuisances were abundant. One such was ticks — Old World cattle ticks. They had come in with cattle imported to the island from Europe. Unlike North American ticks, they carried no dangerous diseases that we needed to worry about, but the Jamaicans did tell us that a "tick fever" could develop if one were bitten by a lot of ticks simultaneously. When we worked on former cattle grazing lands, we had to do a total body check for "grass lice," juvenile ticks, each night. So, along with washing one's teeth, a tick check became part of the ritual. Fortunately, we did not work often in those sites.

On our trip in 1975–1976, we rented a tiny cottage that was a renovated garden house of a landowner near Retreat. Our bedroom was so small that the mattress fit against three walls, leaving little room for clutter. One night, we came home exhausted and noted a column of large ants moving down the door jamb into our bedroom — then we saw another column moving over our bed. I proclaimed to my husband that they were not biting and that we should ignore them and go to bed, pulling the sheet over our heads. George is a mathematician and is not so tolerant of creepy crawlies. Regardless, my solution didn't work for long, for the numbers of ants kept increasing. We finally gave in and tried to sleep in our small Toyota Corolla. The next day, I alerted the owners, who called in the anti-pest people. Although no ants were marching when we got home the next night, the whole place smelled heavily of insect repellant. Knowing that the chemicals could be more harmful than the ants, we had taken the mattresses out on the verandah, but the table could hold only one mattress. My assistant John got the table, and it was another night in the car and very little sleep for George and me.

The next problem with the cottage was the rat. I can sleep with ants, bats, cockroaches, but rats I cannot abide. One moved in with us. After we were in bed, it would wander along the open rafters of our bedroom and gnaw at the walls. We set rat traps, but rats are often clever and refuse to get caught in traps. Our rat was a clever one. One evening, when we came home, the rat was on top of the refrigerator. We armed ourselves with brooms, dustpans, and whatever heavy items we could find and waited for the rat to emerge from under the fridge. I was surprised when my husband began singing the national anthem as loud as he could, for he can't carry a tune and does not sing. John and I were beating the rat into oblivion at the time. I asked George why the singing. He said it was to keep the people in the nearby big house from thinking we were committing homicide in the garden house.

Husk Pile Adventures

The strangest thing happened during one of our husk pile demolitions, a process that meant taking apart a large husk pile to see what kinds and numbers of frogs and nests were inside. The local villagers often came out to watch our strange antics, and sometimes helped us in the process. In one particular coconut grove, a villager and his dog would always appear. The man carried his machete, *de rigueur* for most rural male Jamaicans. He wanted to help us, and did so in his unusual style. He never touched a frog. Instead, as we removed a coconut husk, he would pin down with the tip of his machete any frog that jumped out, surely a most specialized art. I subsequently caught the frog, which, to my surprise, was never damaged. I expected them to be smashed, but they never were. The man's dog was a little black-and-white terrier type that was much interested in whatever would jump out of the husk pile. I called him Spot. One morning, we were proceeding through our regular routine and things were getting exciting as we got near the bottom of the pile and creatures were jumping in all directions. Suddenly, a rat jumped out and Spot went for it. The rat must have bitten Spot on the nose, for it escaped, and Spot dropped to the ground and lay still, as if dead. When Spot did not get up, I got worried. Knowing how much the Jamaicans valued their special pets, even a chicken, I was seriously worried about the consequences. I could be blamed for Spot's death and even taken to court. We four stood and stared, but nothing happened.

The dog was surely dead. Finally, the man said, "Get up, Spot." The dog got up, shook itself, and continued as if nothing had happened. Although I have asked several dog people to explain the Spot story to me, no one has done so to my satisfaction. I could not imagine how the man could have trained his dog to do such a trick; how would they have rehearsed? Unlike most of our "helpers," the man never asked for money.

We once found a small black snake, *Arrhyton funereus*, near the bottom of a husk pile. This time the machete came down with the sharp edge of the blade first. Al quickly jumped in to rescue the snake. At that, the Jamaicans, who have a great fear of snakes even though Jamaica has no venomous species, stepped back. After Al placed the injured snake in a container, one of the fellows said, "You some person!" Sadly, the snake did not survive. In another husk pile, we captured a live, six-inch centipede (called "forty-legs" by the Jamaicans), so we could take it back to the laboratory for photographs. This really bothered our helpers. They were convinced we would use it for black magic. We had to repeatedly assure them that we would photograph it and then kill it for a scientific specimen.

In another coconut grove near the White River, we had a local helper, Lester, an older man. As we worked, we talked about a variety of topics, often having to do with why we were there and what we did at home. One morning, Lester asked me how old I was. I was a little surprised, for at home that is an unusual question to get from a near stranger. So, I returned the question: "How old do you think I am?" Lester looked me up and down slowly and then replied with my exact age. I was dumfounded by his accuracy.

Working on husk piles, we spent a lot of time in coconut groves, but we did not see coconuts in the markets. While eating lunch one day we asked some young boys who were walking by where we could buy a couple of coconuts. With that, one of the boys, probably not yet in his teens, climbed up the branch less trunk of a 30-ft. palm and began kicking down coconuts until about two dozen lay on the ground. The other boys gathered them while another showed us how to chop a hole in the nut with a machete so we could drink the milk. Then, using a discarded piece of the shell as a spoon, one could scoop out the soft "meat." After eating our fill, we piled the rest of the coconuts in the back seat of our VW bug, leaving barely enough room for Al to crawl in.

Southern Charm

We had been warned that foreigners were not always welcome, and we might encounter hostility to tourists in some areas. We were considered to be rich American tourists, even though our work clothes were often in worse shape than those of the locals. Orange VW beetles were used by rental car agencies, so we were immediately identifiable. We did find hostility, but not where we expected it. We were warned not to go into the John Crow Mountains unless we were about six strong. We were three and without even a machete. However, I was in search of *Eleutherodactylus orcutti* (Figure 9), an aquatic species that I suspected might be a livebearer. Although we had found them in streams in the Blue Mountains, we had seen no sign of reproductive activity. So, a search along the north side of the Blue Mountains through the John Crows might reveal some *E. orcutti*.

We found good roads and friendly people going south from Port Antonio into the John Crows, and we stopped at every workable stream from Alligator Church to Bowden, as far as the road went. From there, one had to walk over Corn Puss Gap, and we were not prepared for that at the time. Near the end of the road, we were working in a stream gully when I looked up and saw five men approaching us, machetes in hands. Needless to say, I was apprehensive, but decided I should use whatever southern charm I could muster. After all, it had worked a couple times before. I climbed out of the stream and greeted them with a big smile and explained what we were doing and why. In minutes, they were in the stream looking for frogs with us. They offered to help us the next time we came, for they were sure they could find frogs for us. George, Al, and I were greatly relieved. We shook hands all around and went on our way.

FIGURE 9. The aquatic *Eleutherodactylus orcutti* with its partially webbed hind feet, Sally River near Whitfield Hall, May 1973 (photograph by A. Breisch).

At another streamside location in the John Crows, the three of us were deeply engrossed in our frog search when several children approached us cautiously. When we said we were searching for frogs, they immediately wanted to help. A short while later, the kids had multiplied until they outnumbered the trees. Each had an eleuth or two in each hand offering to sell them to us. Needless to say, these "collections" were without detailed habitat descriptions and some had been "euthanized" by the rough handling. However, the important lesson in these encounters was that we were generally safe in reportedly hostile areas because the Jamaicans we met must surely have thought we were no threat to them, and perhaps a bit crazy to come to their island to look for frogs.

Nocturnal Light Show

Because we spent so much time working at night, we soon learned about some of the nocturnal creatures other than frogs. We encountered the Jamaican Owl, and rats frightened us by rustling about until we recognized their sounds, but the most fun things were the fireflies. Several different kinds were about, and, following the Jamaican custom, each had its own rhythm to its flashes. The most amazing of these turned out not to be a firefly at all. Instead, it was a large click beetle. The "Kittyboo" or "Pina" (*Pyrophorus* sp.) is over an inch long and has two large luminous spots on the thorax. They glow constantly green-yellow when the beetles are in flight, then turn off when they land. Another taillight, this one orange, glows when the beetle is flying. These beetles speed around in the forest, lights on. We often watched two moving lights that would circle about, fly toward each other, then both lights would go out. We never determined whether they were mating or fighting. Much to our surprise, biting insects were not a problem. We soon forgot all about insect repellants, for we seldom needed them, especially at higher elevations.

Mystery Birds and Whooshes

During the day, we had a hard time keeping our eyes on the ground because of the many fascinating birds that distracted us. The hummingbirds were great fun to watch, especially the Doctor Bird, or Streamer Tail, *Trochilus polytmus*. Males have two six-inch-long tail feathers, striking for a three-inch bird. This is one of the flagship species of the island, and its picture shows up everywhere.

We felt right at home with "Hopping Dick," the White-chinned Thrush, *Turdus aurantius*. It acts very much like our own robin, even though it is mostly black. The absolute charmer, however, is the Jamaican Tody, *Todus todus*, locally called the Robin Red-breast. It is a tiny bird with a bright green back and a brilliant red throat and bill. They were curious, with an odd squeaky call, and seemed to follow us about. The family Todidae, related to Kingfishers, is endemic to the Antilles, and each of the Greater Antilles has its own species. When we were working in Puerto Rico, we became very well acquainted with the larger Puerto Rican species.

We were fascinated by the nocturnal Potoo, *Nyctibius griseus*, and we were lucky enough to photograph one at Marshall's Pen, near Mandeville (Figure 10). My photograph even made the cover illustration for the journal *BioScience* (February 1981). Potoos are related to Nightjars. During the day they sit upright on the end of a dead branch, eyes closed, looking ever so much like part of the branch. They even incubate there, with one egg balanced on the end of the snag. We never saw a nest, however.

Our big thrill was in the Blue Mountains, when we heard the Mountain Whistler, the Jamaican Solitaire, *Myadestes genibarbis*. When the fog rolled in among the huge eucalyptus trees and the bell-like varied calls and whistles of the solitaires rang out, the effect was ethereal and other-worldly. They are thrushes, but look nothing like "Hopping Dick." They turn their heads in different directions when they call, so are hard to locate until one gets used to the ventriloquist effect.

The mystery birds puzzled us for quite a while before we realized they were birds at all. When we were on the open slopes during the day near Mossman's Peak and Penlyne Castle, we thought we were being attacked by a miniature fast-flying military jet. A great "whoosh" would pass over our heads, we would look, and see nothing. This continued for quite some time until we realized that our "jets" were large swifts, most likely the Collared Swifts, *Streptoprocne zonaris*. These birds nest on rocky ledges. Whether they were dive-bombing us because we were getting too close to their nests or just doing their usual aerial acrobatics, we never knew. We found the whole thing to be rather alarming until we realized the source of the "whooshes." I later read that a swift holds the speed record for birds, flying at 200 miles/hr.

Plants Make the Garden

In a former British colony, much evidence remains of the past traditions of the Empire. The national botanic gardens are arguably the most beautiful traces of that past. The English have a great love of nature and an appreciation for the floral beauty of the world. The establishment of formal gardens in the colonies was part of that tradition. The gardens were for both beauty and horticultural purposes. Plants were brought from similar climates all over the world. Four such gardens still exist in Jamaica: Bath, Hope, Castleton, and Cinchona. They were built at different elevations, all within reach of Kingston, the capital and major city. We visited all of them and some repeatedly during my trips. Bath, the oldest, opened in 1779. It is in the southeastern "corner" of the island at about 250 ft. elevation. It is the second oldest in the New World (the oldest is in St. Vincent) and was associated with sulfur springs and baths. Its springs are highly radioactive. It contained plants brought by Captain Bligh from the South Pacific. Flooding of the Sulphur River kept doing serious damage to the gardens. Castleton Gardens, north of Kingston at about 1000 ft. elevation, was built in 1862 to escape such flooding, but alas, the Wag Water River on which Castleton is located also flooded. The Royal Hope Gardens, established in 1881, lie at 1000 ft. in the outskirts of Kingston near the University of the West Indies at Mona, so it can be used by students as well as horticulturists and visitors to Kingston. My favorite is the tiny but derelict Cinchona Gardens in the Blue Mountains at 4500–5500 ft., where the air is cool and fresh and where the views are breathtaking. Near Westphalia and Strawberry Hill, it was established in 1868 as a mountainous site to grow Cinchona, the source of quinine. The Cinchona scheme never materialized, and roads now have deteriorated and one would need a guide to get there on the tracks that still exist. I had my eye on Cinchona as a possible work site, since cottages were present, but they were in such bad condition that living there would have been impossible.

FIGURE 10. The Potoo at dusk at the edge of a pasture, Marshall's Pen, Mandeville, January 1976.

In these public gardens, one can see some of the most impressive trees, palms, and flowering shrubs from all over the world's tropics. Many are novelty trees with unusual fruits such as the Cannonball Tree, *Couroupita guinensis*, from South America. At least when we were there, a knowledgeable and loving guide would escort visitors around the gardens with delightful stories of the plants. Unfortunately, some of the oldest trees in the outlying gardens are gone. Hope Gardens are the largest in the West Indies, and the best kept of the four. Even there, periods of neglect and unusual droughts have taken their toll. Since many tour groups include Castleton Gardens in their excursions, it continues to be a beautiful place for tourists to visit, especially from north-shore hotels. In addition to the public gardens, many private gardens are carefully tended by families of the British ex-colonials.

Some of the trees scattered around the landscape were just as magnificent as the ones in the gardens. Humans were dwarfed standing next to huge Bullet Trees, with their impressive buttresses (Figure

FIGURE 11. A buttressed Bullet Tree by Windsor Cave, April 1973.

11). An occasional Silk Cotton, or Kapok Tree (*Ceiba pentandra*), also had broad buttresses that could easily hide a human. The Saman, or Guango Trees, *Pithecellobium saman*, were not just trees but entire communities. These were often in dryer parts of the island in semi-open situations that allowed their broad crowns to develop unhindered. They may have been survivors from coffee plantations where they were planted as shade for the coffee, but cattle also used them for shade. We spent a lot of time observing one on the Ridge, above Discovery Bay, on a former coffee estate. The huge tree was covered with large bromeliads (Figure 12), a mistletoe-like cactus, *Rhipsalis baccata*, several kinds of orchids, other epiphytes of all sorts, and large brown termite nests. These huge trees with their myriad epiphytes create their own climatic conditions. They are called rain trees, because the grass underneath is sometimes wet. Such trees are good homes for many insects and for treefrogs (Hylidae) that live in the large bromeliads (*Hohenbergia* spp.).

FIGURE 12. Bromeliad covered limbs of a Guango Tree near Retreat, December 1975.

Many of the plants had religious or black magical significance to the Jamaicans. The gardener at a hotel near Port Antonio told us how one stem of "Dumbcane" crushed in the swimming pool would make the swimmers blind. We did not take a dip there. He showed us the Judas Tree, and told us that, if the tree is cut on Good Friday, it will bleed red. Of course, with red sap it will "bleed" on any day it is cut, but that would detract from the story.

Other Frogs and Taped Jamaican Music

The frog world of Jamaica is not limited to the ground- or near-ground-dwelling *Eleutherodactylus*, which have direct development. *Bufo marinus*, a large toad known locally as the "Bullfrog," was introduced in 1844 to feed on small rats in the cane fields. However, the species does not like the cane field habitat, so instead lives almost everywhere else. It, like the mongoose, which was also introduced to control rats, has been ineffective in controlling rats.

More fun are the four species of hylids that are native to Jamaica. They live in the bromeliads, the "wild pines," in the trees or on the ground. Searching the bromeliads (Figure 13) was almost as exciting as searching coconut husk piles. We often heard the small *Hyla wilderi* and the larger *Osteopilus brunneus* (Figure 13), which are widespread. Unlike the eleuths which lay terrestrial eggs, the hylids lay their eggs in the water-filled tanks of the bromeliads (Figure 13). Bromeliad tanks hold water, thereby creating a miniature pond for a myriad of creatures. Since there is limited open water on the island, competition for these tiny ponds is keen, each with its own miniature food web. In addition to developing hylid tadpoles, we found red worms, blue annelids, insect larvae of all sorts, and small brown crabs with red-legs. Tree leaves drifted into the bromeliad axils and provided cover for a grand array of creatures. Slugs, beautifully striped snails

FIGURE 13. The author (top) searching a terrestrial bromeliad for frogs in karst country above Runaway Bay, April 1973 (photograph by A. Breisch). A bromeliad tank (*Hohenbergia polycephala*) (middle) containing a tadpole of *Osteopilus brunneus*, Mason River bog, May 1973. *Osteopilus brunneus* (bottom) peeking through leaves of a bromeliad, east bank of White River, near Retreat, December 1975.

(Figure 14), several kinds of spiders, millipedes, and yellow centipedes were abundant, and we also stirred up an occasional "Pawli" lizard (*Sphaerodactylus* spp.) or an anole. We had to watch out for the large black ants with a fierce bite.

The largest of the hylids, the Snoring Frog, *Osteopilus crucialis*, lives high in holes in tall forest trees and lays its eggs either in water-filled tree holes or bromeliad tanks. Little is known about its natural history. We recorded its calls, but never saw one, for its distribution is limited and climbing trees was not our specialty. Its calls sounded just like someone snoring, but these frogs have another call, possibly an aggressive call, which sounded like a woodpecker tapping on a tree trunk. Our attempt at recording their calls was disappointing. Since we had no parabola or dish to focus the calls, we had to do the best we could. We rigged up a long bamboo pole, tied the microphone to the end, and lifted it as high as we could near the tree where the frog was calling. Much to our surprise, we got Jamaican music when we played back the tape. Somehow our microphone had picked up music from afar that had not been audible to us.

Travelers

With excellent support from Director C. Bernard Lewis, George Proctor, and other staff of the Institute of Jamaica, I had accomplished most of my goals for the 1973 trip, and I made plans to return to Jamaica. I had concentrated my work on species occurring in coconut husk piles, for there I could obtain enough data for analysis and see how introduced species interacted with native species at various elevations. We found 0–159 frogs per husk pile, and estimated combined densities of 300–21,400 frogs per hectare of all common species of *Eleutherodactylus* in lowland habitats of Jamaica. These were some of the highest recorded densities of frogs anywhere. We suspected that much had to do with physiological features of the different species involved, so I invoked the aid of physiological ecologist Harvey Pough to help answer that question. Sure enough, species that were introduced had broader physiological tolerances to temperature and desiccation than did native species. They had to be in order to be good travelers.

A Switch in Islands

When we returned in 1975–1976, many of my field

FIGURE 14. A striped snail from a bromeliad, Retreat, December 1975.

sites had been altered by cultivation. I realized that continued work in Jamaica could be difficult without protected field sites for study. Another problem with observing eleuths in Jamaica was their skittish nature. They obviously had adapted to heavy predation by creatures, possibly mammals. I could never see how or where mating took place. People kept telling me I should work on *Eleutherodactylus coqui* in Puerto Rico, for that species was almost unconcerned with lights and handling. It, too, occurred in high densities and a field station at El Verde in the forest of eastern Puerto Rico would provide lodging.

Friends have asked me why I stopped working in Jamaica, which I liked so much. In addition to concerns over the problem of disturbed field sites, fortuitous events intervened. Harvey Pough asked me to collaborate on a grant proposal to study frogs in Puerto Rico. Reluctantly, I agreed. We were awarded the grant. So, in the summer of 1978, I began studying the *Eleutherodactylus* in eastern Puerto Rico and, with my students, continued those field studies until 1989 when Hurricane Hugo altered the habitat dramatically in my study sites. Sure enough, Puerto Rican eleuths were much more cooperative research subjects. They did not mind being looked at or even handled. The forests were beautiful and we had the El Verde Field Station, where we stayed on our many research trips. But the island, reaching only 4,398 ft., and being much farther out in the Atlantic, lacked the incredible biodiversity that we had enjoyed so much in Jamaica.

I have learned over the years that it is not a good idea to look back. Change can sometimes be very depressing — but I had to find out whether the recent conservation measures in Jamaica were working. Reports by naturalists studying there were discouraging. Deforestation continues at one of the highest rates anywhere, with probably only about 6% of the forest cover remaining. The aquatic *E. orcutti*, which we tried so hard to observe, is now extinct, having apparently disappeared in the late 1980s (S. B. Hedges, S. Michael, pers. comms.). On the other hand, additional species have been described from Jamaica, so the island now supports 17 native plus two introduced species of *Eleutherodactylus*.

Christopher Columbus, after arriving in "Xamayca, land of woods and streams," in 1494, described it as "the fairest island that eyes have beheld; mountainous, and the land seems to touch the sky." I agree.

Acknowledgments

George Martin and Al Breisch were excellent company and help in the field. Their memories of Jamaica, comments on this sketch, and several of Al's photographs have improved it considerably.

Suggested Reading

Dunn, E. R. 1926. The frogs of Jamaica. *Proc. Boston Soc. Nat. Hist.* 38:111–130.

Goin, C. J. 1954. Remarks on evolution of color pattern in the *gossei* group of the frog genus *Eleutherodactylus*. *Ann. Carnegie Mus.* 33:185–195.

Hedges, S. B. 1989. An island radiation: allozyme evolution in Jamaican frogs of the genus *Eleutherodactylus* (Leptodactylidae). *Carib. Jour. Sci.* 25:123–147.

Lynn, W. G. 1940. Amphibians, p. 1–60. *In* W. G. Lynn and C. Grant, The Herpetology of Jamaica. *Bull. Inst. Jamaica* (Sci. ser.) 1:1–148.

Pough, F. H., M. M. Stewart, and R. G. Thomas. 1977. Physiological basis of habitat selection in Jamaican *Eleutherodactylus*. *Oecologia* 27:285–293.

Schwartz, A. and D. Fowler. 1973. The Anura of Jamaica: a progress report. *Stud. Fauna Curaçao Carib. Isl.* 43:50–142.

Stewart, M. M. 1979. The role of introduced species in a Jamaican frog community, p. 113–146. *In* H. Wolda (ed.), *IV Symposium Internacional de Ecología Tropical. Proceedings of the 4th International Symposium on Tropical Ecology*. Panama City.

Stewart, M. M. and G. E. Martin. 1980. Coconut husk piles — a unique habitat for Jamaican terrestrial frogs. *Biotropica* 12: 107–116.

Trueb, L. and M. J. Tyler. 1974. Systematics and evolution of the Greater Antillean hylid frogs. *Occ. Papers Univ. Kansas Mus. Nat. Hist.* 24:1–60.

Stewart, M. M. 2003. Recollections of Jamaica, p. 121–132. *In* R. W. Henderson and R. Powell (eds.), Islands and the Sea: Essays on Herpetological Exploration in the West Indies. Society for the Study of Amphibians and Reptiles, Ithaca (New York). Contributions to Herpetology, volume 20.

Hispaniola

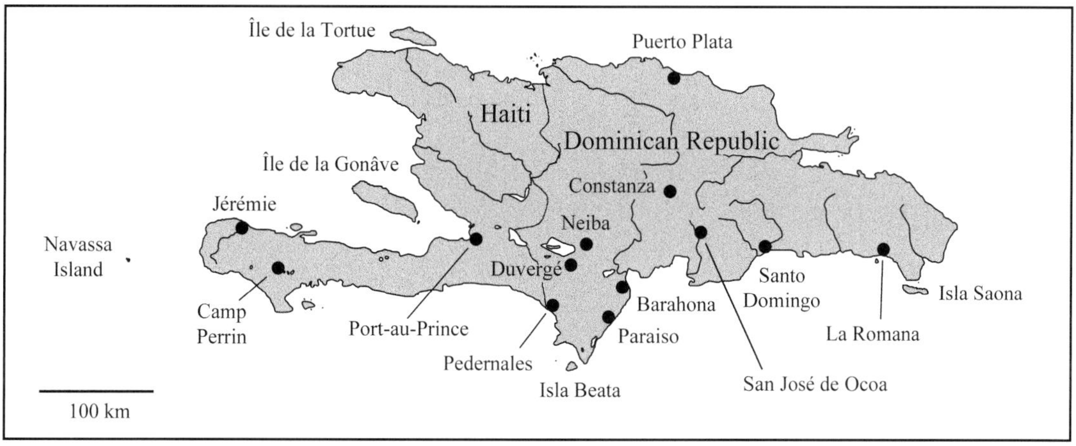

Île de la Tortue
Puerto Plata
Haiti
Dominican Republic
Île de la Gonâve
Constanza
Jérémie
Neiba
Navassa Island
Duvergé
Santo Domingo
Isla Saona
Barahona
Camp Perrin
Port-au-Prince
Paraiso
La Romana
Pedernales
Isla Beata
San José de Ocoa
100 km

A Solenodon for Harvard

A. Stanley Rand[1]

I worked in the West Indies off and on for years after my first trip, with Clayton Ray, July–September 1958, to Puerto Rico and the Dominican Republic (DR). However, in recent times, I have abandoned the islands, but not the tropics. This essay was first drafted in a wood frame house on the edge of the lowland tropical forest in Panamá.

In addition to exploring my own memories, I have asked a couple of other people about the trip. I called Clayton Ray, who also ended up at the Smithsonian, although he works in the Natural History Museum, whereas I joined the Tropical Research Institute. Clayton lent me a copy of his field notes that I have mined unscrupulously for this essay. He read a draft and helped correct both grammatical and factual errors. He also recalled several stories on the phone and over lunch, none of which I remembered. He remembered that money was hard to find at the Museum of Comparative Zoology at Harvard (MCZ), but that we got $50 from Bill Clench in the mollusk department to pick up snails. He also remembered that, when we got to Puerto Rico, we caught anoles at the airport and put them in barf bags, because our collecting bags were still packed — I learned quickly from that experience, and I still always carry a couple of plastic bags folded in my hip pocket.

I also asked my brother what stories he remembered me telling about that first trip. His favorite was of our arriving at an isolated farm at lunch time. We were invited to lunch. The lady of the house apologized for not serving chicken. If she had known the night before, she said, she could have caught a chicken while it was asleep, but during the day, the chickens were too wary. The story is true — but it happened in the back country in São Paulo, 10 years later and 3000 miles away, but it is illustrative of the generosity of country people throughout Latin America, including those we met on our trip to the Dominican Republic.

Cold Stone Steps and Peculiar Teeth

The beginning of my involvement with the West Indies (WI) started under the stairs by the basement door of the MCZ. The need for caffeine, nicotine, and conversation drove the vertebrate paleontologists and herpetologists to gather each afternoon to drink coffee and smoke cigarettes, which was then (as now) prohibited inside the building. The group included curators and whatever graduate students and visitors were about. Regulars included Al Romer, the museum director, Ernest Williams, the curator of herpetology, a turtle anatomist and paleontologist, who was developing an interest in West Indian anoles, and Brian Patterson, a vertebrate paleontologist, whose father had shot the lions of Tsavo. Williams's Tibetan prayer flag flapped above. Most of what I learned in graduate school at Harvard, I learned on those cold stone steps.

Williams was already deeply involved in West Indian zoogeography. He had been to Cuba with Rudy Ruibal and to Jamaica with Karl Koopman and Max and Bessie Hecht. The MCZ knew about the theory of continental drift, but didn't believe it because no mechanism was known. Phil Darlington's *Zoogeography* had just been published and argued strongly that the Greater Antilles had never been connected to each other or to the mainland. However, scattered mammal fossils, mostly Miocene, had to be explained: ground sloths, a monkey on Cuba, and giant rodents on Puerto Rico; but these were scant, and more vertebrate fossils, particularly from the Dominican Republic, were needed. Patterson also was interested in the very peculiar zalambdodont molar teeth of *Solenodon*, a giant insectivore (Lepotyphla) endemic to the Greater Antilles

[1] Smithsonian Tropical Research Institute, MRC 705, Washington, DC 20560, USA.

(Figure 1). The evolution of mammalian molars was a big thing in those days; "protocone, hypocone, and paracone" are still familiar words, although I could no longer draw the appropriate shapes and spatial relationships, much less explain their evolutionary and functional roles.

Clayton wanted to find fossils and I wanted to look at herps in the field in the tropics and, if I was lucky, find a thesis topic. A small amount of money had been found, partly from Sigma Xi, partly from other sources around the museum, to send one of Patterson's graduate students (Clayton) to Hispaniola to look for fossils, and I was offered the chance to go as field assistant and collect herps on the side. My bag was packed and ready before the details were fully explained to me.

Literature and a Dictator

Actually, we spent a lot of time before we left accumulating, packing, and repacking things for the trip. We were told that we would not need the United States postage stamps we were packing. I got copies of the latest books on herps, Doris Cochran's 1941 *Herpetology of Hispaniola*, and K. P. Schmidt's 1924 *Herpetology of Puerto Rico*. At this time, Juan Rivero was still working on his Puerto Rico book, Williams had published only one anole paper, though his "Anoles of Jamaica" with Garth Underwood was to be published the next year. Tom Schoener was still a scrawny high school student interested in birds. Al Schwartz had just started his field work in Cuba in 1954.

Permits were not necessary. We did not need them for collecting or exporting anything from the Dominican Republic or for importing anything into the United States. We did need a zoo vet's certificate of good health before the airline would carry our solenodons, but that was all. Today, a large part of organizing a research trip involves arrangements for permits and complaints about the process — some things were easier in the old days.

FIGURE 1. *Solenodon paradoxus* (drawing by Laurie Walz for F. E. Sergile, C. A. Woods, and P. E. Paryski. 1992. Final report of the Macaya Biosphere Reserve Project. Florida Museum of Natural History, Gainesville).

We took a Greyhound bus from Boston to New York and then flew to Puerto Rico, where we spent a few days fruitlessly searching in caves and trying to locate the quarry where, years before, H. E. Anthony had found a handful of old fossils in a fissure filling. The collecting locality was poorly described, as was the custom in those days, and we were never sure we were in the right place. Regardless, we were sure we didn't find any fossils.

From San Juan, we flew to Ciudad Trujillo in the Dominican Republic. The capitol had been renamed Ciudad Trujillo for (and by) the dictator, Rafael Trujillo, who had been installed by the United States Marines as head of the army when they pulled out of the country in 1924. Using that power base, he had become president, dictator, and "El Jefe." His name and his picture, in a resplendent uniform, were everywhere: cities, provinces, highways, even the local beer was called "El Presidente." The wooden whisky boxes mounted on wheels that small boys pushed through the city were labeled "todos por Trujillo." The front room of every house in the country that we saw had two pictures on the wall: the Virgin Mary and El Jefe. Trujillo was to survive until 1961, when he was assassinated.

A Limited Vocabulary and an Elegant Hat

In the Dominican Republic, we were met by Eugenio de Jesus Marcano, professor of botany at the University of Santo Domingo, which provided us with a car and a driver. We paid $2–3 per day for car expenses and another $3 per day for the driver, Luis Garcia. Garcia started as our driver, but he ended up cooking, carrying gear to the caves, catching bats, entertaining our local guides so that we could get on with fossil hunting, and sharing my beer. Marcano was a wonderful person and incredibly valuable. Williams had recruited him through Dr. Howard at Harvard, who was in correspondence with Dr. José Jimenez, a medical doctor and enthusiastic amateur botanist, who lived in Santiago, near Marcano's home town. Ernest Williams later named an anole and Brian Patterson a fossil *Solenodon* after him. I hope those tributes and the stories about Clayton and myself that he must have told for years recompensed him for his efforts that summer. Marcano was a teetotaler who tried to block my search for beer, and who tried to persuade me to stop smoking cigarettes. He was temporarily successful in lowering my consumption of beer, but more than 20 years passed before I quit smoking.

Marcano introduced us to Prof. Boyrie de Moya. He had established and designed the zoo in Ciudad Trujillo, and was instrumental in arranging a gift of two solenodons for us to take to Harvard.

Marcano spoke very little English, and Clayton no Spanish when we started the trip. With my broken Spanish, I became the "official" translator. By the end of the trip, I was quite fluent, though completely agrammatical and with a very limited vocabulary mostly limited to words pertaining to caves, bats, and lizards. In the last days of the trip, we were completely confused when told excitedly that Marcano had just had his first "baron." Only when we did not react appropriately were we told that his first son had just been born.

Among the field supplies we bought in the market were straw hats. Clayton shaped his immediately. It took me longer since I wanted mine to look different from his, so that we could easily tell them apart. I remember being frustrated that I could not get a shape that was as elegant as his. I never did.

Bones and Decrepit Guides

We had come to look for fossils in caves and, by the end of the trip, we must have looked in several dozen. Marcano had sent a mimeographed letter to public officials around the country, asking them about caves in their areas. He had collated the answers and outlined a series of field trips. Piling our gear into the university's Land Rover, we visited those areas that seemed most promising. We were looking for high, dry caves, for these preserve bones best. We hoped to find bones of mammals no longer present on the islands, like those that had been found elsewhere, from the subrecent past back to the Miocene or even older. A rhinocerotoid ungulate jaw has recently been found in Jamaican Eocene deposits — and that was exactly the sort of fossil we were hoping to find — if we were incredibly lucky. We were not incredibly lucky, but we did find bones of some extinct mammals.

At each site, we hired local guides and set out for the caves. Clayton claimed that he would go to fewer caves if he didn't know it would be downhill on the way back. We became highly selective, carefully picking the most decrepit looking guides, but we were still out-walked by a series of old men and young boys, who kept assuring us, for what seemed like hours, that the cave was very close.

We found lots of bones in some of the caves. Most were from bats living in the caves or from owl pellets, some with introduced *Rattus* and some with equally fresh looking *Nesophontes*. *Nesophontes* was an insectivore (Lepotyphla). Like *Solenodon*, it was unique to the West Indies and was believed to be extinct. Karl Koopman, mammalogist at the American Museum of Natural History, had contributed $100 to the trip — so, we used some of our precious peanut butter as *Nesophontes* bait on mouse traps, but to no avail. On a later trip to Haiti, on the other end of Hispaniola, Skip Lazell and I carried a pack full of mouse traps to the far corners of that country in hopes of catching a *Nesophontes*, but had no luck there, either. The species either really was extinct or barn owls were better at catching small mammals than we were. I had better luck with anoles, but I must admit it is hard not to find anoles in the West Indies, even if you are as naïve as we were.

We did find "hutia" bones, several genera of endemic large rodents, some extant forms of which are the size of woodchucks. We found none of the older and more exotic fossils we had hoped for, although we did find the bones of both native insectivores and a few scraps of ground sloth bone.

Slingshots and Lizard Markets

Clayton also was interested in collecting the bats that were roosting in the caves. The risks of contracting histoplasmosis from the guano (which we probably did) and rabies from bat bites (which we did not) were ignored. Field work seems to be riskier now than it was then, certainly I am more aware of the risks these days. But then, as now, young men in their 20s feel immortal, and Clayton and I were no different. We seized bats from cave walls, grabbed others as they flew by, shot some with a slingshot (although a ricochet broke Clayton's glasses), or knocked them down with a stick. This was a new technique for me. One of us, standing in the cave passageway, whipped a long stick back and forth rapidly as the bats, panicked by our presence, flew past. Most escaped safely, but some unfortunates were struck by the stick and we pounced on them as they hit the floor. Mist nets as a way of catching bats were beyond our purse, but we caught a lot of bats and made a lot of notes on where they roosted and how they behaved when disturbed.

Anoles were easier than bats to catch, and I became quite skilled with a running noose of nylon monofilament on the end of a slender stick. Some lizards, like *Ameiva*, are hard to noose, and I would have liked my .22-pistol loaded with dust shot — but even we were not naïve enough to try to bring a gun into a Latin dictatorship. In fact, I could not even get a permit for my lizard pistol in Cambridge (Massachusetts). The police claimed, with some justice, that there were no lizards to shoot in Cambridge. We had brought a length of aluminum conduit to use as a blowgun. With it, we could hit lizards, but even small ones kept running away with the darts. A slingshot worked well on bats.

Although we caught a lot of anoles, we bought even more in the markets that we set up at each village. After we returned from our caving, we passed the word that we would buy lizards, snakes, and frogs. Only live lizards with complete tails would be bought, we explained. The initial response was incredulity, but when we paid for the first lizard, the village kids began to line up with lizards and frogs and an occasional snake. A lizard with a broken tail was rejected, and sometimes turned up later in the hands of the youngest and cutest in the village. This ploy didn't work, we recognized the lizards. One of the tricks of running a successful lizard market is to adjust the prices so more is offered for the rare than for the common species. It was quickly apparent that the kids knew what lizards lived in and around the villages and where to find them.

"Water Good, Beer Better"

Mostly we set up our cots in rural school houses. School was out for the summer and we had Marcano to explain our needs. So, we had no competition for space. The schools were not elaborate and some were quite simple. One had a thatched roof and a dirt floor. It was built near a steep hillside and had only three walls, the hill providing the fourth. The hill may have kept out the wind, but it did not keep out the mule that wandered into our classroom one night and awoke us by stumbling over the desks. The schools were dry, but even with newspapers from Marcano's plant press spread over them, our canvas cots were cold. I have added a substantial layer of subcutaneous insulation in subsequent years, and would now be less bothered by the cold.

We ate a lot of rice boiled over a wood fire with tuna fish stirred in for flavor. As a treat, we got a West Indian stew of starches: platanos and three or four different roots all boiled together. Boiled with some kind of meat this made a good stew, without the meat it was tasteless, but filling. Restaurants in the country were rustic; one had no running water and the patrons washed in a small basin filled from a pitcher. I was delighted when I poured out my basinfull and got tadpoles. After lunch we explored the well and found the adult treefrogs hiding under the wooden cover. In the restaurant, I splurged on a bottle of beer, El Presidente in a 1-liter bottle. Marcano was concerned about my consuming this quantity of beer and insisted that I share with the driver. At one stop, he assured me that I did not need to drink beer, the water was good. I still feel, as I did then, "Water good, beer better."

The country villages had no TV, and of course no boom boxes or VCRs. We provided a great deal of entertainment as people crowded around to watch us examine the day's fossils, pickle the bats, kill, label, and fix the day's herps, and, of course, write up the day's notes. When we got back to Cambridge, our notes were typed up by Clayton's wife, Donna. In writing this, I worked from a carbon copy of Clayton's notes. We had no access to any kind of electronic storage, not even photocopies. We both grew beards and postponed haircuts. In the country, because of our long hair and beards, and because we were so dirty from digging in guano-floored bat caves, we were thought to be Spaniards. Clayton shaved his beard when we got home; so did I, although I kept my mustache as a memento of the trip, and have it still — even though it has since turned white and hides in my current beard.

We weren't always mistaken for Spaniards. One dark night, we drove around a curve in the mountains near the Haitian border and into an army camp. The soldier who stopped us and peered into the car saw four bearded men and jumped back exclaiming, "Padres!" The days when bearded men in field clothes were assumed to be priests didn't last much longer. The Castro revolution was in full swing in Cuba, Jamaica was still a British colony, Papa Doc Duvalier had just taken over in Haiti.

A Red-hot Needle

Collecting bones, bats, lizards, and snails took most of our time, but we did make some natural history observations. I was tearing open a rotten log to see what might be living inside when I saw two small white ovoids. As I picked up the first, a red-hot needle was plunged into my hand. I jumped back tossing the anole eggs into the air. The tactic of burying her eggs in the nest of a ponerine ant with a fierce sting had saved the egg from me, at least momentarily. How the anole laid her eggs in the ant nest without herself getting stung, I still don't know. We even collected some information on the habits and natural food of hutias when we visited a hutia trapper in Sabana de la Mar. We bought several endemic rodents from him, which we took back alive with the solenodons.

It's Easier To Be Cheerful If You Aren't Hungry

We were frugal in our expenditures, camping in schools was cheap, but, as Clayton wrote in his notes, gas was 50¢ a gallon, and the Land Rover's mileage was not good. On 11 August, he noted that we had $450 and hoped to stretch it for five more weeks. We almost made it. We left on 11 September. A $200 check from Patterson arrived at the end of August. It was most welcome, since it paid for shipping back our 287 pounds of specimens on Pan American World Airways.

We traveled over much of the country, at least those parts with limestone and caves. We also made one trip up into the central mountains to visit the bromeliad-draped pine forests at the high elevations that have a distinctive herpetofauna. In a 1962 paper, I reported on observing anoles in "Santo Domingo and vicinity, Santiago and the nearby Sierra Septentrional, Padre Las Casas, Sabana de la Mar and the mountains south of there, and the Sierra de Neiba." Clayton noted that the country people were much more energetic and more cheerful in the more fertile and better-watered part of the country. We decided that it is easier to be cheerful and energetic if you aren't hungry. The Dominican Republic seemed wonderful to us, but it certainly was no paradise for many of the people living there.

Our Land Rover, the classic model, green and square, took us and our gear over some roads that, on the Esso map, were marked "summer roads" and over others that were not marked at all. The 4-wheel drive of the Land Rover and the skill of the driver were defeated only once. We encountered a place where the road turned into a large lake of mud. We were creeping around the edge when we slid sideways into the morass and sat, wheels spinning, while mud oozed in under the door. We might be there yet, except for a yoke of oxen that came plodding slowly up the trail. They were hooked to the Land Rover and they plodded on, pulling the Land Rover behind them. The oxen moved slowly, but not perceptibly slower than they had without a load.

We made one trip by mule-back, after taking the Land Rover as far as it would go. The first 15 minutes were exhilarating, I looked down on everyone I passed. By the end of the first hour, riding was uncomfortable, and, by end of the day, torturous. I was assured that the third day was the hardest, but, since the trip only lasted three days, I did not find out. Marcano asked each person we met: "Where are you going? Where are you from?" When I asked him about his interest in the activities of the country people, he explained that a rider had to say something, it was not polite to just ride by in silence.

On this trip we stopped at the Marcano family farm and rope factory. In a long, thatch-roofed shed with open walls, rope was twisted. First, the sisal fibers were spun into three independent threads, each the length of the building. Each was separately twisted in one direction and then allowed to untwist together to form a rope. The resultant rope was far superior to the army surplus product we had brought with us from the United States. Marcano's father was obviously pleased by this, but far too polite to point it out.

The main house was quite grand. We were told that the kitchen was in a separate building in case of fire from the wood stove over which everything was cooked. This custom was unfamiliar to me then, but, as I have discovered since, was common in colonial America. Mount Vernon, for example, was built that way. On our last night, the Marcano kitchen produced a feast, quite the best meal we had all summer. It included, among many other things, guinea pigs and squab, both raised on the farm.

Deforestation was evident everywhere we went. Western Europeans had farmed on the island for almost 500 years and most of the forest had been cut for farm fields and pastures and for firewood and charcoal. I remember Marcano telling us that an area of scrub pasture had been wet forest, and explaining that he knew this because one of the spindly shoots in a weedy fence row was a West Indian mahogany and that mahogany only grew in wet forests. Some of the original flora was still hanging on. Perhaps, if this area were protected for a century, it would regenerate — but, even in 1958, I knew that the farmers planting fields and grazing cattle and goats, the women cutting wood for their cooking fires, were not going to let it happen.

Between trips into the field, we stayed in the Hotel República in Ciudad Trujillo, where $5.50 each got us a shared room and "pretty good food." We complained that the city shut down between noon and 3 PM and that there was nothing to do. I have since discovered that that time can be pleasantly filled with a siesta.

A Thesis Topic and a Bad-tempered "Opossum"

By the end of the trip, we had seen a lot of the country, collected a lot of bones, caught a lot of bats, picked up a bunch of snail shells (one a new species), pickled a lot of herps, and I had found a thesis topic. In those days, a guiding principle of ecology stated that no two species could occupy the same niche in the same place, one species must outcompete and displace the other. In the Dominican Republic, I saw commonly three different species of anoles in the same tree. This seemed like an ideal place to look for resource partitioning and possibly interspecific competition. After I got back to Harvard, I organized my notes, and got Marcano to send me three species of live anoles. I used these to test structural niche preferences in a small room that I redecorated as a very minimalist representation of a Dominican landscape. My starting hypothesis was that sympatric species, even though they might live on the same tree, used different parts of it. My results were concordant with this.

I was starting from two anole papers. Jim Oliver in 1948 had described microhabitat differences among three species on Bimini and Bruce Collette in 1961 would describe the habitat use by anoles in a park in Havana that he saw when he visited his family on vacations. These descriptions were accurate, but qualitative. I wanted to quantify the differences and extend the single-site descriptions that Collette and Oliver provided to a large island. Tom Schoener later went on to extend the quantification and explore its ecological importance.

I had hoped to go back to the Dominican Republic to do the research for my thesis, but logistical considerations made me decide on Puerto Rico instead. The latter supported a smaller anole fauna, had a much simpler geography, and Juan Rivero, an old friend of Williams, was willing to provide some logistical support. I was successful in that I got my degree and published the core of my thesis — but that was all still in the future during the summer of 1958.

In the excitement of leaving I forgot my airline ticket. I somehow find it reassuring that my present forgetfulness is not entirely a recent phenomenon.

The *Solenodon*? We received a pair of them from the national zoo in Ciudad Trujillo. We escorted the animals safely back to the United States. In New York, Clayton took a bus to Indianapolis to collect his family. I flew on to Boston with the live animals, where we were met at the airport by a contingent of 11 from Harvard, the largest welcoming party I have ever seen. It was a bit deflating when I realized that they had come for the solenodons, not me.

The solenodons were filmed and photographed eating a variety of things. What the paleontologists learned about the function of zalambdodont teeth from all this, I don't remember — but I do remember the *Solenodon*. A large black and white photograph of the beast, looking like a scruffy and bad-tempered opossum with a long pointy nose hung on my wall for years, a reminder of my introduction to the West Indies. I wish I knew what happened to it.

Since 1958, I have collected, watched, and filmed anoles on all of the Greater Antilles and most of the Lesser Antilles as well. All of the trips were fun, and some were truly great, but nothing matched the excitement of that first trip to the Dominican Republic that introduced me to the West Indies.

Suggested Reading

Rand, A. S. 1962. Notes on Hispaniolan herpetology. 5. The natural history of three sympatric species of *Anolis*. *Breviora* (154):1–15.

Rand, A. S. 1964. Ecological distribution in anoline lizards of Puerto Rico. *Ecology* 45:745–752.

Rand, A. S. 2003. A *Solendon* for Harvard, p. 135–140. *In* R. W. Henderson and R. Powell (eds.), Islands and the Sea: Essays on Herpetological Exploration in the West Indies. Society for the Study of Amphibians and Reptiles, Ithaca (New York). Contributions to Herpetology, volume 20.

Rediscovering the Diversity of Dominican Anoles

Richard E. Glor[1]

Herpetological fieldwork has touched nearly every remote corner of the Caribbean, from the peaks of the highest mountains to the tiniest atolls. Although this might lead one to believe that nothing is left to discover, striking new species continue to be described every year — for example, the recent discovery of what may be the world's smallest lizard on a small Hispaniolan satellite island. Perhaps even more exciting than these remarkable new species is the discovery of new layers of biological diversity among well-studied species that have been revealed by applications of molecular biology. In particular, molecular analyses have found that many widespread species are actually comprised of multiple independent evolutionary lineages, the presence of which had been concealed by a lack of morphological differentiation, but have nevertheless been isolated from one another for millions of years. In addition to suggesting that species diversity has been vastly underestimated, this new level of diversity offers a rare opportunity to examine the evolutionary processes that have contributed to the biological diversity of the Caribbean herpetofauna.

An examination of these processes is at the heart of my thesis studies under Jonathan Losos and Allan Larson at Washington University in St. Louis. In the summer of 2001, those studies took me to Hispaniola with Steve Poe, who was at the time a Miller post-doctoral fellow at the University of California at Berkeley, and Paul Hime, an eager young undergraduate from Washington University with a lifelong interest in reptiles. Both Steve and I have long focused our studies on *Anolis* lizards, the world's most species-rich amniote genus (Paul remains enamored with snakes, but I harbor some hope that he will grow out of it). Known to anyone who has visited a pet shop in the United States as "American chameleons," only one species (*A. carolinensis*) is native to the continental United States. However, just a few miles off the Florida coast, anoles dominate forest ecosystems on the islands of the West Indies, where the genus is represented by over 150 species, many of which occur in such high numbers that every tree in a forest will be cohabited by numerous individuals. Although each anole species shares a set of distinctive features — perhaps the most remarkable being their adhesive toe pads and an extensible throat fan or dewlap — they have evolved an astonishing array of morphological, physiological, and behavioral adaptations that allow finely-tuned specialist species to occupy nearly every available habitat in the Caribbean. As a consequence, anoles may be found everywhere from the driest forests to the wettest mountaintops — and several species are even known to be aquatic.

The anoline radiation is particularly impressive on the four Greater Antillean islands (Cuba, Jamaica, Hispaniola, and Puerto Rico), where anoles exist in communities of up to 11 sympatric congeners. Even more remarkably, similar anoline communities have evolved independently on each island. In particular, strikingly similar ecological specialists known as "ecomorphs" have evolved separately on multiple islands. The trunk-ground ecomorph, for example, has evolved similar adaptations on all four islands for a life spent pouncing on leaf-litter prey from perches low on the trunks of trees. The trunk-crown ecomorph also has evolved on all four islands, but it is adapted to a life high in the leafy canopy.

The Challenge

Hispaniola, the second largest Caribbean island, is home to over 40 species of anoles, despite having an area smaller than the state of Maine. These spe-

[1] Department of Biology, Washington University, St. Louis, Missouri 63130, USA.

cies represent at least six ecomorphs and a host of ecologically and morphologically unique forms. Paul and I had traveled to the Dominican Republic, which occupies the eastern two-thirds of Hispaniola, to sample material for my thesis work investigating the evolutionary processes that have contributed to morphological and molecular diversification in the Hispaniolan Green Anoles (the *A. chlorocyanus* group). Preliminary analyses of mitochondrial DNA sequences already had indicated that two subspecies (*A. c. chlorcyanus* and *A. c. cyanostictus*), which had been distinguished by subtle differences in dewlap color, have actually been diverging from one another for as long as ten million years. My purpose for this trip was to crisscross the island and obtain a more complete sample of genetic material from this species. Steve was primarily interested in documenting variation in crown-giant anoles. However, a new goal emerged shortly into our trip — a goal that appealed to us mostly as lifelong herpers: we were going to catch every species of Dominican anole before we left the island.

We knew that catching all of the Dominican anoles would not be a trivial challenge. A number of the island's species are known only from a handful of specimens collected at remote localities. Furthermore, we would need to visit each of these localities without sacrificing any of the sampling required by our studies of geographic variation in the green and crown-giant anoles. However, our new goal represented a personal challenge through which I hoped to relive the sense of adventure and the thrill of the hunt that is sometimes lost in strictly academic fieldwork.

This goal was not without academic justification. Personally, I felt like I needed to understand the diversity that already existed before I set out to document new levels of diversity with my molecular analyses. Perhaps more importantly, our efforts could provide invaluable material for other biologists studying the evolution of the Dominican herpetofauna. Although our predecessors had amassed remarkable collections of museum specimens, these have traditionally been preserved in formalin, which destroys DNA and other important molecules. As a result, even species that are well-represented in museums are often unavailable for molecular studies. These studies require the collection of new material and, in particular, tissues preserved in alcohol or frozen in liquid nitrogen. Comprehensive tissue collections have already proven essential to answering a number of key questions in Caribbean herpetology. For example, Blair Hedges at Pennsylvania State University has used molecular analyses to determine that most of the Caribbean herpetofauna arrived via over-water dispersal rather than merely riding the land masses that now comprise the islands after they broke away from Central and South America.

Our First Real Challenge

We arrived in the capital city of Santo Domingo in mid-July, but fled the insanity of the city as soon as our permits were approved. After checking off a few common species along the east coast, we were ready for our first real challenge — Fowler's Anole (*Anolis fowleri*). Described in 1973 and still known only from a few individuals, this curiosity among anoles doesn't fit into any of the recognized categories of anoline ecomorphs. Although occasionally described as a twig anole because of its habit of sleeping on twigs and vines, *A. fowleri* is clearly an ecologically unique species. However, like many other rare anoles, its natural history is completely unknown because it has never been observed active in nature — all of the known specimens have been collected while sleeping at night. Adding to its mystery, *A. fowleri* is restricted to a few pockets of hardwood forest along a narrow altitudinal zone in Hispaniola's impressive central mountain range, the Cordillera Central, which boasts the tallest peak in the Caribbean.

Indeed, the species is known only from a few specific patches of forest on the northern and southern faces of the mountains along the notorious dirt road leading north from San José de Ocoa to the mountain valley town of Constanza. On a good day, this road is manageable with four-wheel drive, but after a bit of rain (which comes frequently in the mountains), deep puddles and the danger of mudslides render the road totally impassible. Steve and I had each made attempts on previous trips without success. I had tried in the summer of 2000 driving a rented Chevy Swing, perhaps the smallest four-door car ever made, and had to turn back when I bottomed out after about a half hour of driving. This time though, we had pooled our resources to rent a new Nissan 4 x 4 pick-up with four good tires, a spare, and plenty of ground clearance. We could avoid the worst of the road by foregoing the trip over the mountain and approaching the localities on the northern face from Constanza, but we de-

FIGURE 1. The author trying to catch a male *Anolis distichus* spotted displaying on a log in a stream running through La Palma.

cided to risk starting in San José de Ocoa and driving over the mountain from the south so we could scout for new localities and perhaps even spot some *A. fowleri* active by daylight.

We departed with every confidence that we would have *A. fowleri* by nightfall; in the worst case we would get them that night from one of Albert Schwartz's published localities. As we approached the known altitudinal range of *A. fowleri*, Steve started to take careful notes on habitat and mileage, while simultaneously keeping an eye out for new *A. fowleri* localities. In a way we were lucky, the road was in better shape than usual; we even saw a four-door sedan that had made the journey. We had time to stop several times for lizard hunting and still arrived in Constanza before nightfall. We didn't find any anoles along the way, but we were still confident that success was imminent. So, after a short stop in Constanza to find lodging, we were on our way back up the mountain just after nightfall. As we approached the edge of town, I reset the odometer and we began to log our distance to the type locality, just as Albert Schwartz and his companions had done nearly 30 years previously when they initially discovered *A. fowleri* (what an exciting night that must have been!). Reading and re-reading his dog-eared copy of Schwartz's 1973 paper describing the species, Steve guided us as we retraced Schwartz's directions. By nightfall, we were convinced we had found the spot. It was a small patch of hardwood forest where the road had been cut into the side of the mountain, revealing a wall of deep red earth beneath the forest — exactly as Schwartz had described it. We even found the same small stream running across the road that Schwartz had noted. The mileage didn't exactly match, but that was probably due to us setting our odometer in a slightly different part of town than Schwartz did.

That first night we spent hours bush-whacking in the cold of the mountainside forest by flashlight and headlamp — but we found only a single twig anole, *A. insolitus*. Although it was no *A. fowleri*, it was a new species for us, so when we decided to end our search in the wee hours of the morning, we were content with our one new species and the knowledge that we could afford at least two more nights of searching for *A. fowleri*.

The next day, we took a break from our *A. fowleri* search to visit La Palma (Figure 1), the locality just northeast of Constanza that was made famous by Stan Rand and Ernest Williams's classic ecological study. We spent the day there wandering through

FIGURE 2. *Anolis aliniger*, one of the most colorful Dominican anoles, from the famous La Palma locality in the Cordillera Central. Note the black axillary patch.

FIGURE 3. A male *Anolis distichus* displaying at La Palma.

small mixed-crop farms and wading up a stream. We were rewarded with some of the world's most magnificent anoles and we checked two more species off our list. *Anolis aliniger*'s (Figure 2) striking emerald green color made this trunk-crown species, which is found only in high altitude forests, one of the most beautiful anoles I've seen. We also spotted a few *A. christophei* along the stream. This long-legged species is rather drab until it opens its dewlap, which is pale purple and remarkably large for such a slender-bodied lizard. After catching one of these animals, we noticed that the dewlap is so large that it extends all the way to the abdomen.

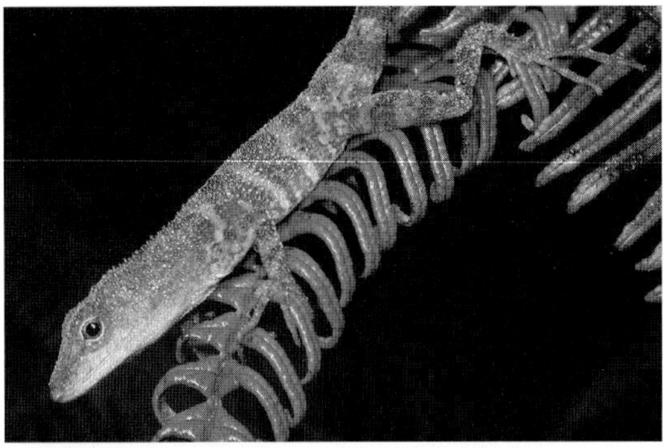

FIGURE 4. Dew-covered *Anolis etheridgei* that were sleeping on ferns at La Palma when we found them.

After a great day at La Palma (Figure 3), we spent a second night of exhaustive (and exhausting) searching at the same locality as the previous night — again without any sign of *A. fowleri*. In fact, we didn't find a single lizard and were beginning to doubt ourselves. With only one night left in the area, we returned with our goal on the line.

We were on the mountain all night long, checking every branch, twig, and vine in the area (Figure 4). We ordinarily wouldn't have spent so much time at the same place with such poor results, but we were convinced that we were at Schwartz's locality. In field herpetology, established localities are like tried and true fishing holes; why risk wasting time at a new spot when you have a proven winner? We all agreed it was too risky to go exploring when we were confident we had the right spot.

As the night wore on, we started getting more desperate and adventurous. We had already checked the easy spots, so we started sliding down steep embankments and climbing up stony ridges to check unexplored patches. Sometime in the middle of the night, I started to work my way up the stream that cut across the road. It quickly became a steeply sloping, overgrown rivulet winding its way to the top of a nearby peak. In the streambed, the soft layer of topsoil had been washed away to reveal a substrate of slick, gray rocks. Clambering along while maneuvering through the dense vegetation that had sprouted around the stream wasn't easy, but I was desperate. Somewhere near the summit, I lost my balance trying to jump up to the top of a small boulder and ended up somersaulting backwards down the mountainside. After bouncing off a few rocks, I ended up on my back with my head in small pool of remarkably chilly water. As I regained my composure and glanced to my right, I noticed a jagged rock in the pool that surely would have ended my night (and maybe my life) if I had landed on it. As a consequence of this rather rude interruption of my search, I began to question the risks and the logic of continuing this apparently futile search and headed back toward the truck to see if the others were having any more luck.

My watch read four AM when I arrived at the truck. I found that I wasn't the only one who had reached the breaking point. All of us were cold, wet, scratched, bug-bitten, and generally exhausted from hours in the brush. Adding insult to injury, Paul's stomach, which entered the trip a virgin to Dominican cuisine, was beginning to stage a full-blown

FIGURE 5. The locality on the road heading south from Constanza where we finally found *Anolis fowleri*.

rebellion. After some painful deliberation, we decided to cut our losses and head back to Constanza. I was convinced that we had failed at our first big challenge and our goal of catching all the Dominican anoles was over before it had really begun, but Steve insisted we make one more stop at a patch of forest he had noted on the way up the mountain, so I reluctantly pulled over next to an unimpressive patch of forest flanking a tiny stream (Figure 5).

In no mood to repeat my earlier brush with disaster, I checked the roadside vegetation while Steve headed up the stream. After a few minutes of searching, I heard Steve yelling from a distance. Had he found *A. fowleri*? If not, I was going to kill him for making me trudge up another overgrown stream. I arrived to find him straining his eyes at something high above his head in an impossible tangle of vines that looked like a baby twig anole, *A. insolitus*, or perhaps just a twig. If it was *A. insolitus*, it would be only our second one, so I commuted Steve's death sentence and started thinking about how we were going to catch it. Steve extended his 14-foot collapsible fishing rod and attempted to lasso the lizard along with the twig to which it was clinging. Unfortunately, the lasso snagged and the twig snapped, sending the lizard tumbling into the thicket

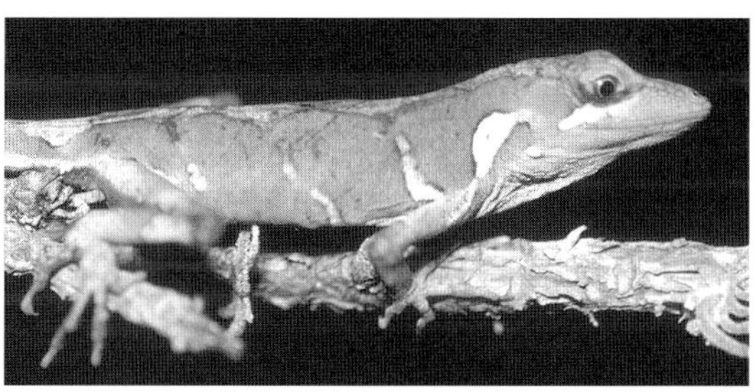

FIGURE 6. The first *Anolis fowleri* we found sleeping on a fern.

in front of us. With the energy of collective frustration, we started sifting through the densely matted vegetation and leaf litter for the tiny lizard. As he searched, Steve broke through a cluster of vines into the open forest understory that the dense vegetation around the stream had concealed. There, directly in front of his face, he saw a perfect little green anole resting peacefully atop a fern. As I looked over his shoulder, I thought it must be a baby crown giant (*A. baleatus*), but how strange to see one so near the ground on a fern. As Steve approached the sleeping lizard, he noticed the telltale white neck markings he had read about in Schwartz's description and shouted: "*fowleri*!" I stared in disbelief as he scrambled through the brush to grab the animal (Figure 6).

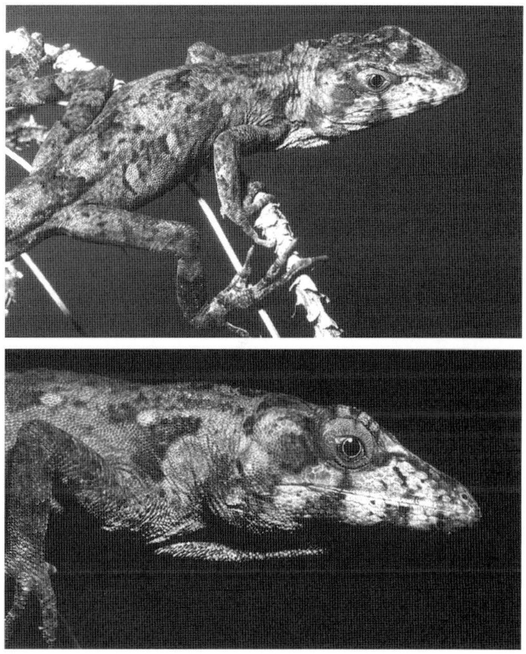

FIGURE 7. An adult male *Anolis fowleri*, which was found sleeping on a tree fern.

After all this time scanning the canopy and tracing the outline of every twig and vine with our lights and eyes, I couldn't believe that we could find one of these beauties exposed on a fern — especially after we had come within moments of giving up. With renewed energy, we scrambled through the understory hoping to see a few more. Within minutes, we had found another juvenile sleeping on a fern and an adult male hanging from the frond of a tree fern about two meters off the ground (Figure 7). It seemed more like a fern anole than a twig

FIGURE 8. Steve Poe (right), Paul Hime (left), and the author, each holding an *Anolis fowleri*.

anole to us, but the true secrets of its natural history still await a naturalist with the skill and patience to make some daytime observations. Back at the car, we shared our treasures with Paul (Figure 8), whose response provided the six letter code we needed to label the locality in our Global Positioning Unit: "Hell ya!"

Needle in a Haystack on Top of a Mountain

Buoyed by our success with *Anolis fowleri*, we headed to the north coast where we traveled for several weeks, collecting tissue samples for our molecular studies and checking off a few other common anoles along the way. We didn't meet our next real challenge until we headed back south toward the second largest of Hispaniola's main mountain ranges — the Sierra de Neiba. Roads into this range are in even worse condition than those in the Cordillera Central, but the Neiba is the only place on earth where you can find the rare twig anole, *Anolis placidus*, which was not formally described until 1989. We knew of only one way to see this species — driving straight up and over one of the weathered dirt roads straddling the mountains. We brushed off warnings that the roads over the mountain were impassible and drove to the small town of Hondo Valle near the Haitian border. From there we would attack the mountain's steep northern face.

We found the road without much trouble and were beginning to enjoy the view from the steeply ascending dirt track after about an hour of driving. Then we rounded a corner and heard a sudden hissing sound from under the truck. I looked at Steve: "Is that our tire?" He glanced out the window and saw that we had badly blown our front right tire, which was completely flat in a matter of seconds.

Flat tires were nothing new — this was our fifth of the trip — but this one came at a really bad time. We were already riding on three other surgically repaired tires; two of them had plastic strips stuffed in puncture wounds, and the third was filled with a cheap inner tube after a bad tear rendered the original shell irreparable. We still had a spare, but we were just beginning the drive. Could we risk getting stuck in the mountains if we lost another tire? As we changed tires in front of an audience of over a dozen curious locals, I was convinced that our chances for *A. placidus* had vanished as rapidly as the stale air from our tire. We decided to err on the side of caution and headed back to town to have our tire patched by one of the local gomeros. These young tire repair entrepreneurs have shops in nearly every town across the country, regardless of size. All you need to start a basic gomero shop is a screwdriver, a few strips of plastic, and a bucket of water. Although I'm generally amazed at their resourcefulness and efficiency, gomeros are not risk free — some will do more damage than good.

On our heart-breaking ride down in search of the local tire man, we struggled with the logistics of making another attempt at the mountain. Darkness would arrive soon, and driving at night would be risky, but we had only a week left, and we would need at least that long to collect the samples we needed in the south. Returning for *A. placidus* on another day was out of the question.

Although the biggest risk of driving at night was handling the truck on a narrow and unknown one-lane road where a wrong move could leave us rolling down the mountainside (guard rails in the Dominican Republic are a rare sight indeed), I was equally worried about dealing with the Dominican military outposts that occur regularly along remote roads near the Haitian border. The guards are typically very understanding, but a truck full of gringos passing after midnight looking for lizards would probably test their limits. I was beginning to think we should abandon *A. placidus* and head south before we got ourselves into trouble.

When we reached the gomero, the decision was taken out of our hands. He told us he couldn't repair our tire; it was too badly damaged to fix without an inner tube, and he didn't have any. We piled back into the truck to plan our route south, but just as we were pulling out we got another lucky break, a woman started yelling from inside the repair shop, she'd managed to scrounge up an inner tube for us!

FIGURE 9. *Anolis placidus* from the Sierra de Neiba.

Taking this as a sign that good fortune was smiling on us, we headed back up the mountain, this time driving on two inner tubes and two patched tires. Although it was pitch black before we reached even the base, we all agreed that *A. placidus* was worth the risk. However, several hours were needed before we would arrive in *A. placidus* habitat and we would need to pass a few checkpoints first. After about two hours, we approached the first checkpoint, where my hopes that the guards would have dozed off for the night were quickly dashed. Two soldiers with machine guns and one with a shotgun, all of whom were surprised to see anyone passing at this hour, quickly surrounded the truck and ordered us to turn on the cab light so they could see our faces. Then the commander of the outpost approached and asked us why we were driving on the mountain this late. He was unconvinced when I explained that we were biologists looking for lizards. Understandably, he wondered why we should be out at night when everybody knows it's easier to see lizards by day. After a few tense moments, I convinced him that nighttime hunts were perfect for catching sleeping lizards. Although I still don't think he really believed us, he warned us to be careful and sent us on our way. Thankfully, by the time we reached the next checkpoint all the guards were asleep, so we passed without incident.

Unfortunately though, the first *A. placidus* locality on our map was right next to that outpost. To avoid waking the guards, I drove over the next rise before shutting off the engine. We then gently shut our doors before slinking off into the night to begin our nocturnal hunt. Twig anoles are rarely ever observed by day and are among the treasures that make nighttime herping a necessary and worthwhile risk. Unlike daytime herping, where half the challenge is catching the animal after you've spotted it, the biggest challenge to hunting twig anoles at night is merely finding them. These slender-bodied and well-camouflaged anoles typically occur in dense mats of twigs and vines, so its not too grand an exaggeration to say that finding them is like finding a needle in a haystack, in a tree, in the dark, on top of a mountain. The best approach is often to go to an area where twigs and vines are abundant and start tracing everything in sight with your flashlight or headlamp, scanning for the silhouette of a sleeping lizard. Even for a trained eye, this can lead to long, tedious nights, full of more false alarms than actual lizard sightings. On a night like that my heart raced with every lizard-like bump on a twig.

After you manage to spot them, catching twig anoles is often anti-climactic. Though highly skilled at traversing narrow perches, they are not particularly fast, and their hopes of evading capture hinge primarily on crypsis. Their immediate response to disturbance is typically just to cling more tightly to their perch. Sadly for the anoles, this strategy is ineffective against humans, who can simply snap off their perches if they're within an arm's reach.

As soon as we turned on our lights, the troubles on the road were forgotten and we were filled with the sense of excitement shared by every naturalist searching for a rare species. It wasn't long before our efforts were rewarded. After less than an hour of searching, Steve found our first *A. placidus* in the forest along the road (Figure 9). Once in hand, I found it difficult to believe that we had considered skipping this species — it was magnificent. Like a tiny dragon, it had an impressive spiked crest and it was not difficult to imagine this paragon of adaptation scaling twigs with the perfection that came with millions of years of evolution by natural selection.

We found a few more specimens over the next few hours before heading back down the south face, still not knowing how far we would have to drive to get off the mountain. After about four more hours of driving and just as the sun was rising, we reached the main highway that cuts across the valley between the Sierra de Neiba and the Sierra de Bahoruco to the south. As we pulled off to the side of the deserted road (Figure 10) to figure out where we

FIGURE 10. How far are we from Barahona? Signage in the Dominican Republic is rare and, when present, is often unreliable.

FIGURE 11. The famous "Green Chapel," which marks the entrance to the road leading to a well-known *Anolis barbouri* locality.

were going to stay, Steve glanced out the window and spotted a sleeping anole. It turned out to be a grass anole, *Anolis olssoni*, another new species for our list. This had indeed been a good day. With our rewards in hand, we headed toward the town of Neiba with the hopes of finding a cheap hotel and get some well-deserved sleep.

A Haitian Village and the Green Chapel Road

Unfortunately, Paul had to be back for the start of classes in late August, so Steve and I were on our own to catch the last five species on our list. After dropping Paul off at the airport, we set out to find the Dominican Republic's strangest anole, *Anolis* (= *Chamaelinorops*) *barbouri*. This species lives in the mountains of the Sierra de Bahoruco and the Cordillera Central, and for unknown reasons has an extraordinarily wide vertebral column.

Just before we began our search for this species, I called one of my thesis advisers, Jonathan Losos, to give him a progress report and to get some pointers on finding *A. barbouri*. When I called, I rattled off a list of the species we had yet to find: *A. barbouri, A. armouri, A. whitemani, A. singularis,* and *A. strahmi*. His immediate response was: "Well you shouldn't have any trouble with *barbouri*, just head north from Barahona and turn left on the road marked by the green chapel. At the end of the road, you'll find a Haitian village where the residents can help you find the lizards for a small finder's fee." Steve had received exactly the same advice from Ernest Williams and José Rosado at Harvard and from Kevin de Queiroz at the Smithsonian. This was going to be a piece of cake, we just needed to find the Green Chapel Road, and how many green chapels could there be on the road heading north from Barahona?

We soon found out, however, that the "green chapel" was neither a chapel nor was it green (Figure 11). It is perhaps more accurately described as a roadside monument comprised of a concrete cross surrounded by a low fence, all of which has been painted blue and white — apparently, the folks responsible for maintaining this monument are unaware of its importance as a landmark for visiting herpetologists! Thankfully, Steve had thought ahead and noted the names of some towns along the road on the basis of specimen labels in the Museum of Comparative Zoology, so we were able to navigate with those landmarks. Once we found the "White and Blue Concrete Cross Road," we figured we were set, all we needed to do was drive straight ahead until we reached the Haitian village.

This too would not prove so simple. We went as straight as we could, and nearly drove right off a cliff. Some locals who observed with amazement our attempt to drive down a steep donkey trail told us we had missed the turn to the Haitian village. We thought: "What turn, nobody said anything about a turn." After a bit of backtracking, we made the turn and found the Haitian village at the end of the road. Although we had some difficulty communicating with the Creole-speaking Haitian immigrants, they were adamant that no strange lizards lived in the area. This must be the place, though, for how many Haitian villages could there be at the end of the "White and Blue Concrete Cross Road?"

As it turns out, there are four, and we visited every one of them. Some new roads apparently have cropped up since our mentors and colleagues last visited the nameless Haitian village of lore. It really didn't help that no matter which route you took, the maze of poorly maintained dirt roads dead-ended in a Haitian village. When we reached the third village, I asked if there were any more, and we were given instructions to one last village, this one much farther up the mountain.

The next 20 km required several hours of driving along an apparently abandoned track that only with great charity could have been called a road. Just as we approached an area where it appeared that dense growth in the middle of the road would mark the end of our drive, we spotted a man wearing shorts and an ancient gray sport jacket walking along the road with his machete. He asked what we were looking for and I explained that we were biologists in search of a unique lizard native to the region. To our great delight, he claimed to know what we were

FIGURE 12. The Haitian village where we finally found *Anolis barbouri*.

talking about, and hopped into the truck to guide us to his village, which was just a bit farther up the road. We seemed to have finally found our spot, but as we struggled to drive over the dense growth in the road, I couldn't shake the feeling that this guy was just taking us to an even more remote spot where he could rob us at machete-point.

Thankfully, my fears were completely unfounded and, after a bit more driving, we reached a small village (Figure 12) consisting of ramshackle huts occupied by Haitian immigrants and their families. Our guide took us to a small stream cutting through a coffee grove where he claimed the lizards were common. We spent most of the afternoon clambering down this steep stream, but after hours of searching, we had only seen a few individuals of the colorful bush anole, *A. bahorucoensis*. Our guide assured us that our failure was due to the overcast skies and that if we came back the next day we would be sure to find some. Unfortunately, we hadn't budgeted for another trip up the mountain and, at that point, we weren't even sure if this guy understood what we wanted, but we were too close to our goal to give up now.

Early the next morning, we drove back up the mountain where we found our guide waiting patiently for our return. He came through. By the end of the morning, we had four *A. barbouri*. Because we needed to get back down the mountain and all the way across the Barahona Peninsula before nightfall to stay on schedule, we didn't stick around for long. Before leaving the village, though, we gave all our food to the villagers and compensated our guide for his help. The children (Figure 13) seemed particularly delighted with the Captain Crunch cereal we had picked up for them in Barahona. We now had found all but four species, but time was short and getting these last few would prove a real challenge.

Herping with an Audience

Our experiences with the desperately poor residents of that isolated immigrant village were typical of my interactions with the people of the West Indies — namely, they know their natural history and they are always willing to help. This is a real blessing, because interacting with local residents is unavoidable. On the relatively small islands of the Caribbean, the days of working in pristine forests untouched by humans are long past. Today, people are everywhere, from the peaks of the region's highest mountains to the hearts of the best-protected parks. In the Dominican Republic, in particular, people are so ubiquitous that most herping is done in front of an audience.

When I first started working there, this was a source of endless frustration. In addition to the logistical impossibility of catching animals with spectators scaring off our prey, the unrelenting stares of the locals could be disconcerting. A big lanky gringo like me simply cannot blend in — and, of course, the fact that I'm generally walking around in remote areas picking lizards off trees with a lasso tied to the end of a fishing pole doesn't help.

Eventually, I was able to overcome the frustration by putting myself in the shoes of the locals. I realized that to them, I wasn't just a biologist trying to do my job, I was an foreigner engaging in remarkably odd behaviors in their neighborhood. I tried to come up with an appropriate analogy in the United States. How would I react if I was an Iowa farmer who looked out my window one day to see an obvious foreigner picking animals out of the trees in my front yard with a twenty-foot-long pole? I would probably greet him with a shotgun in hand.

FIGURE 13. A young girl in the Haitian village enjoying some Captain Crunch.

Thankfully, most Dominicans are content just to stare. The solution to my discomfort was simple, rather than treating people like an annoyance, I began to engage them. I quickly learned that a smile, a wave, and a quick hello were the surest way of turning a stare into a smile. Nowadays, I make it a point to speak with everyone I come across in the field, and I can count on one hand the number of times I've been told to take a hike.

Close Calls While Closing In

After getting *Anolis barbouri*, we entered a mad dash to the finish. We only had two days left to get the final four species: *A. whitemani, A. strahmi, A. singularis,* and *A. armouri*. I knew we could get these species because all occur on the Barahona Peninsula and I had seen them on previous trips — but a single day of bad weather could ruin our best-laid plans. Thankfully, the hurricane that was supposed to pound the Barahona Peninsula skirted the coast and only caused a single night of downpours.

As soon as we bagged *A. barbouri* on the eastern edge of the peninsula, we raced down the mountain and drove all the way across to the western side, toward the small fishing town of Pedernales. We had planned a day-long trip up and over the Sierra de Baoruco the next day, but we still had enough daylight left for a quick late-afternoon trip in the Pedernales area. We decided to drive up the old Alcoa mining road to the Aceitillar region to look for *A. armouri* and *A. singularis*. This road is another famous herping locality in the Dominican Republic. Originally built by Alcoa to bring bauxite down from their mountaintop strip mining operation to the docks at Cabo Rojo, this impressive three-lane road was abandoned when the mines were closed years ago. However, it now provides a unique access point to the region's incredible montane pine forests. The Dominican government eventually hopes to use this road to shuttle ecotourists and birdwatchers to a new national park surrounding the old mines, but facilities are not yet completed.

FIGURE 14. The rare Dominican snake, *Ialtris doralis*, found on a trail in the Sierra Martín García.

Because we had to drive all the way across the peninsula to get there, we didn't arrive until late in the afternoon when the sun was already setting — a bad time to arrive in the mountains where lizard activity periods are typically short. Our first goal was to find the montane green anole, *A. singularis,* which is endemic to the Sierra de Bahoruco and the Sierra Martín García (Figure 14). This species has a reputation of being one of the country's most difficult anoles to capture and was given the name *A. singularis* because only a single individual is generally collected from any locality. In addition to occurring only at high altitudes, they are found only high in the forest canopy. To further complicate things, they're among the most wary of anoles, typically reacting to disturbance by scampering up a trunk and farther into the canopy.

After we checked in with the soldiers guarding the developing tourist attraction from poachers, we began scanning the pine trunks for the remarkable emerald-green-and-blue *A. singularis*. While Steve was struggling over a barrier in the road, I spotted one high on a trunk, well out of the range of my 20-foot pole. Against my protests, the curious soldier who was tagging along chucked a few rocks at it, barely missing and sending the lizard scampering even higher into the canopy. Given this trauma, I doubted this lizard would be coming back down anytime soon. I quickly turned my attention to other trunks in the area and, after another half hour, spotted another anole high in the canopy. The soldier had lost interest by this time, so I was able to proceed more deliberately with this one. I was lucky because it was within lasso range on one of the smaller trees in the area, but it scurried out of reach as soon as I began extending my pole.

Realizing that I might not get another chance, I gently shook the tree with the hope of scaring the lizard back down into lasso range — but, when I started shaking, the lizard just scurried up even farther into a cluster of pine needles. I started shaking more vigorously, trying to jar it loose. After a few minutes of violence, it started to lose its grip. With a few more solid jars, the lizard came parachuting down to me. I reached to snatch him from the air, but he fell though my hands onto the ground and without missing a beat turned around and ran back up the trunk. As soon as I saw him jump back onto the trunk, I swung my arms around the tree, hugging with the hope that my arms would create enough of a roadblock to prevent him from climbing farther. Since he ran up the other side of the trunk I wasn't even sure he hadn't already dashed back up into the canopy, so I peeked around the trunk — and he was still there, only about a meter off the ground! Keeping one hand above to prevent him from climbing, I reminded myself of the most important rule of lizard catching as I moved into position — the lizard is always quicker. No matter how quick you are, you're bound to end up with a handful of trunk if you don't anticipate where the lizard is going to run and catch it on the move. Just as a quarterback leads his receiver, the herpetologist leads his lizard. Knowing this canopy-dwelling species would try to sprint up the trunk, I brought my hand down a few inches above his current location and, to my great delight, I ended up with the surprised but unharmed lizard in my hand.

Steve and I celebrated for a minute before heading off to see if we could also check the pine forest trunk-ground species, *A. armouri*, off our list before darkness descended. After turning a few stones around one of the old mines, we had our *A. armouri*. Now we had only two species left, *A. strahmi* and *A. whitemani*, but they would need to wait until the next day. Both of these species are trunk-ground anoles, closely related to the nearly ubiquitous *A. cybotes*, but *A. strahmi* and *A. whitemani* each have unique ecological requirements. *Anolis strahmi* is a rock-dwelling species found only on outcrops and roadcuts on the slopes of the Sierra de Bahoruco, whereas *A. whitemani* is a dry forest specialist found only in the most desolate portions of the Valle de Neiba and surrounding areas. However, I knew we could get both of the species in one day if we drove across the badly worn road leading from Pedernales to Puerto Escondito. This road would take us from dry coastal forests up through the pine forests of the mountaintops and back down into the dry forests of the Valle de Neiba.

Early the next morning, we started the long drive. On previous trips, I had found *A. strahmi* in road cuts along the south face, so we stopped a few times on the way up to look, but without any success. The sun was out and the weather seemed perfect for anoles, but we encountered a surprising amount of traffic on the first stretch of this one-lane dirt road, mostly in the form of badly overloaded pick-ups teetering on the edge of control, so I thought it would be best to postpone our search for *A. strahmi* until we got to the outcrops on the other side of the mountain. I wasn't thinking that the weather might turn by the time we got to the other side.

By the time we reached the top and spent some time looking for more *A. singularis* and *A. armouri* in the pine forest, clouds had moved in and were threatening our chances of finding *A. strahmi* and *A. whitemani* on the other side. The sun also was beginning to set, so we hurried down the north face, heading directly to the exposed road cuts just past the military outpost at El Aguacate. After more than an hour of scanning the rocks there, we had yet to see a single *A. strahmi*. We were just about out of daylight when I spotted a lizard silhouette high up on the rock face. Typical of *A. strahmi*, it was perched next to a crevice, meaning I would only have one chance to catch it before it dove into the cliff for safety. This was a high risk, high-pressure attempt; Steve was standing next to me and we both knew our hopes of completing our checklist rested on my catching this lizard. As I inched my pole toward the lizard, it attacked my dental floss lasso, probably mistaking it for a flying insect. Fortunately, I had a second chance because the lizard didn't dive for cover after I succeeded in wrestling the string from its mouth. As I moved my lasso into place for a second attempt, the lizard broke for the crevice, but my lasso was already in position and he ran straight into my trap.

We had no time to celebrate, since we still had one more species to go and the sun was already on the horizon. We sped toward Puerto Escondito, to the exact spot where I had caught *A. whitemani* on a previous trip. Slowing down as we reached the locality, we started talking about our plan of attack, it was already dusk and we might have to wait until it got dark when we could find the animals sleeping. Just as I was stopping the truck, I glanced out my window and saw an anole perched on a rock next to the road — and it was an *A. whitemani*! Mission accomplished. We enjoyed a moment of celebration and snapped a photo just as the sun faded below the horizon (Figure 15).

Acknowledgments

I want to take this opportunity to thank two mentors who introduced me to academic herpetology

FIGURE 15. Steve Poe (left), holding an *Anolis whitemani*, and the author, holding an *A. strahmi*, as the sun sets immediately after catching the final two species on our list.

and helped me make the most of my undergraduate years. All of the life-changing academic experiences I had as an undergraduate I can attribute to Kraig Adler — including my first trip to the West Indies with Bob Powell, who has provided me and many other undergraduates with our first experiences in tropical field biology through his REU programs.

I also thank José Ottenwalder, whose hospitality has made my trips to the Dominican Republic possible. Finally, none of my work in the Dominican Republic would be possible without the logistical support of the Dirección General de Vida Silvestre y Biodiversidad and the Dirección de Areas Protegidas y Biodiversidad. Gloria Santana in Vida Silvestre and Andreas Schubert and Matilda Mota in Areas Protegidas have been particularly helpful.

Financial support for the trip described in this essay was provided by a grant from the National Science Foundation to Jonathan Losos.

Suggested Reading

Flores, G., J. H. Lenzychi, and J. Palumbo, Jr. 1994. An ecological study of the endemic Hispaniolan anoline lizard, *Chamaelinorops barbouri* (Lacertilia: Iguanidae). *Breviora* 499:1–23.

Hedges, S. B. 1996. The origin of West Indian amphibians and reptiles, p. 95–128. *In* R. Powell and R. W. Henderson (eds.), Contributions to West Indian Herpetology: A Tribute to Albert Schwartz. Society for the Study of Amphibians and Reptiles, Ithaca (New York). Contributions to Herpetology, volume 12.

Hedges, S. B. and R. Thomas. 2001. At the lower size limit in amniote vertebrates: a new diminutive lizard from the West Indies. *Carib. Jour. Sci.* 37:168–173.

Hedges, S. B. and R. Thomas. 1989. A new species of *Anolis* (Sauria: Iguanidae) from the Sierra de Neiba, Hispaniola. *Herpetologica* 45:330–336.

Howard, A. K., J. D. Forester, J. M. Ruder, J. S. Parmerlee, Jr., and R. Powell. 1999. Natural history of a terrestrial Hispaniolan anole: *Anolis barbouri*. *Jour. Herpetol.* 33:702–706.

Rand, A. S. and E. E. Williams. 1969. The anoles of La Palma: aspects of their ecological relationships. *Breviora* 327:1–17.

Schwartz, A. 1973. A new species of montane *Anolis* (Sauria, Iguanidae) from Hispaniola. *Ann. Carnegie Mus.* 44:183–195.

Williams, E. E. 1965. The species of Hispaniolan green anoles (Sauria, Iguanidae). *Breviora* 227:1–16.

Glor, R. E. 2003. Rediscovering the diversity of Dominican anoles, p. 141–152. *In* R. W. Henderson and R. Powell (eds.), Islands and the Sea: Essays on Herpetological Exploration in the West Indies. Society for the Study of Amphibians and Reptiles, Ithaca (New York). Contributions to Herpetology, volume 20.

Dewlaps and the Lack Thereof; or, Lizard Studies in the Neotropics

Henry S. Fitch[1]

Long before I ever saw the tropics, I had become very familiar with several regions of the United States, and had marked and studied populations of many species. My interest in animals had always centered on snakes. However, at the San Joaquin Experimental Range in the Sierra Nevada foothills, where I worked for several years while employed by the United States Bureau of Biological Survey and its successor, the United States Fish and Wildlife Service, my official duties involved studies of mammals that were rangeland pests. Later, at the University of Kansas Natural History Reservation under Prof. E. Raymond Hall, mammal study was "encouraged." Lizards, raptors, even spiders and mollusks also have been subjects of my research. As decades passed, many of my colleagues at the University of Kansas made field trips to the tropics, but my lot seemed that of a stay-at-home, looking out for my family and my marked populations of animals. However, in 1965, at age 55, I enrolled in the Organization for Tropical Studies' six-week course in tropical ecology in Costa Rica. The course was interesting and profitable.

I took a LACSA plane from Miami and, as the trip ended, with the plane swooping south over the continental divide to a landing near San José, I was overwhelmed with a feeling of elation that, at last, I was in the tropics and would have the chance to study populations of tropical animals in the field. Coursework took me to most parts of Costa Rica, and I became acquainted with the country's diverse climates and fauna. I returned with my family after successfully applying for a National Science Foundation (NSF) sabbatical grant in 1967 to study lizard and snake populations under the different climatic regimes in Costa Rica. The lizard project developed as planned, but the snake project was a total loss. Despite my best efforts, no Costa Rican snake species could be found in the numbers needed for a population study.

The Mainland

I have done many kinds of field work in Central America, most often alone, sometimes with a crew of family members, at other times with a crew of wildlife technicians employed by the local government. Each type of activity had its benefits and frustrations. While on sabbatical from the University of Kansas in 1967–1968, my main activity was lizard-catching at each of 16 study areas along transects that divided Costa Rica into east-west and north-south quarters. Our living quarters consisted of a camper for four on the back of a truck. My wife, Virginia, did not do field work, but was the boss, cook, washerwoman, and public relations contact. Also, over time, I had come to realize that stressful contacts with local people, such as at border crossings, could best be handled by turning the expedition over to my daughter, Alice. The severe and punitive attitude that I often encountered was immediately transformed into one of fawning cooperation. Our crew on this particular trip consisted essentially of my younger son and daughter. My daughter was then 19, a junior at the university. She combined beauty and intellect and the insight to see potential research projects everywhere, and she tended to give these priority over lizard catching. My son, Chester, then 14 and in junior high, liked to do field work, and was endowed with the split-second reflexes that it takes to be an expert lizard catcher. In addition, at several of our study areas,

[1] Fitch Natural History Reservation, 2060 East 1600 Road, Lawrence, Kansas 66044, USA.

we acquired a young local man (different at each site), who attached himself to us and worked for free just to be near my daughter. Our older son, John, had completed school, left home, and was working for the Smithsonian Institution.

Work on a field project usually involved interactions with local people, and those interactions, positive or negative, can make or break a project. At worst, the people may be kleptomaniacs who will whisk away anything that is left lying loose, even momentarily. When they find your car locked they will at least unscrew and appropriate rear-view mirrors. They view visiting field workers in terms of what can be begged or stolen from them. Fortunately, others are curious about the field work, eager to learn about its purpose and results, and anxious to participate in any way possible.

I sometimes recruited locals to help in my work, even when rather precise teamwork and timing was necessary, such as in the "ghost pen" maneuver we used to catch *Cnemidophorus* and *Ameiva* on sandy, level ground. Noosing was not very effective for catching these kinds of lizards, and the ghost pen constituted a break-through — but it required concentration. Three or four or five people made up the team, each member carrying an aluminum sheet about 6 feet long and 18 inches high. With careful maneuvering the team would surround a lizard, and at a given signal all would advance and set down their sheets on edge, and hopefully the series of arcs, with ends meeting or overlapping, would form a circle with the lizard inside.

During our Costa Rican field work, at each of our many study areas, the awareness that several or many kinds of venomous snakes might be lurking nearby, ready to bite any intruder, was a constant nagging worry — especially as I urged my helpers to intensify their efforts. For the record, we never encountered a venomous snake on any of our 16 study areas; but they were undoubtedly there.

After my sabbatical, when I returned to Costa Rica alone to check on the lizard populations at the study areas, labor that was free or almost so was sorely missed. Earlier NSF funds were not available for salaries, so I paid my helpers out-of-pocket, but I was already on half-salary for sabbatical and had to pay them much less than the salary they would have liked. To bolster morale, I assured them that lizard-catching was fun, but I have to admit that because of hot, humid weather and stinging and biting insects, it was less so than it might have been. Lack of enthusiasm was chronic. My helpers were often reduced to a trance-like state, staring out over the ocean toward China, and I would have to remind them that lizard-catching was the order of the day.

To the Islands

My first glimpse of the Antilles came in 1985 after extended field work in Nicaragua. I was 75 years old. I caught a plane from Managua to Miami and after a brief stopover flew to Santo Domingo, where by pre-arrangement I met Bob Henderson and Nate Kraucunas of the Milwaukee Public Museum. They had given me the address of the hotel where they planned to stay, and where I subsequently found them. They told me about their recent, rather harrowing trip through Haiti, and we talked about the field work that we planned to do.

The next morning, we set out in a rented car and drove west to the Barahona Peninsula, not far from the Haitian border. Natural communities and the influence of humans were of great interest to me, especially since this island was settled very early by immigrating Europeans. As we drove west, the country became quite dry and barren. A type of cactus with an erect stem that branched several feet above the ground was common on the hillsides, and I wondered if it evolved over millions of years through the selective pressures imposed by the giant land tortoises that constituted the "megafauna" of this island before the coming of humans (Amerindians), who hunted and succeeded in exterminating them.

In the little town of Paraiso, south of the city of Barahona, we found a hotel of sorts, acceptable primarily because it was near the site of our anticipated field work (Figure 1). A fifteen-year-old boy was in charge of the hotel, and it seemed that we were the only guests. Facilities were primitive but adequate, at least at first; but before we were ready to leave, the drains of the showers and toilets had become plugged. Our youthful host was not quite sure how to deal with this and flooding became imminent — but on that first night it was not a problem, and we went to sleep lulled by the distant booming of the surf.

The next morning we drove for several miles up a narrow road in the bottom of a canyon. The hillsides were dry and fairly barren, but the riverbottom was wooded and jungle-like in the few places where it was not much disturbed by human

FIGURE 1. The author (standing) watching Bob Henderson process an *Epicrates striatus* near Barahona.

activities. This was where *Anolis bahorucoensis* (Figure 2) could be found, and little time had passed before we first encountered one of these beautiful little lizards. Bob and I had hunted anoles together at many places in Mexico and Central America, but he explained to me that *A. bahorucoensis* is unusual and exceedingly special.

In male anoles, a dewlap or throat fan is a prominent feature of the anatomy. It is mobile and brightly colored, and every species has its own dewlap display. The lizard tips its head back to allow space for the dewlap to be fully expanded by erection of the hyoid cartilage. The display is a highly stereotyped sequence of movements as the lizard alternately spreads and folds its dewlap, sometimes with nodding or bowing movements of the head and neck, and sometimes accompanied by pushups as the lizard raises and lowers its forequarters in rhythmic sequence. The display may last several seconds and may have several parts, performed with exceeding precision and timing.

Anolis bahorucoensis and its near relatives in the *A. hendersoni* group (*A. hendersoni* and *A. dolichocephalus*) are peculiar in having a relatively small yellow or whitish dewlap, and it does not spread widely because the hyoid cartilage, essentially a rod hinged on the chin and embedded in the skin, is not as long or erectile as it is in most other species. Thus this species poses a minor mystery: why is its dewlap so poorly developed? How is its ecology different from those of other species in which the evolutionary trends seemed to focus on developing large and colorful dewlaps?

Despite its reduced dewlap, *A. bahorucoensis* is more conspicuous than most other anoles. In contrast to the dull and uniform brown or olive color of most species, it has a bold, disruptive pattern with stark contrasts and bright colors, almost as if its striking body markings were compensating for the lack of a conspicuous dewlap. Schwartz and Henderson (1991) described it with the following terms: ". . . 4–5 bright green diamonds alternating with purplish brown constricting dumbbells; white to cream flank stripe, coarse brown and white stippling; short, often dark-edged lateral bars; nuchal scrolling blue-green, green or gray, or (2) dorsal diamonds dull, grayish-tan, alternating with rich brown dumbbells; flank stripes tan to cream outlined above and below by dark pigment; short, dark-edged lateral bars, nuchal scrolling pale green to gray; fine substripe scrolling, head dark green to tan; eye skin pale orange anteriorly, eye followed by black patch in turn followed by pale or bright blue or tan; . . . vertical crescent, interparietal pale blue to white; neck blue-green, greenish or gray, venter green; throat yellow, cream or bright orange."

To solve the enigma of the rudimentary dewlap, we decided that an experimental approach was needed, with observations of how members of this species behaved toward each other in their natural surroundings. Unfortunately, those we saw in the field were either singles or male-female pairs, and their distribution was such that the chances of seeing two males interact seemed poor. We decided to speed things up by introducing two males to one another. We chose two very nice and large representatives of the species. Immediately after capture, we dropped them into a large, clear plastic bag, which we inflated, sealed with a twisty, and placed in the shade in the crotch of a tree about 1.5 m above ground level. We then retired to the car, about 12 m away, to have our lunch while watching expectantly.

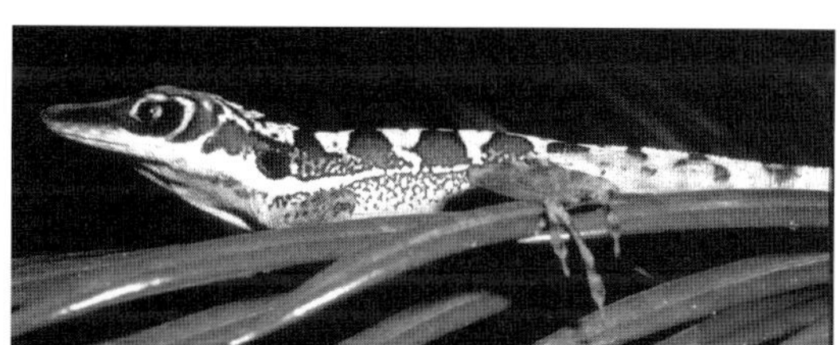

FIGURE 2. *Anolis bahorucoensis*, bearer of an enigmatic rudimentary dewlap (photograph by Robert W. Henderson).

The two males had retired to opposite corners of the bag, but their frantic scrambling was ineffective on the slippery plastic. We had barely retired to the car when an unexpected attack came from the outside. A large anole that must have been watching us all along rushed at the bag from a distance of several meters and lunged at one of the captives — but its attack was foiled by the slippery plastic, which prevented it from obtaining a hold on its intended victim. Both of the *A. bahorucoensis* cowered inside the bag, obviously terrified by the attack of their much larger congener, *A. cybotes*, which could have easily killed them but for the invisible barrier.

Anolis cybotes (Figure 3) is a common species occurring islandwide on Hispaniola. It reaches 77 mm SVL (males) and 66 (females), versus 51 and 44 mm for *A. bahorucoensis*. In later experiments with tethered *A. bahorucoensis* and other small anoles, we verified that *A. cybotes* is an aggressive predator, even on the young of its own species, and smaller anoles undoubtedly constitute an important part of its diet. Upon seeing one, the larger lizard will rush at it, attempt to seize it, subdue it with a killing bite, and swallow it with a few gulps.

We also were able to verify that *A. bahorucoensis* does have a dewlap display. The display is infrequent and involves relatively small movements, mere vibratory nods with no accompanying motion of the dewlap or torso. It is fast and simple and on a very small scale, intended for rival males at close range, but not showy enough to be spotted by ever-watchful *A. cybotes*.

Cybotoid anoles have lived on Hispaniola since the Miocene, at least, and *A. bahorucoensis* must have evolved along with them, no doubt under intense selective pressure to escape or avoid predation. *Anolis bahorucoensis* would seem to be extremely vulnerable, and the full extent of its adaptations for *A. cybotes* avoidance have not yet been studied. Still, we do not understand why large and aggressive *A. cybotes* do not hunt down, catch, and eat every last one of the smaller species. *Anolis bahorucoensis* is a little more terrestrial and that may somehow be linked with its survival. Similarly, *A. distichus*, another smaller species that occurs commonly in the same places as *A. cybotes*, is a little more arboreal and an alert and skillful climber that may be able to out-dodge and outwit the larger species, even while living in close association with it, perhaps on the same tree trunk. *Anolis cybotes* occurs sympatrically with several other kinds of smaller anoles, and all must have been subjected to intensive selection pressures to escape predation. In *A. bahorucoensis*, one such adaptation is to simplify and shorten territorial and courtship signals, so that they are not easily seen from a distance. In addition, these lizards are extremely wary; those we encountered usually fled over leaf litter with a series of rapid hops, but other escape tactics included hiding in leaf litter, running up a stem and flattening against it with legs flexed, jumping from bush to bush, and running along the upper or lower side of a stick into the shelter of dead brush, vine tangles, or dense ferns. Just how these tactics permit *A. bahorucoensis* to escape a pursuing *A. cybotes* is not clear, but some aspect of its escape behavior must be very effective to prevent it from being caught and eaten.

Anolis coelestinus is a medium-large (to 84 mm SVL in males, 60 mm in females) arboreal anole that occurs everywhere that *A. bahorucoensis* is found in southern Hispaniola. The former is an accomplished climber, and often lives on trees with smooth and slippery bark. Using tethered *A. bahorucoensis*, we learned that *A. coelestinus* also would catch and eat them, but it lacked the speed and decisiveness of *A. cybotes*, and I doubt that it is a major predator on *A. bahorucoensis*. The latter species interacts with a host of others in its range. On the whole it is successful and has survived, but

FIGURE 3. *Anolis cybotes*, an aggressive predator of *A. bahorucoensis* (photograph by Robert Powell).

with a modification of its courtship and territorial display, and no doubt other adaptations that are as yet undetected.

Mainland versus the Islands

Coming to Hispaniola from Central America after extended field work on an unrelated project, I was tempted to make comparisons of the climate, people, and the animals. Compared with their mainland relatives, West Indian lizards and snakes, in general, attain high population densities and are relatively tame (depending on whether or not they have evolved in a predator-free environment). Some of the largest kinds of island anoles are relatively predator-free and are rather tame, whereas smaller species of anoles are kept alert and wary by the larger species that prey on them.

In Central and South America, on our lizard study areas where concentrated field work was needed, we were always subject to the gnawing uncertainty that we might step on a venomous snake without seeing it, or might even brush our faces against it on a tree branch while stalking a climbing anole. These unknowns caused stress and worry, much of which may have been unnecessary, and may have caused avoidance behavior when we had nothing to avoid. Undoubtedly, the possibility that dangerous snakes might be present affects lizard-hunting negatively, beside detracting from the pleasure of field work. However, venomous snakes are absent from most of the Antilles, and that is generally considered a benefit for anyone who does field work. Personally, I like snakes and my usual attitude is "the more of them, the better."

Almost anywhere in the tropics, heat can make fieldwork uncomfortable, but, in the Antilles (Figure 4), the proximity of the ocean often keeps temperatures within tolerable bounds — and the pleasant sea breezes tend to alleviate the attention of ants, mosquitoes, and biting flies, which, though present, seem less abundant than on the mainland.

If I had it to do over again, and had known then what I know now, I might have chosen the Dominican Republic instead of Costa Rica as a place to do my sabbatical research. If, at an early stage, I had decided to do my tropical research in the West Indies, I might have avoided times when it was so hot that I was gasping for breath; I might have avoided theft of equipment, cached nearby while I was hunting lizards, and I might have avoided the worry about my crew or me being bitten by a venomous snake — and I surely would have caught more lizards. If I had a choice as to where I would do more field work, I would choose the islands, but as I look back over my long career, I'm happy that my tropical experiences included field work on the Central and South American mainland, too.

FIGURE 4. Where the Sierra de Baoruco meets the Caribbean south of Barahona (photograph by Robert Powell).

Suggested Reading

Fitch, H. S. and R. W. Henderson. 1987. Ecological and ethological parameters in *Anolis bahorucoensis*, a species having rudimentary development of the dewlap. *Amphibia-Reptilia* 8:69–80.

Fitch, H. S. and D. M. Hillis. 1984. The *Anolis* dewlap: interspecific variability and morphological associations with habitat. *Copeia* 1984:315–323.

Lenart, L. A., R. Powell, J. S. Parmerlee, Jr., A. Lathrop, and D. D. Smith. 1997. Anoline diversity in three differentially altered habitats in the Sierra de Baoruco, República Dominicana, Hispaniola. *Biotropica* 29:117–123.

Orrell, K. S. and T. A. Jenssen. 1998. Display behavior of *Anolis bahorucoensis*: an anole with a diminutive dewlap. *Carib. Jour. Sci.* 34:113–124.

Fitch, H. S. 2003. Dewlaps and the lack thereof: or, lizard studies in the Neotropics, p. 153–157. *In* R. W. Henderson and R. Powell (eds.), Islands and the Sea: Essays on Herpetological Exploration in the West Indies. Society for the Study of Amphibians and Reptiles, Ithaca (New York). Contributions to Herpetology, volume 20.

Wet Mountains and Mountain Frogs

Richard Franz[1]

Most herpetological exploration on Hispaniola had been completed by the mid-1980s. Albert Schwartz and his collectors had scoured most places that were accessible by road or boat. Only a few rugged montane areas remained unexplored. One of these remote mountainous areas, the Massif de la Hotte, became the target of the Charles A. Woods–Florida Museum of Natural History expeditions in 1984. Our mission was to explore this high mountain area for new plants and animals and identify sensitive habitats that could be included in a newly-created Haitian national park in the Massif de la Hotte. Explorations were supported by the Haitian government and a United States Agency for International Development (USAID) grant (to Charles A. Woods and the University of Florida).

The Massif de la Hotte is a rugged montane area, located on the western end of the Tiburon Peninsula in southern Haiti (Figure 1). It includes an extensive limestone plateau (Les Platons) and the Formon and Macaya ridges. Pic Formon, at 2200 m, and Pic le Ciel (= Pic 2170), at 2170 m elevation, on the Formon Ridge and Pic Macaya, at 2340 m, on the Macaya Ridge dominate the rugged scenery. The towns of Les Cayes in the south, Jeremie in the north, and Aquin and Miragoane in the east are the nearest large settlements.

The expeditionary assaults on the Massif de la Hotte always originated from Camp Perrin near Les Cayes (Figure 2). Dan and Tia Cordier had been sent by Charles (Woody) Woods to this small mountain village on the cobble-strewn banks of the Riviere du Sud in February 1983 to set up a base from which future penetrations would operate. Woody rented a large concrete-block house on the main (gravel) road that ran through the middle of Camp Perrin. Dan and Tia had to move in, estab-

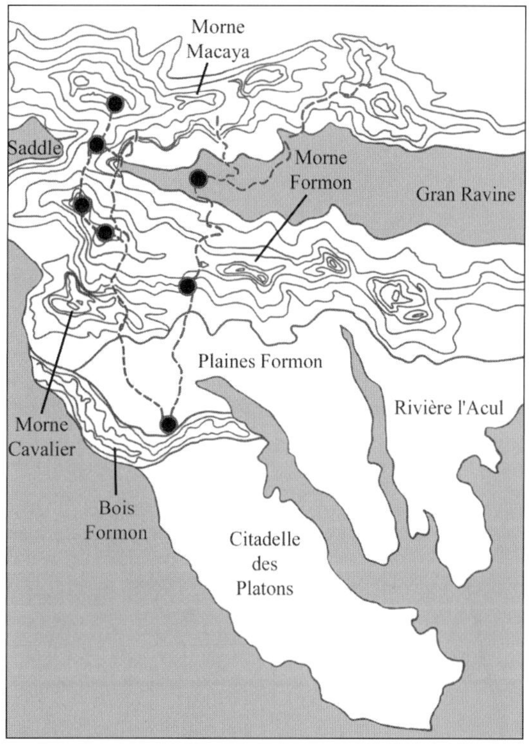

Figure 1. Map of the collecting area in the Haitian Massif de la Hotte. The dashed line represents our route and dots indicate camping sites. Gray areas are lowlands.

Figure 2. Street scene near our base camp in Camp Perrin, southern Haiti.

[1] Florida Museum of Natural History, University of Florida, Gainesville, Florida 32611, USA.

FIGURE 3. Madame Robert's house, our mountain base camp on Plaine Formon.

lish cordial relations with the townspeople and the local military contingent, learn Haitian Creole, and find routes into the massif, all before the first expedition arrived in January 1984. Lebien Barbeau (Babbo for short), his wife, Aline, and their eight children became Dan and Tia's mentors and their path into Haitian society.

An Old Fort, Mountains, and a Fossil Lizard

Three major expeditions that yielded herpetological materials, each several weeks in duration, were launched into the massif between January and August 1984 (Appendix 1). Expeditions progressed by road along the Riviere L'Acul to a point where the road left the river and ascended the steep escarpment via a donkey-wide switch-back trail, cresting the plateau at the ruins of the southern Citadel (Citadelle des Platons). This stone fort was built by the new Haitian military after the defeat of French forces in Haiti and the declaration of Haitian independence in January 1804.

Expedition vehicles were left at the old fort, and expeditionary members made a strenuous two-hour march across the plateau through Sou Bois (= "under trees") and Bois Formon (Forest of Formon) to the main expedition camp at Mme. Despagne's (Madame Robert for short) house on Plaine Formon at the base of Morne Formon (Figure 3). This small Haitian house with its tin roof sat at the junction of two well-traveled footpaths that originated in the lowlands and led into the proposed park area. The house became the center of expedition operations and a secure place to store equipment and supplies, including extra food. The stored food became a test of wills between expedition members and the local black rat population. "Museum Specials" (large rodent snap traps) brought a quick end to the mentally-challenged segment of the rat population, but were ineffectual against the smarter survivors. We carried our homes on our backs. When we worked the plateau, expeditionary tents were lined up in rows on the large flat field at the base of the mountain in front of Madame Robert's house. The tents traveled with us on the mountain trails to other camps as work progressed into other regions of the proposed park. Haitian drums beat almost every evening and the smell of wood smoke flavored the night air.

A mixture of Barbancourt rum and warm Gatorade (no ice) became the evening relaxer, and the local yam and corn field next to Madame Robert's was our toilet the next morning. All supplies (including rum) came up the mountain on the heads of Haitian porters. Drinking water was retrieved from high mountain spring seeps on the slope of Morne Formon. Dan and Tia kept meticulous track of everyone. They established a network of runners that radiated out each day in all directions to bring us supplies and news on slips of Haitian newsprint usually signed NEPH.

Trails led out from Madame Robert's toward the park's boundaries on the south slope of Morne Formon. Two Haitian trails traversed Morne Formon, one leading through our Formon Ridge Camp (1650 m) and cresting the Formon Ridge at an elevation of 1900 m, and the other passing through the Jeremie Sinkhole Plain and the slopes of Morne Cavalier, cresting Formon east of Pic le Ciel (Pic 2170). Both trails crossed into the Grande Ravine du Sud. The west trail ended in the ravine; the east trail ascended the east shoulder of the Macaya Ridge. A third trail, cut by an expedition member, Brent Mitchell, and a Haitian work force in January 1984, branched from the Cavalier Trail and followed the crest of Formon, accessing Pic le Ciel and the Saddle between Formon and Macaya. The climb to Pic Macaya originated from the Saddle Camp. We may have been the first people to climb this peak. Surveyors had made it to the summit in the past and left a concrete marker. However, we suspect that they had been airlifted to the top, since we were unable to find any obvious signs of their route up the mountain. The collective memory of the local Haitians also did not remember anyone having been there. We established several expeditionary camps along these routes, including Formon Ridge Camp on the south slope of Morne Formon, Brent's Camp along Brent's Trail near Pic le Ciel, Ravine Camp and Mud Camp in the Grande

FIGURE 4. Morne Macaya viewed from the north side of Morne Formon.

Ravine du Sud (1338 m), Saddle Camp (1900 m), and Pic Macaya Camp at the summit of Morne Macaya (Figure 4). Many of these camps became important collecting stations, e.g., the Formon Ridge Camp was the site where the paratypic series of *Eleutherodactylus amadeus* was collected. These frogs were so numerous on this narrow perch of land that each morning we had to sweep dozens of them from each of our tents.

Several thousand herpetological specimens were collected in the vicinity of the park during January and early February, June, and late July and early August 1984 by expedition members and local collectors. The majority of these specimens are in the collection of the Florida Museum of Natural History in Gainesville to serve as a permanent witness to the amphibian and reptile fauna that existed there in the mid-1980s. Two new frogs were discovered during the various penetrations into the park (*E. amadeus*, Figure 5; *E. thorectes*, Figure 12).

One of the most spectacular finds during the survey work was the fossil remains of a giant diploglossid lizard found in the sediments at the bottom of several of the deep sinkhole caves on Plaine Formon. The bones appeared as fresh as if the creatures had died yesterday. We advertised a reward of one hundred dollars for the first live specimen among the local populace, but we had no takers. The lizard must be extinct. It may have achieved total lengths in excess of 45 cm and probably represented an undescribed species. Its bony morphology fits well with that of other giant diploglossid lizards known from northern Hispaniola. This new fossil lizard still awaits a scientific name and a formal description.

Rain, Drizzle, and More Rain

Weather played an important role in determining the success of the Haitian expeditions and the comfort level experienced by its members. Too little rainfall meant little activity on the part of the frogs we had come to see, and too much rain made tent-living, cooking, and trips to the yam field hell. More importantly, heavy and continuous rains, such as those experienced in June, were a considerable threat to the safety of expedition members as they attempted to traverse steep mountain trails or ford flooded streams and rivers. Avalanches in this precipitous country were common. People who wore glasses were even more disadvantaged as they were blinded with their glasses awash with rain drops or fogged-over from the high humidity. Raincoats were never efficient enough to keep you dry, and they were either so hot that you sweated to death or so cumbersome that you constantly tripped on their hems. You might as well been standing directly in the rain. One tip I learned from Donald Dodd, the Puerto Rican orchidologist, to mitigate raincoat and eyeglass problems was to carry an umbrella with a bandolier strap. When in use, the umbrella shielded my face, hair, and upper body parts from the rain; when the rain stopped, the umbrella could be carried attached to your pack or across your shoulder like a rifle. The hooked portion of the umbrella handle was also useful in hooking flexible overhanging branches to bring lizards closer to hand or within range of rubber bands or nooses.

Based on recorded rainfall patterns, the lowlands of the Tiburon Peninsula show two well-defined wet seasons, one in the April–June period, and the other in August–December. Les Cayes experiences more than 210 cm of rainfall annually. Little data, however, are available for the Massif de la Hotte, but it probably exceeds that of the lowlands.

FIGURE 5. Mozart's Eleuth, *Eleutherodactylus amadeus*, from northeastern slope of Formon Ridge, Macaya National Park, Haiti, 1700 m elevation (photograph by S. Blair Hedges).

FIGURE 6. Hispaniolan Brown Snake, *Darlingtonia haetiana*, from 13 kilometers SSW La Guazara, Dominican Republic, 975 m elevation (photograph by S. Blair Hedges).

It rained little during our stays at Madame Robert's in January, February, and July, and August 1984. However, it rained horribly in June when expedition members were nearly washed off the mountainside. In one instance, Dan Cordier, Kurt Auffenberg, and Roger Portell chose a hillside tent location (Mud Camp) above the ravine and dug in for the night. Following a meal of gorp, granola bars, and Lipton's "Cup of Soup," they turned in. The incessant rain continued. In the middle of night, they were awakened by the movement of softened mud under their tents. They were slipping into the ravine! To make matters worse, rocks and plants were giving way around them. Somehow they managed to hang on and survive the night.

Precipitation, either as rain or mist, occurred practically every day during the study on the ridges of Formon and Macaya. Switch-back trails up Formon and Macaya were often muddy and with standing water collecting in the low spots. Boots and feet were always wet. Clouds blew through the montane forests daily, and dripping water was a constant companion on the long walks between mountain camps and collecting spots. Evenings were usually clear and cool, but by morning the clouds had usually reformed. The Saddle showed the most severe weather conditions as a result of localized adiabatic cooling and local wind patterns, which passed rapidly up the ravine from the sea and up and over the ridge.

FIGURE 7. Tiburon Curlytail, *Leiocephalus melanochlorus*, from Bois Sec, Haiti, 360 m elevation (photograph by S. Blair Hedges).

Herpetological Adventures and Falling Rocks

Les Platons.—Remnant forests, named "Wet Forests on Limestone" by the expedition botanists, occurred on more rugged limestone hills, known as Bois Formon, and the lower slopes of Cavalier, Formon, and Macaya, below elevations of 1281 m. This forest is characterized by a wide diversity of trees and shrubs, particularly melastomes (*Macranium*, *Calcygonium*, *Miconia*, and *Ossaea*).

Most of the original forest trees had been cut, deep organic soils were burnt to expose more alkaline "terra rosa" soils close to the limestone, and the areas were planted in gardens or converted to pastures. The alkaline soils are more favorable for the growth of starchy tubers (sweet potatoes, yams, and taro) that make up the bulk of the Haitian peasant diet. Peasant homes dotted the landscape, and gardens often went right down into the throats of deep sinkholes. I sometimes wondered how farmers managed to dig these patches, some so steep that the farmers must have tied ropes around themselves to keep from falling into the pits. Locals often told us that falling out of their gardens was a real threat. Cows and goats were tethered on small pastures. Scattered bread fruit, mango, and other fruit-bearing trees had replaced the native forest trees. No wonder the giant diploglossid lizards had disappeared.

The more hardy members of the local amphibian and reptile fauna continued to inhabit the decimated plateau landscape. Hispaniolan Brown Snakes (*Darlingtonia haetiana*; Figure 6) were common under rocks in the pastures. This small snake with its brown back and orange-red or black belly reminded me of the North American Red-bellied Snake (*Storeria occipitomaculata*). About the same size as *Storeria*, *Darlingtonia* feeds on eleutherodactylid frogs, including their eggs, and small lizards. It has been reported to lay up to seven eggs under rocks in the late summer. Other herps common in pastures and gardens of the plateau included the skink-like anguids (*Celestus costatus* and *C. stenurus*),

FIGURE 8. Hispaniolan Giant Treefrog, *Osteopilus vastus*, collected from rolled up green banana leaves on Plaine Formon. This is one of the largest treefrogs in the world.

Curly-tail Lizards (*Leiocephalus melanochlorus*; Figure 7), Haitian Racers (*Antillophis parvifrons*), and numerous frogs (*Eleutherodacylus* of various species, *Osteopilus pulchrilineata*, *O. vastus* [Figure 8], and *O. dominicensis*) (Appendix 2). I suspect that these species normally lived in areas of natural disturbance in the forest and probably were much less common when the original forests covered the plateau. One frog (*E. lamprotes*), a bromeliad-dweller, was found only in undisturbed segments of forest remnants at Bois Formon.

Sinkholes and Caves.—The plateau and the eastern shoulder of the Macaya Ridge form extensive karst regions. These limestone areas are characterized by hundreds of sinks and vertical-walled solution pits. A dramatic pit, locally known as Trou (= "cave") Zonbi, near Madame Robert's, is more than 100 m deep. These sinks and pits were important sources of fossilized bones of amphibians, reptiles, birds, and mammals. They also were sources for rare live frogs. One of the deeper pits contained human bones, and rumors intimated that this was a place where criminals were punished.

Dan Cordier, Ada Fowler, and some Belgian cavers spent much of their time descending into many of the pits. The few pits I tried scared the "zombies" out of me. One particular incident in Trouing (= sinkhole) Jeremie #2 nearly ended my life on an otherwise perfectly beautiful day. Our guide, Babbo, dislodged a *very* large rock that fell at Newtonic speed toward my quivering mass, which was suspended from a rappel rig and ropes some 8 m below him and 13 m above the floor. Have you ever tried melting into a solid limestone wall? Fortunately, the rock flew safely by me, barely skimming my coveralls, and landed harmlessly on the bottom with a huge hollow-sounding thud. Hearing my warning call, Dan had somehow managed to secure himself behind some rocks in a small alcove at the bottom of the pit before the rock hit. Seventeen years later, I still cringe at that memory.

Frogs were my draw to these caves; otherwise, I saw no redeeming value in jumping into deep, dark pits. The fact is that it was a frog that stopped me at a small ledge at the 8-m point in that fateful cave. I was examining an *E. glandulifer* with eggs that had positioned itself with its bounty in a small crevice in the limestone wall as that projectile was launched. I do remember, just prior to that fearsome moment, pondering how this frog had managed to get so far underground to attend its eggs.

Several other frogs, including *E. chlorophenax* (Figure 9) and *E. nortoni*, were recovered from pit caves on Plaine Formon. We also collected a small series of *E. counouspeus* in the bottom of the deep Trouing Jhing near the Citadel. This spotted cave frog was not found in other caves on the plateau, although it is well known from caves near the base of the plateau.

The Formon Ridge.—We spent a lot of our time exploring the trails across the Formon Ridge. The mountainside was so steep, though, that you rarely had opportunities to leave the vicinity of the trails except at the Formon Ridge Camp and on top. Montane Pine Forests or Cloud Forests replaced the Wet Limestone Forests of Plaine Formon at elevations greater than 1280 m. West Indian Pines (*Pinus occidentalis*) grew on the slopes and crests of Formon and Macaya and near the summit of Morne Cavalier. This forest had three recognizable plant associations that graded into one another: (1) moist pinelands, including "cloud forests," had scattered pines and a dense layer of woody shrubs (*Didymo-*

FIGURE 9. Tiburon Peninsula Crevice Frog, *Eleutherodactylus chlorophenax*.

FIGURE 10. A contemplative Charles Woods on Pic Macaya. Dr. Woods was the expedition leader of the Haitian project.

panax, *Persea*, *Myrsine*, *Brunellia*, *Garrya*, and *Miconia*), tree ferns (*Cynathea* and *Alsophila*), bromeliads, and Climbing Bamboo (*Arthrostylidium*); (2) dry pinelands included pines, blackberries (*Rubus* sp.), and ferns (*Pteridium*, *Histiopteris*, and *Blechnum*); and (3) open balds, such as found on Pic le Ciel, were dominated by low, dense stands of wax myrtles and hollies (*Ilex*). The last two plant associations probably resulted from fires or other natural disturbances.

FIGURE 11. Walter Judd in the pine forest and fern carpet on the summit of Morne Macaya.

The herpetofauna changed with the vegetation. Different species of eleutherodactylid frogs and anoles replaced the plateau species (Appendix 2). One of the most unusual was the toad-like *E. ventrilineatus*. We found this species also on Pic Macaya. The underside and the bottoms of the feet of this species were a brilliant sulfur color. We may have been the first to recollect this species since the original description by Doris Cochran of the Smithsonian Institution in 1941. Another unusual species we encountered was an odd little lizard, *Anolis* (also called *Chamaelinorops*) *barbouri*, named by Karl P. Schmidt in 1919, in honor of Thomas Barbour, then the Director of the Museum of Comparative Zoology at Harvard, in recognition of his contributions to West Indian herpetology. This strange brown lizard with its shield-like back was found among sprays of dried pine needles still attached to branches of fallen pine trees close to the ground.

My son, Jeff (at that time 13 years old), spent considerable time climbing bromeliad-festooned trees near the Formon summit. He would break air-plants free with a machete and toss them to me on the ground, so we could sample bromeliad-inhabiting frogs. We tried to collect in as many microhabitats as possible to insure a thorough sampling of the area's fauna.

This was Jeff's first trip out of the country. He and Bryan Woods (Woody's son), also 13, gained life experiences on that trip that they could not otherwise have had. Learning to communicate with locals who spoke a very different language and lived a very different life style, on top of having a true wilderness experience, provided them with opportunities to test their own mettle.

Pic Macaya.—The traverse up the side of Macaya was a serious climb and at times very scary. One particular 30-m segment perched on a near vertical slope 300 m above the Grande Ravine left many of us fretting the return trip. The top of Macaya was a wonderful experience (Figure 10). Huge Caribbean Pines, 1 m in diameter, laced with floating mists and luxuriant growths of ferns, gave the appearance of some prehistoric garden (Figure 11). Here, Walter Judd and I wandered the primeval forest, he was picking up new plants and I was catching frogs. One find was a series of tiny frogs that I thought at the time must represent the immatures of some larger unknown species. Later, upon closer examination of the preserved material, Blair Hedges discovered

FIGURE 12. Macaya Breast-spot Eleuth, *Eleutherodactylus thorectes*, from the southern slope of Peak Formon, Macaya National Park, Haiti, 1920 m elevation (photograph by S. Blair Hedges).

that this frog was an undescribed species, which he named *E. thorectes* (Figure 12). This miniature frog is one of the smallest anurans in the world. Its appearance and behavior were similar to those of the North American frog, *Pseudacris ocularis*, hopping elfin-like among decaying leaves, pine needles, shrubs, bromeliads, and ferns. One of the females carried three eggs inside her body.

The trip back to the Saddle was a piece of cake. But, the adventure did not end there. Walter and I, camping on the Saddle that night, were awakened by the sounds of boulders tumbling down the mountainside. We both sat straight up out of a dead sleep, as one of these boulders came crashing within feet of our tent before veering off over the edge and plunging into the Grande Ravine below.

The June trip to Pic Macaya was also jam-packed with adventures. Kurt Auffenberg, Roger Portell, Dan Skean, and two local Haitians climbed to the summit for a one-day stint; Dan Skean became lost at the top, and Kurt stayed on the mountain looking for Skean, while Roger and the guides climbed back down to the Saddle. Kurt finally met up with the wanderer, and they made that horrible descent without flashlights as dark was closing in. I can't imagine it. Kurt, Dan, Roger, and the two guides spent that night in the rain on the Saddle, sharing the only food they had — a can of smoked oysters; one of the porters with lightning-fast reflexes snatched the abandoned can from the flames and drained the remaining hot oil and juice from the can, which completely grossed Kurt out. Sitting at the Formon Ridge Camp the next night, Kurt figured that they had traversed nearly 4700 vertical meters in the four days since leaving Camp Perrin.

Grande Ravine du Sud.—The Riviere du Sud is the dominant stream in the Macaya Park area. This easterly-flowing stream originates from several springs at the base of the Saddle between the Formon and Macaya ridges. Shelley Etheridge, later to become Shelley Franz, and I traversed the eastern Formon Trail, first stopping at the Formon Ridge Camp, and then proceeding to the Ravine Camp, where we camped for several days along the plunging Riviere du Sud (Figure 13). Several Haitian porters traveled with us. They helped us collect specimens and cook our meals, mostly the tubers of the Elephant-ear Plants (*Colocasia esculenta* and *Xanthosoma sagittifolium*) they called Karaib, Taro, Malanga, and Mazonbèl, but, for the most part, they spent their time playing cards with a stack of yellowed leaves during the long sunny days.

The mountain torrent was absolutely gorgeous and, in Shelley's and my opinions, the most splendid part of our trip. Shelley and I were able to find refuge among the boulders to bathe for the first time since leaving Camp Perrin. The cold

FIGURE 13. Riviere du Sud in the Grande Ravine below the Saddle.

FIGURE 14. Hispaniolan Green Treefrog, *Hyla heilprini*, from a streamside along the Riviere du Sud.

water helped mend those scrapes, bruises, and bug bites that had accumulated over the previous weeks of intense field work.

Spindly pines grow in profusion in the alluvial cobble and boulder bars along the stream. Although close to water, the forest appeared arid with ferns (particularly *Pteridium*), blackberries, and a few woody shrubs (*Heterotrichum*, *Gyrotaenia*, *Senecio*, *Garraya*, and *Cestrum*) forming a dense layer under the pines. Walter Judd and his botanical colleagues recognized this streamside flora as unique, naming it the "Ravine Pine Forest." This peculiar forest ranges up the inner gorge of the Grande Ravine to elevations of at least 1300 m.

The herpetofauna of the ravine offered many rare species, some of which we had not encountered before (Appendix 2). We were fascinated by the bright green treefrogs, *Hyla heilprini* (Figure 14), that we encountered among the streamside rocks. We also found their tadpoles in eddies away from the main current. Several of these tadpoles were in the process of transforming, hence we were able to associate the tadpoles with the adult frogs.

We collected several bizarre anoles in the vegetation that stretched over the stream. The bright blue and green male *Anolis monticola* often displayed with their yellow throat fans from the leaves of low shrubs; pursuit often caused them to drop from their perches into the stream below. The elongate-headed stick anole, *A. dolichocephalus* (Figure 15), lacks the characteristic dewlap used by most male anoles to communicate with conspecifics; their slender, brown-striped bodies and tails were seen to snugly snake along the midribs of bracken ferns. They are reported to sleep on vines at night. Leaving the ravine was difficult, particularly knowing that we had to climb over 600 m out of the ravine and another 600-m climb down to Madame Robert's house. Shelley's knees finally gave out on her on the trip down, and she ended up backing down the mountain side. The exodus took us nearly eight hours, but we did it.

Going Home

I always had mixed emotions leaving Formon and Camp Perrin. The herpetofauna was fantastic, the experiences wild, and the hardships incredible. Leaving Madame Robert's meant a long walk back to the vehicles parked at the Citadel. Then, we had to face that scary trip back down the donkey switchback. Arriving in Camp Perrin meant the preparation of specimens for transport back to Florida and filling out United States Fish and Wildlife Service importation declarations. It also meant getting our gear cleaned sufficiently to get through Miami customs and Florida agricultural inspections — but the worst was leaving our friends. Following the last trip down from Formon in August 1984, Dan and Tia Cordier, with some tears and final good-byes, put Shelley and me on a gaudily-painted Tap-Tap diesel-powered truck (tap-tap means "fast, with no problems") headed for Port-au-Prince and the international airport. The August trip was my last excursion into the real Haitian wilderness. I wonder how our Haitian friends have fared. So much has happened to Haiti since that last trip.

Acknowledgments

The people who participated in the expeditions are listed in Appendix 1. To these compatriots, I render my thanks for their help in collecting important material and for their companionship. I specifically thank Charles Woods for the opportunities to travel with the expeditions to the Massif de la Hotte and later to the Massif de la Selle; Dan and Tia Cordier, both good friends, for their logistical and emotional support; Kurt Auffenberg and Roger Portell for their June tales; Jeff Franz, Shelley Franz, and Walter Judd for the wonderful times we spent wandering around the Haitian landscape; Babbo and our many

FIGURE 15. Tiburon Peninsula Stick Anole, *Anolis dolichocephalus*.

Haitian friends who worked their hearts out to help us have successful trips; Charles Woods, Dan and Tia Cordier, Shelley Franz, Blair Hedges, Walt Judd, and Roger Portell for reviewing this manuscript; Blair Hedges for the use of excellent photographs of frogs and reptiles; and finally the Haitian government and USAID for financial support.

Suggested Reading

Franz, R. and D. Cordier. 1986. Herpetofaunas of the proposed national parks in southern Haiti. Report to USAID/Haiti. Florida State Museum, Gainesville.

Hedges, S. B. 1988. A new diminutive frog from Hispaniola (Leptodactylidae: *Eleutherodactylus*). *Copeia* 1988:636–641.

Hedges, S. B., R. Thomas, and R. Franz. 1987. A new species of *Eleutherodactylus* (Anura, Leptodactylidae) from the Massif de la Hotte, Haiti. *Copeia* 1987:943–949.

Appendix 1. Members of the Charles A. Woods–Florida Museum of Natural History Haitian Expeditions to the Massif de la Hotte, Haiti.

January–February 1984 Expedition: Charles A. Woods, Project Director, mammalogist, Florida Museum; Dan and Tia Cordier, in-country field coordinators, Florida Museum; Ray Cleeland, volunteer; Donald Dodd, botanist and orchidologist, Puerto Rico; Ada Fowler, volunteer; Richard Franz, herpetologist, Florida Museum; Jeffrey Franz, volunteer; Walter Judd, botanist, Department of Botany, University of Florida; Bruce MacFadden, geologist, Florida Museum; Brent Mitchell, Peace Corps volunteer; Paul Perski, volunteer; Fred G. Thompson, malacologist, Florida Museum; Bryan Woods, volunteer; and Scott Yocum, entomologist, Florida Department of Agriculture.

Specific Goals: collect herpetological materials during the dry season on the plateau, Morne Cavalier, Morne Formon; cut a trail to the saddle and make the initial ascent to the summit of Morne Macaya, collect on the saddle and Pic Macaya.

June 1984 Expedition: Dan and Tia Cordier; Kurt Auffenberg, malacologist, Florida Museum; Ada Fowler; Dan Skean, botanist, Department of Botany, University of Florida; and Mike Thomas, entomologist, Division of Plant Industry, Florida Department of Agriculture.

Specific Goals: collect materials during the rainy season on the plateau and mountain slopes.

July–August 1984 Expedition: Dan and Tia Cordier; Ada Fowler; Richard Franz; Shelley Etheridge (Franz), volunteer; Stott Woods, volunteer; and John Yarington, volunteer.

Specific Goals: collect materials on the plateau and in the Grande Ravine du Sud; special efforts were made to collect amphibians in the pit caves on Plaines Formon, Morne Cavalier, and the east shoulder of the Macaya ridge.

Appendix 2. Altitudinal distributions of herpetological species collected during field work in the Massif de la Hotte, Tiburon Peninsula, in southern Haiti in 1984. Upper limits in elevations for each species are listed in parentheses.

Below 700 m (outside the study area).

700–1030 m (wet forest on limestone and lower portion of ravine pine forest): *Eleutherodactylus chlorophenax*, *E. counouspeus* (730 m), *E. glaphycompus*, *E. lamprotes*, *E. nortoni*, *E. wetmorei*, *Hyla heilprini*, *Osteopilus* (= *Hyla*) *pulchrilineata* (1000 m), *O.* (= *Hyla*) *vasta* (900 m), *O. dominicensis* (1000 m), *Celestus costatus*, *C. stenurus* (1030 m), *Anolis armouri*, *A. coelestinus*, *A. cybotes*, *A. distichus*, *A. dolichocephalus* (1030 m), *A. monticola*, *A. ricordii* (800 m), *Leiocephalus melanochlorus*, *Epicrates gracilus*, *E. striatus* (970 m), *Antillophis parvifrons*, *Darlingtonia haetiana*, and *Uromacer catesbyi* (1020 m).

1030–1327 m (wet forest on limestone and the upper portions of ravine pine forest): *Eleutherodactylus chlorophenax*, *E. glandulifer*, *E. glaphycompus*, *E. lamprotes*, *E. nortoni*, *Hyla heilprini*, (1300 m), *Celestus costatus*, *Anolis armouri*, *A. coelestinus*, *A. cybotes* (1260 m), *A. distichus*, *A. monticola* (1150 m), *Anolis* (= *Chamaelinorops*) *barbouri*, *Leiocephalus melanochlorus*, *Antillophis parvifrons*, and *Darlingtonia haetiana*.

1327–1600 m (transition between wet forest on limestone and montane pine forests): *Eleutherodactylus amadeus*, *E. brevirostris*, *E. glandulifer*, *E. glaphycompus*, *E. nortoni* (1365 m), *Celestus costatus* (1365 m), *Anolis armouri*, *A. coelestinus* (1530 m), *A. distichus*, *Anolis* (= *Chamaelinorops*) *barbouri* (1530 m), *Leiocephalus melanochlorus* (1500 m), *Antillophis parvifrons* (1500 m), and *Darlingtonia haetiana* (1420 m).

1600–1900 m (montane pine forest): *Eleutherodactylus amadeus*, *E. bakeri*, *E. brevirostris*, *E. glandulifer* (1900 m), *E. ventrilineatus*, *Anolis armouri* (1650 m), and *A. distichus* (1800 m).

2340 m (Pic Macaya): *Eleutherodactylus bakeri*, *E. brevirostris*, *E. thorectes*, and *E. ventrilineatus*.

Franz, R. 2003. Wet mountains and mountain frogs, p. 159–167. *In* R. W. Henderson and R. Powell (eds.), Islands and the Sea: Essays on Herpetological Exploration in the West Indies. Society for the Study of Amphibians and Reptiles, Ithaca (New York). Contributions to Herpetology, volume 20.

As Good As It Gets

Robert Powell[1]

Alec Waugh, in *Hot Countries*, noted that "the reward of travel is not the seeing of certain accepted beauties but the discovery of a series of special and particular sensations . . ." and "it is [not] through any conscious seeking of accepted beauties that you come across those moments of sheer rapture that leave life permanently enriched . . ., a chance turning your head will bring you one of those moments whose beauty is so complete that it seems possessed of eternal properties."

I initially came to the West Indies by pursuing logical answers to purely pragmatic questions (and a fortuitous decision to ask Bob Henderson for advice). However, pragmatism quickly flew out of the window once I was subjected to the island experience. Lizards were everywhere. Although I quickly became accustomed to anoles on every tree, ameivas running down the busy streets of Santo Domingo's colonial sector, and curlytails (*Leiocephalus* spp.) sitting on the berms of provincial roadsides (Figure 1), I am no less enchanted by those same experiences today. However, the broader appeal of the West Indies quickly became obvious even to someone as one-dimensionally focused as I was (and still am), and this wider perspective quickly and substantially supplemented in a most delightful way the obviously wonderful opportunities to do a little herpetology.

Figure 1. Curlytailed lizards often sit on roadside berms; this is a female *Leiocephalus personatus* from the eastern coast of the Dominican Republic near Juanillo.

An Island of Islands

The Dominican Republic (DR) was the best possible introduction to the West Indies. To paraphrase Al Schwartz, Hispaniola is an "island of islands," with an incredibly diverse herpetofauna in an equally incredible diversity of habitats (Figure 2). Since that first trip in 1986, I've revisited the island over 30 times and each time have experienced something new. It will always have a special place in my memories, but that's not to say that other islands don't have a lot to offer.

In the last decade and a half, I've spent varying amounts of time on Grenada, St. Martin/St. Maarten, Antigua, Anguilla, Saba, Puerto Rico, Mona, Grand and Little Cayman, and Navassa. In the context of my trip to Navassa, I even managed very brief visits to Great Exuma in the Bahamas and to the United States Naval Base at Guantanamo Bay, Cuba. The latter has all the charm of any military base — until one notices the "iguana crossing" signs (Figure 3)

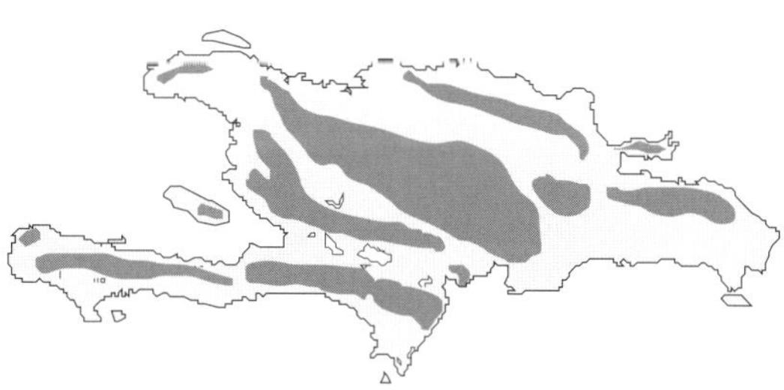

Figure 2. Hispaniola is an "island of islands;" each mountain range (indicated in dark gray) and many of the intervening lowlands have their own endemic species.

[1] Department of Biology, Avila University, Kansas City, Missouri 64145, USA.

FIGURE 3. An iguana crossing sign at the United States Naval Base at Guantanamo Bay, Cuba.

or wanders but a short distance into the surrounding scrub. In each instance, the dichotomy of the island experience was apparent, with critters and people invariably winning over the negatives associated with development and poverty.

Heat, Sand, and Salt

When I tell folks that I do research in the Caribbean, they roll their eyes knowingly and assume that I've perpetrated a clever scam that allows me to spend time on the many sun-drenched beaches for which the islands are so deservedly famous.

The reality is that I hate spending time on beaches and I've avoided them on nearly every trip. In my mind, they have three strikes against them: (1) they're hot, intensely so in most instances; (2) they're covered with sand, insidiously invasive stuff that finds its way into even the most private nooks and crannies of your body; and (3) they're fronted by salt water, an odious concoction obviously intended to test the patience of even the most tolerant person. Now, don't get me wrong, I don't mind looking at beaches. Many are beautiful, especially when framed by palms gently waving in the warm breezes. As a matter of fact, I have a photo of a particularly wonderful beach in Barahona (Dominican Republic) hanging in my office (Figure 4). I stare at it longingly during the long months between trips (particularly during Missouri winters when herps are not to be found and venturing outside invites frostbite). Also, just so you understand that I'm not a fanatic about this beach thing, I have been known to wander onto one now and then — when it's absolutely necessary. Sometimes lizards are found on beaches. On small islands without airports or docks, one has to cross a beach to reach the interior. Also, because I think every biologist should experience the wonders of a coral reef at least once, I often accompany my students across beaches on the way to snorkeling over nearby reefs.

However, I've become quite adept at minimizing my exposure to beaches. Unfortunately, many of the other less appealing features of the islands defy my best efforts to avoid them. I hate heat, and the islands are hot. Poisonous plants and nasty blood-sucking insects are ubiquitous, and a phenomenally high percentage of the latter are convinced that I'm the tastiest morsel they've ever encountered. Drinking the water on many islands is very risky, and ice is another unknown. At a restaurant one evening, my students and I happened to notice that our waiter was rinsing a block of "safe" ice with tap water (we had warm drinks that night). Almost everyone has problems on their first trip or two, but I've found that much of the unpleasantness can be avoided by

FIGURE 4. A beach in Barahona, Dominican Republic; I stare longingly at this picture during cold winter months in Missouri.

chewing a Pepto Bismol tablet with each meal (a little bismuth poisoning is far preferable to worshipping at the porcelain alter). I also hate the high prices for essentials (especially on the smaller islands), and I become quite irate when I have trouble finding the necessities of life. Although I enjoy many of the local brews and soft drinks (some are of exceptional quality), I'm all but addicted to Mountain Dew. Only in recent years have I managed to stumble across this vital fluid from time to time, a benefit primarily attributable to the continuing growth of tourism (a decidedly mixed blessing).

FIGURE 5. Traffic in Santo Domingo is always an adventure (photograph by Donald D. Smith).

Air Travel, Customs, and Car Rentals

Because I can't drive to the island, I'm exposed to the uncertainties and frustrations of air travel, the inevitable hassles with customs (especially when returning to the States with specimens), and I have to rent a car. The latter isn't always so bad (actually went very smoothly on a 2000 trip to Anguilla), but, generally speaking, renting a vehicle is a risky proposition. On my first trip to the Dominican Republic in 1986, I had reserved a compact car, but ended up with a Russian-made Lada with broken door handles, headlights that worked intermittently, and windshield wipers that didn't work at all. I exchanged it (a four-hour process) for a decrepit Nissan that ran well, but had a suspension incapable of handling smoothly paved roads, much less the treacherous tracks encountered in the campo. I managed OK, but my wife, who was pregnant at the time, suffered immensely. On another trip, I had confirmed reservations for two vans, but only one was available (I suspect that a cash-paying customer arrived on an earlier flight). Fortuitously, the agent at the rental desk happened to have a boyfriend who would rent us his van. Lacking a viable alternative, I agreed. A broken drive train while high in the mountains, a missing second gear, four flat tires, and sundry minor breakdowns provided numerous opportunities to explore the universal utility of duct tape (without which one should never travel in the tropics).

Renting a car is but the first step in the adventure called "driving" in the islands. Lane markers, red lights, and other vehicles are minor inconveniences, and defensive driving is incompatible with survival. If you hesitate for even a second, every driver in the vicinity will immediately sense your reticence and move in for the kill, exploiting any vacillation on your part in order to progress a few feet at your expense (Figure 5). When combined with the overall quality of rental cars, city driving is akin to skydiving without a parachute. Some of my students still speak in awe of my expletive-laden tirade when, in the midst of a major Santo Domingo traffic jam (a redundancy), the gearshift lever of the rental van broke off in my hand, leaving me at the less-than-tender mercies of surrounding drivers while incapable of shifting out of second gear.

The Goat Will Probably Die

Leaving the city behind helps immensely, cutting the frequency of life-threatening encounters by at least half. Of course, this is more than offset by the abject quality of the so-called roads. Unless one happens to be visiting during an election campaign and driving through a region in which

FIGURE 6. Roads in the campo (top) are seldom what they appear to be on maps (photograph by Donald D. Smith). Without the help of this grader (bottom), we might still be trying to climb out of the valley.

FIGURE 7. These warning signs were erected the year after our unfortunate encounter with a cow on the road.

the voters' loyalties had been bought by capital improvements, bone-jarring potholes, missing shoulders or bridges, and other minor annoyances are to be expected — and these are the major highways. Back roads always present challenges (not to mention that maps often show them going places they might someday go, but never have or, at the very least, haven't gone in decades; Figure 6). Additional hazards occur when roads pass through small villages, whose inhabitants, for lack of sidewalks, walk along the road itself. Combined with a lack of lighting at night, it's a wonder that island populations continue to grow.

Animals on the road constitute another threat. Many a chicken has bitten the dust beneath the wheels of our vehicles. In contrast, we've managed to avoid dogs, cats, goats, donkeys, horses, and all but one cow (Figure 7). That such hazards are not limited to any one island was reinforced when, concerned about possible legal consequences of a collision with one of Anguilla's ubiquitous goats, I asked a friend what would occur if we happened to hit one. The immediate response: "The goat will probably die."

Mañana

Also exasperating is the need to adjust one's pace. In the tropics, life simply doesn't progress quickly. Even on the French- and English-speaking islands, the "mañana principle" applies. Folks don't hurry, even when we expect it. After a while, adjustments are possible. I've long since ceased being upset when forced to wait. After ordering a meal, I go watch lizards, take a walk, or merely soak in the ambiance. If a promised document doesn't materialize on day one, I come back the next day and, if necessary, the day after that. The failure to adjust to the slower pace simply generates stress, and I've learned (albeit reluctantly in the early days) that stress detracts immeasurably from the island experience. I'm no longer willing to pay that price (and things don't move a bit faster when the Norte-americano gets upset).

Too Many People

Growing human populations on islands with no room for expansion results in the displacement of natural areas, with a generally negative impact on native plants and animals. Additonal problems can be attributed to political corruption and the spoils systems that prevail in most island governments. After a recent election, a friend who works for a federal agency in the DR commented dryly that he had "many new colleagues."

Worse yet is an exploitive attitude toward natural resources. Even when laws exist to conserve a resource or to define a national park, enforcement frequently is in the hands of individuals who see their authority primarily as a means of exploiting that resource or area for their own amusement or profit. Cutting trees is forbidden in the DR, and, even when it's essential, the paperwork is daunting. However, when a local military commander re-

FIGURE 8. Sometimes a roof to keep the rain out is all it takes to have a schoolhouse.

FIGURE 9. The author exploring the clearcut on Loma Remigio where Gloria Santana stood up to machete-wielding campesinos (photograph by James D. Forester).

decorated his country home a couple of years ago, he thought nothing of cutting two magnificent mahogany trees that had shaded one of our study sites.

Even worse are the tragedies that occur when rapid population growth is combined with the extreme poverty that is all too evident on many islands (Figure 8). One becomes sympathetic to the plight that forces the poor to exploit whatever resource is available in order to survive another day. Still, I shall always remember the day that Gloria Santana, a very classy, but somewhat diminutive lady who works for the Dominican Departamento de Vida Silvestre, stood up to some machete-wielding campesinos, telling them in no uncertain terms that they must cease immediately all efforts to clear plots at a cloud forest site on Loma Remigio (Figure 9).

However, the most discouraging aspect of the problem is the apparent lack of ability to address it. We frequently see "children" bearing children, which results in slums and villages overrun with hordes of naked youngsters, most of whom have no hope of improving their lot in life — and appear to be completely resigned to their fate.

A Mixed Blessing

I keep hearing about the potentially positive impact of increased ecotourism, but what I see are hotels and resorts for environmentally conscious tourists being built on sea turtle nesting beaches or along scenic streams. In both instances, the damage far exceeds any gain beyond the purely fiscal. As a consequence, the potential benefits of ecotourism are, at the very best, a mixed blessing. Some areas may be spared from destruction if the income from tourism exceeds the potential gain from exploiting the area. The temptation is to blame the islanders, but they're merely victims of competitive market demands. The crime is that developers cater to the ignorance of tourists who want to "experience" the tropics while sleeping in air-conditioned comfort and taking a break by playing a round of golf or lounging about a pool.

Why, Then, the West Indies?

With so much to dislike, why do I keep going back? Why do I question my sanity when I step from a plane only to be hit by the intense tropical heat (if arriving by day) or an oppressive mustiness (when arriving at night), and then begin planning my next visit before I depart?

I know some colleagues have been attracted by the diversity and/or incredible abundance of frogs, lizards, and snakes, whereas others were taken in by the beaches, palm trees, or even the general allure of the islands. In my case, the decision was made initially for purely pragmatic reasons. I wanted to study herps; to do so, I needed a place where my students and I could collect meaningful data in short periods of time.

Although I was largely ignorant of West Indian realities, a cursory exploration of the literature suggested to me that nearly all of the work on the herpetofauna had been systematic in nature — with the notable exception of some fascinating studies conducted at the El Verde Field Station in the Luquillo Experimental Forest on Puerto Rico. Consequently, I assumed that Puerto Rico had been worked intensely — and categorically eliminated it as a suitable destination. Cuba was intriguing, but politically unfeasible. I subsequently excluded the smaller islands (Bahamas and Lesser Antilles) because I feared that the lower species diversity on each would render them uninteresting (talk about simplistic!). So, I ended up looking more and more to Jamaica or Hispaniola (specifically the Dominican Republic; Haiti, like Cuba, was intriguing but logistically unfeasible).

Because of the responsibilities associated with taking students along, I wanted some first-hand input regarding the suitability of either destination. At the time, I knew no one who had worked in Jamaica, but several years before, after a visit to the Yucatan Peninsula, I had contacted Bob Henderson with questions I hoped he could answer based on his work in Belize. More recently, Bob had worked in the DR. When I called this time, he not only remembered me (or so he claimed), he responded with enthusiasm to my query regarding the feasibility of taking students to the DR and promised to send some information. Two days later, I received a bulging manila envelope packed with copies of papers, lists of names and addresses, and receipts for everything from lodging to car rentals. I immediately began writing to some of the names on his list and simultaneously planned a short visit with my wife, Beverly. In January 1986, she and I spent a largely delightful week (we did have a rental car, after all) visiting Santo Domingo, Barahona, and La Romana — and I took my first group of students there in March of the same year.

Now, looking back to that first trip and over the many subsequent trips to the DR and other islands, I'd have to say that four major reasons are responsible for my returning again and again: (1) I can do the kind of science I enjoy and make a meaningful contribution; (2) I really, really like the critters; (3) I have come to respect and like most West Indians and nearly all of my colleagues who work in the region (collectively they've enhanced my experiences immeasurably); and (4), although I hate to admit it, I, too, have become enamored by the islands' irrepressible ambiance.

My Kind of Science

I like to think of myself as a naturalist, a noble profession with a long history — not all of it inspiring. Martin Gorst, in *Measuring Eternity: The Search for the Beginning of Time*, described the travails of the earliest natural historians:

"At the time, [John] Ray and [Edward] Lhuyd's pursuits were quite new and in some ways appeared rather strange to contemporaries. They were among the first naturalists to examine nature closely for its own sake. As they traveled from village to village, their peculiar activities — collecting stones, drawing flowers and asking questions about the landscape — drew suspicious stares from the locals, or in some cases outright hostility. In Wales, Lhuyd and his party were considered conjurors; in Cornwall they were arrested as thieves; and on a trip to Brittany in northern France, Lhuyd was accused of being a spy and thrown into jail for two weeks. It obviously wasn't easy being a naturalist at this time."

Obviously, the situation improved with time, ultimately reaching its apex, inspired by Charles Darwin's theories, in the second half of the 19th century. However, Edward Larson, in *Evolution's Workshop: God and Science on the Galápagos Islands*, made the following observation regarding biology at the turn of the 20th century:

"By 1900, laboratory experiments had displaced fieldwork as the preferred means to study life... Genetics, embryology, cytology, microbiology, and the like came into vogue among scientists, and the best work in these departments was confined to research universities in the United States and Europe. "Naturalists" of the order of Cuvier, Agassiz and even Darwin no longer defined the cutting edge, and their fieldwork was derided as akin to stamp collecting. Fast-breeding laboratory fruit flies displaced Galápagos finches as the favored source of information about evolution. Field studies were left largely to museum curators and amateur naturalists, who were expelled to the periphery of the scientific enterprise."

Unfortunately, little has changed in the intervening 100 years. We now express concern about the loss of biodiversity, but funding for survey work or basic systematic studies is almost non-existent. Consequently, if one wishes to be a naturalist, especially in the tropics, out-of-pocket funding is a necessity. Fortunately, I have been able to make a

FIGURE 10. One of our long-term study sites near Baní became a dumping ground for the city.

FIGURE 11. The curlytail lizard, *Leiocephalus semilineatus*, has an inexplicably disjunct distribution in the Llanos de Azua.

case for using low-tech natural history studies to teach science (although I am routinely criticized by reviewers of my funding proposals because my students will be engaged only in "descriptive ecological studies" — as if this was a bad thing). Unfortunately, even the most gifted undergraduates are available for only brief windows of time, from the acquisition of essential knowledge (usually in introductory courses) until they matriculate to post-graduate endeavors. This reality places a premium on work for which ample and relevant data can be collected during short periods, criteria ideally suited for work in the West Indies. So, I keep going back to the isles in large part because research in the region allows me to be a naturalist while fulfilling my goals as an educator — and both allow me and my students to make meaningful contributions to a growing body of knowledge about one of the most fascinating faunas in the world.

Ever-changing Questions

Along these same lines, and when answering colleagues unfamiliar with the area and students who often ask why I keep returning to the Caribbean instead of foraging farther afield to South America or even to the other hemisphere, I acknowledge the spectacular settings, the abundant and wonderful critters, and the nearly always friendly and helpful people. However, one principal reason for returning again and again is that every trip raises a plethora of never-ending and ever-changing questions that cry out for someone foolish enough to seek some answers — and I, of course, am willing to repeatedly make that sacrifice. In other words, I want to tackle at least a substantial subset of the questions while I am still able to wander afield and catch lizards.

FIGURE 12. *Anolis coelestinus* is very common on Loma Remigio, but is relatively rare only a few kilometers to the west.

Why Here and Not There?

Some of these questions are simple and focused, although they have defied my ability to provide even a speculative answer. At one of our study sites near Baní (Figure 10), *Leiocephalus semilineatus* (Figure 11) is the only representative of its genus. Just down the road through what appears to be suitable and continuous habitat, these lizards are found in sympatry with *L. schreibersii*. A tad farther down the same road, nearer Barahona, only *L. schreibersii* is to be found. Why are these two congeners sympatric in one area and allopatric at both ends? The assumption is that some environmental variable is responsible, but, at least so far, we've been unable to identify the causative agent or entity.

Similar questions abound. Also in the DR, an anoline community composed of four abundant forms (*Anolis bahorucoensis*, *A. coelestinus*, *A. cybotes*, and *A. distichus*), variously supplemented by as many as four other species, is found on the eastern and northern slopes of the Sierra de Baoruco. However, at one site, one of the "big four," *A. distichus*, is relatively rare; only a couple of kilometers away, through essentially uniform habitat, *A. distichus* is abundant, and *A. coelestinus* becomes difficult to find (Figure 12). If, on our next trip, we should succeed in identifying the responsible variable, I'd bet a year's salary that we'd simultaneously discover another apparent anomaly that defied every logical effort to explain it.

Lizards Don't Read Textbooks

Another type of relatively focused question deals with animals that refuse to follow the rules, occupying habitats with which they're not supposed to be associated or behaving in a fashion that contradicts what's been reported in the literature. Students new to research are often troubled, at least initially, by such contradictions, and my reminder that "lizards don't read textbooks" has been voiced at least once on every trip.

FIGURE 13. *Anolis olssoni* is a grass anole; this ecomorph is tiny, slender, and, as the name implies, associated with grasses on which they forage and sleep.

Anolis bahorucoensis is a small, shade-loving anole with a diminutive dewlap. These lizards apparently maintain a low profile largely because they're very susceptible to predation by larger congeners. However, during a study at higher elevations, these attractive little critters behaved quite differently. Not only were they overtly heliophilic, at some locations they basked on the few available perches in the immediate proximity of their purported predators. Now, I have to think that a large lizard with an inclination to eat a smaller one would do so regardless of elevation, but apparently the more temperate climate at higher elevations puts a premium on thermoregulation that exceeds the dangers of associating too closely with someone who might eat you. This assumption raises all sorts of interesting questions. Would a lizard transported to a site at a higher or lower elevation adjust its behavior? Are additional behavioral safeguards against predation operative at the higher elevation sites? Who knows — I don't, but I want to find out, and I plan to revisit the site in the near future.

An Excuse to Watch More Anoles

When Ernest Williams and Stan Rand applied the concept of ecomorphology to Greater Antillean anoles (ecomorphs are species with different phyletic origins but similar morphological adaptations to similar ecological niches; Figure 13), they stimulated an entirely new set of investigations by an entire generation of scientists. Explorations of phylogeny in the context of ecology and morphology created new insights into all of these disciplines — and these, in turn, raised new questions awaiting yet another generation of biologists. My initial forays into this arena were efforts to elucidate ecological relationships of anoline communities that had not been intensely studied, particularly communities in varying habitats, often those subjected to the adverse effects of humans. This alone could keep me going for more careers than I have left (and I'll continue to look into these situations when the opportunities arise), but I've also become fascinated by situations involving ecomorphs in circumstances that differ from those that apparently prevailed during their evolutionary history.

For example, species of Greater Antillean origin and with characteristic ecomorphologies, but isolated from congeners that might restrict them to some subset of available habitats, should exhibit ecological (and possibly morphological) release. This appears to have occurred to some degree in Navassan *Anolis longiceps* (a trunk-crown ecomorph probably of Cuban origin; Figure 14) but not in *A. monensis* from Isla Mona (a trunk-ground ecomorph obviously allied with closely related Puerto Rican species). Is the apparent release in *A. longiceps* real? Is it a reflection of a more versatile ecomorphology (trunk-crown versus others)? Does it reflect a longer period of isolation? How do the niches of these species compare to Lesser Antillean anoles that evolved on one- or two-species islands, but still reflect ecomorphological characters and constraints to varying degrees? I don't know the answers, but I'm excited about the possibility of collecting the data that lead me to at least some provisional conclusions — or maybe it's just an excuse to have fun and watch more anoles.

Two Islands Into One

One of the more fascinating aspects of Hispaniolan zoogeography is the effect of the island's history as

FIGURE 14. *Anolis longiceps* from Navassa Island may exhibit ecological release (a broadening of the niche to take advantage of the absence of competition).

two separate paleoislands now joined. Many species found on either the North or South island have ecological counterparts on the other, presumably a consequence of allopatric speciation. However, other species appear to have island-wide distributions and still others appear to have extended their ranges to the other paleoisland after the merger. Are the former the result of incomplete speciation events or of over-water dispersal from one paleoisland to the other? Are the latter successful in invading the other island because no ecological counterpart existed or are the relevant distribution patterns indicative of events not yet recognized?

Complicating the issue even more are apparently inexplicable distributions of some species that occur in places that should have been inaccessible or are absent from areas where logic would place them. An example of the latter is the apparently disjunct distribution of *Anolis whitemani* (as traditionally defined), a xerophilic species restricted to some very dry habitats in various parts of Hispaniola while absent from others (even some in intervening areas). Were the intermediate populations extirpated at some time in the past or did these disjunct "populations" evolve independently in parallel fashion in response to similar environmental conditions? Recent phyletic data generated by Rich Glor of Washington University in St. Louis suggest the latter, and also raise the question of how many similar situations exist in the complex habitat mosaics of the Greater Antilles. Such questions are limited only by the number of critters we can identify; and, although my example focuses on Hispaniola, similar situations exist on the other large islands as well as on the many archipelagos of the region. I'm pretty sure that I'll never run out of questions.

The Fauna

On that first brief trip with my wife, I managed to collect a small sample of specimens with which I was quite enamoured (I still didn't have a clue just how diverse the herpetofauna really was). Although I readily recognized the genera, I wasn't sure of very many of the species (*Bufo marinus* was a notable exception). I carefully preserved one or two of every species I found and, after returning to Kansas City, sat down with John Parmerlee, a former student and current colleague, to see what I had caught. Armed only with a copy of Henderson and Schwartz's 1984 guide to the identification of Hispaniolan herps, in which ranges are addressed solely by region and only a few taxa are illustrated, we succeeded in identifying all but two troublesome anoles. Only someone who has struggled through the process of getting to know an extremely diverse but unfamiliar fauna without help will understand how thrilled we were when the experiences gained on subsequent trips supported every one of our very tentative identifications.

The Anoles

I dearly love the anoles (my vanity license plate even reads "ANOLIS"). They're ideal models for ecological and evolutionary studies and, additionally, can serve as object lessons in animal (and human) behavior. I've watched a large male displace a smaller individual who had succeeded in mounting a receptive female (Figure 15), and, on many occasions, witnessed two more equally matched males attempting to throw one another from a tree in order to gain an advantage in courting the young female patiently awaiting the outcome of their bout. Most young men pass through a stage (some never leave it) during which they were at least tempted to emulate the very direct course of action taken by these lizards. I also admire the efforts of male anoles to impress the females with colorful dewlaps and absurd headbobs and pushups — and can readily relate these pretentious behaviors to those in which many of my male students routinely engage (I, myself, once owned colorful neckties on the perhaps futile assumption that they would attract the attention of a particular female). When males respond to interlopers and potential competitors by making themselves look as fierce and prodigious as possible, they once again parallel equivalent human behaviors. I've often thought that, in general, the lizards

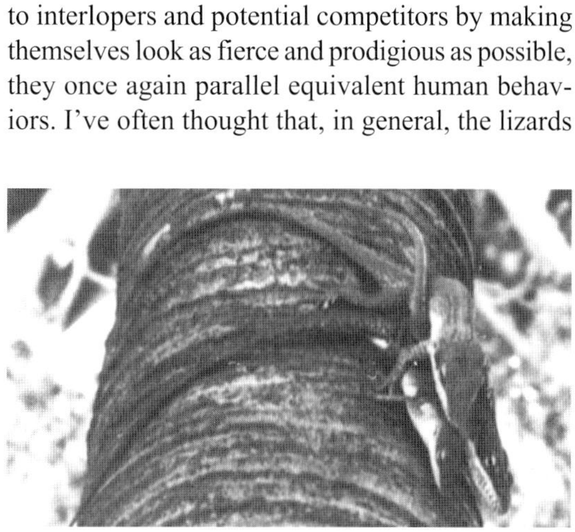

FIGURE 15. Attempted displacement of a mating male *Anolis porcatus* in Santo Domingo, Dominican Republic (photograph by Donald D. Smith).

are more honest; when hormones rage, they respond accordingly. When human hormones rage, we frequently try to deny their influence by playing games that fool no one, and may well obscure the true intent and honest message of the perpetrator. Might we not learn something from the humble anoles?

The "Sosua Method"

I rapidly became enchanted with the curlytails (*Leiocephalus* spp.), although I initially found them very difficult to catch. On my first trip, I had captured a *L. lunatus* near the airport. Later, I noticed that the lizards I was seeing around Altos de Chavon, a scenic little resort village built near La Romana in the style of a 16th-century Spanish town, were different. I couldn't get close enough to grab one and even had trouble reaching them with my little noose (cut short enough to fit in a suitcase). I finally managed to slip the noose over the head of one individual and was delighted to feel its weight when I closed the noose. Joy quickly turned into dismay, however, when the loop slipped off and the lizard went flying away. I swear I looked and listened for the longest time before I realized that the poor little fellow had landed right at my feet, sufficiently dazed that I had no trouble bending over and picking it up. I did this, of course, in the most casual manner, as if I had planned on using this technique all along.

On a later trip, John Parmerlee and I were hunting anoles in the central square of Sosua, wanting a sample from northern populations of several species. Although we didn't particularly need large males, they tend to catch the eye. One exceptionally handsome *Anolis chlorocyanus* had repeatedly defied our best efforts, and we became obsessed with catching him, in spite of the fact that any number of other individuals were obviously much more prone to being noosed. After a number of futile attempts on our part, he finally took refuge in the crown of a small palm. Not to be dissuaded, John grabbed the trunk, warned me to be on the lookout, and began to shake it vigorously. With the first full swing, the lizard became airborne — and flew in a high and graceful arc directly to me. Reaching out with one hand, I easily caught it, responded affirmatively to John's query whether I had seen where it went, and casually dropped it in a collecting bag. Although we now refer to this as the "Sosua method," I doubt that we could repeat the feat in another century of trying. Our audience, however, was most appreciative. They clapped at our performance, and then most walked away shaking their heads in amazement, thinking the show was over. One little boy, obviously in awe of our skills, subsequently pulled on my sleeve and asked just how I knew where to stand in order to catch the flying lizard.

Lip Service

On another trip, John and I were collecting *Anolis cybotes* in Monte Cristi (this population had been associated with *A. whitemani* and we wanted to collect a series in an effort to clarify the situation). Because anoles had been hard to find in the scrub surrounding town, we hadn't expected to find very many and had pocketed only a few tubes in which to keep specimens while on the go. Unexpectedly, anoles were abundant, but, to our dismay, we quickly ran out of tubes. We could easily hold several animals in one hand, but were then unable to remove more recently caught specimens from our nooses. Desperate for any solution, I stuck several anoles in my mouth, holding them firmly with my lips and careful not to decapitate or suffocate them. This technique was sufficiently effective to allow the capture of several additional animals, but walking back to where we had left our car, we overheard several comments questioning the sanity of the Norteamericanos who eat live little lizards.

High Hopes

While waiting for breakfast one morning in Barahona, I stepped outside to enjoy the view and noticed a very small *Leiocephalus* slowly moving toward the edge of the veranda. Upon closer inspection, I ascertained that the little guy was dead and, instead of walking, was in fact being moved by an uncountable number of tiny ants. I called to the students who had remained inside and soon a rather respectable audience had gathered to witness this tribute to the strength, collaboration, and perseverance of social insects. When they finally succeeded in pulling and pushing the carcass to the edge of the patio, we cheered when it fell to the ground below, but figured they'd have a much more difficult time in the sand. Actually, they managed at least equally well (maybe poor traction on loose substrate is less of a factor with six legs?). Interestingly, the ants soon split into two divisions, one continued moving the lizard while the other enlarged the opening into the colony's burrow. By the time our breakfast arrived, these tiny ants had

FIGURE 16. *Epicrates striatus*, world champion stinky snake. This specimen was collected near Copey in the northwestern Dominican Republic.

moved a dead lizard, several orders of magnitude larger than any one of them, over two meters and into an underground chamber deep enough that the carcass disappeared entirely from our view. I don't think the ants moving a rubber tree plant (as in the popular children's song) could have been much more impressive.

A Fortuitous Hand-off

Anyone who has ever collected snakes knows that many species have the unpleasant habit of crapping all over the unfortunate, albeit successful, collector. Here at home, garter snakes (*Thamnophis* spp.) and watersnakes (*Nerodia* spp.) are particularly renowned for the pungent quality of their musk, and I'm sure other species throughout the world are at least equally adept at rendering themselves as unpalatable to a predator as possible. Still, I'm convinced that large boas of the genus *Epicrates* have earned the right to be placed prominently in anyone's top-ten list of stinky reptiles (Figure 16).

We encountered our very first live specimen (having previously seen several roadkills) crossing the road late at night on the return from a very long trip across the Barahona Peninsula. Most of the students were sleeping when I hollered "snake!," slammed on the brakes, and jumped from the van.

Fortunately, one young man had managed to stay awake. Steve, who had been on field trips with me back in the states, knew how single-minded I could be when encountering a snake, especially one that was rapidly disappearing into the roadside vegetation. Sitting right behind me, he ignored the snake, dove over the seat, and managed to set the emergency brake just as the van began rolling toward the seaside cliffs.

In the interim, I had grabbed the snake, a really nice 1.5-m specimen. Martin Jolley, a friend and colleague who had been riding shotgun, had followed me out of the vehicle and, as excited as I was, asked to see the catch. One of the smartest things I've ever done was hand it to him. As soon as the exchange was complete, that sucker dumped on Martin in a manner that would have put the most ambitious watersnake to shame. I must say that both the quantity and quality of the musk was something to behold! We did graciously let Martin ride back with us, but we made sure every window and even the rear hatch was opened as wide as possible.

Bushy Tails and the Lack Thereof

During another trip, we were collecting lizards on the grounds of an upscale tourist resort — and attracting a considerable amount of attention. A couple from New Jersey, obviously fascinated by what we were doing, followed us around for some time before approaching us and asking for an explanation. When we told them we were collecting specimens for a scientific study, they asked if we knew anything about animals other than lizards. When we indicated we might, they asked us about the "darling" squirrel-like creatures they kept seeing in trees around the resort (they knew they weren't squirrels because they lacked bushy tails). I admit to a bit of wicked delight in telling them that the "cute" little critters they were watching with such affection were Black Rats. Although the couple quickly left with no further conversation, I'm pretty sure they were at least slightly disappointed in their brush with the local fauna.

Delight and Dismay

The capriciousness of field biology was amply demonstrated during my visit to Isla Mona. Wandering along one of the many island trails with Carlos, then a student at the University of Puerto Rico–Mayagüez, with whom I had corresponded before the trip, I glanced up in time to see the tail of a Mona Racer (*Alsophis variegatus*) disappearing around the trunk of a tree. Quickly snagging and retrieving it, I proudly displayed my catch. Carlos's reaction was a combination of elation (this was also his very first Mona Racer) and dismay. He explained the latter response by telling me that, in his many visits to the island, he had not previously seen a racer, much less caught one, and here I find one after being on-island for only a few hours.

The People

I'm not sure just why I felt so comfortable in the DR, even before I could navigate the streets of the capital or identify even the most common species of anoles. In retrospect, I might have been more seasoned (and receptive) than when I had first ventured into the Mexican tropics in 1978, or maybe it was because the Dominicans were so openly tolerant of my antics. Most likely, it was the friendly welcome by Sixto Incháustegui, one of the more prominent names on the list Bob Henderson had sent me only a few weeks previously. At the time, Sixto worked at the Museo Nacional de Historia Natural. When I arrived unannounced (I had written him, but had set no specific date or time for a visit), he dropped everything and escorted me to the Departamento de Vida Silvestre to help me acquire a permit. When everyone in the office rose and greeted him with open arms, I knew Bob had steered me in the right direction. This became even more obvious when the staff immediately accepted me, in spite of my almost non-existent Spanish and nearly complete ignorance of the country's biological marvels. Sixto also cushioned my immersion into Dominican society by explaining that the myriad colors decorating everything from rocks and trees to vehicles and entire houses represented the various political parties which were, at that time, engaged in a spirited election campaign. He tempered my exhilaration only a little when he advised me to avoid trucks flying colored flags and over-flowing with armed men, noting that "they can get a little too enthusiastic at times."

Dominicans are great folks. I met Julio Felíz (Figure 17) during my first visit to Barahona. A school teacher turned tourist guide, Julio gives good advice and has proven to be a valuable friend on many a subsequent trip. Once he took the bus to Santo Domingo to purchase a rear window for our van (broken as a consequence of an unfortunate collision with what I swear was a jumping tree). On another occasion, he demonstrated his prowess with a slingshot, helping us quickly collect a sample of *Ameiva chrysolaema* on a trip devoted specifically to that purpose. Yet another time, he imposed on a friend to take us onto private property along the Río San Rafael, where the rainforest follows the river almost to the ocean's edge and where, not coincidentally, *Anolis bahorucoensis*, arguably the most attractive of all anoles, reaches the lowest limits of its elevational range.

When we arrived on one trip with baseball equipment we wanted to donate to a school, Julio introduced us to the local teachers. He's arranged for us to acquire a series of giant anoles (*A. barahonensis*), locally known as Saltacocotes ("throat-grabbers"), and to make connections with a fisherman who ferried us to Cayo Pisaje, and collect some very large specimens of *Aristelliger lar*. He's personally guided my students to the region's larimar mines (a turquoise-like semi-precious stone unique to the island), through the local sugar mill, where we surprisingly found machinery manufactured in St. Louis, and to the market, where the vibrant sights, unfamiliar sounds, and pungent smells immerse even the most reluctant visitor in another world.

No Hablo Español

Although Sixto and Julio facilitated my adjustment to the DR, some encounters defy all efforts of anticipation. On one trip with students into the far reaches of the Río Nizaito valley, a number of local children materialized out of nowhere, apparently attracted by our strange antics. Seeking an explanation, they turned to an African-American student on the assumption that he was Dominican. Obviously having never encountered a dark-skinned North American, they peppered him with questions regarding our activities. Poor Paul, whose Spanish began and ended with a heavily accented "no hablo Español," tried to explain his situation, but the kids, certain that he was a Dominicano, only laughed, convinced that he was perpetrating a terrific practical joke.

La Gringa Loca

On another trip, we were collecting sleeping anoles in a cafetal in the Sierra de Baoruco when we spotted a Vineboa (*Epicrates gracilis*; Figure 18) high in

FIGURE 17. The author and Julio Feliz (photograph by David M. Nieves).

FIGURE 18. A vineboa, *Epicrates gracilis*.This specimen was collected near Bávaro in the eastern Dominican Republic.

branches overhanging the road. Unable to reach it with conventional collecting tools, we constructed a human pyramid with the smallest student on top. Again, we attracted considerable attention from the locals who had gathered quietly to observe our inexplicable behavior. When Debb, from the top of the pyramid, hooked the branch on which the snake was foraging and shook it loose, the audience was entirely too polite to openly criticize the performance, but the term "la gringa loca" was easily discerned from their muted conversation.

Bounties and a Promise

We were collecting blind snakes (*Typhlops*) near Paraiso when a passing resident of a nearby community hollered loudly, charged into our midst, and frantically tried to chase us away. Afraid that we had trespassed, we retreated to the road only to find, when the situation calmed a bit, that he was not at all concerned with where we were but mortified at what we were doing. In all seriousness, he explained patiently that these little snakes would sting with the spiked tips of their tails and could inject a potentially fatal venom. Even when I showed him that the tails were incapable of penetrating skin and explained that the spikes serve merely as anchors when burrowing, he persisted and would not allow us back onto the site. We returned later in the day and successfully collected a nice series of specimens. Again, an audience materialized, although no one offered to help us in our endeavors which was understandable, but unusual.

Dominicans in rural areas are typically quite adept at catching animals, and offering bounties can be an effective means of collecting large numbers of specimens (and was a common practice of many herpetologists who had preceded us in the area). Sixto had asked that I not use this method because it conveyed the idea that animals were merely another exploitable and expendable resource. Although I occasionally pay someone to lend assistance, I always try to emphasize that I'm paying them for their help rather than for the animals they happen to catch. Only rarely, when circumstances rendered the likelihood of catching specific animals impossible without help, have I violated my promise to Sixto.

A Language Barrier

I quickly learned to trust local advice regarding the presence and density of particular species (although we often needed one in-hand to convey exactly in which animal we were interested). Consequently, we often stop and talk to passersby. On one of my earliest trips, before I became aware of the many Haitian seasonal workers who are hired to harvest cane (a particularly nasty job avoided by all but the most poverty stricken Dominicans), I addressed a gentleman encountered along the road to Lago Enriquillo. When he didn't immediately respond, I attributed his reticence to the poor quality of my Spanish, and tried again, speaking more slowly and, to my mind at least, enunciating more clearly. This time he spoke up. Unfortunately, I understood not a word — and was pretty sure he hadn't even been speaking Spanish. Only after thanking him politely and driving on down the road did it occur to me that he was speaking Creole and must have been heading home to the batey (a Haitian barrio) down the road. On another occasion, deep in the forest, we addressed a youngster who replied in perfect Castillian Spanish (even with the lisp that characterizes that dialect). Surprised, we asked about his use of language and were told proudly that: "My mother always insists that her children speak properly."

Reflections on Navassa Island

My memories of Navassa are a microcosm of the entire island experience. As a member of a survey team with the assignment of assessing the island's natural resources, a love/hate relationship developed

FIGURE 19. *Celestus badius* "swimming" in dead Poisonwood leaves on Navassa Island.

rapidly. On the plus side of the ledger, I quickly came to appreciate the "ubiquitous four" (the four extant reptiles: two geckos, *Aristelliger cochranae* and *Sphaerodactylus becki*, an anguid, *Celestus badius* (Figure 19), and an anole, *Anolis longiceps*) — I only wish I had spent more time collecting ecological data instead of searching for species that had been reported from the island in the 19th century and not seen since. For example, I accumulated numerous observations of habitat associations, but I never attempted to quantify what obviously were phenomenally high population densities of all four species. Certainly, their abundance in various habitats constituted an ideal natural laboratory.

I also enjoyed the company of the "dry team" (a "wet team" was synchronously assessing marine habitats around Navassa during much of our stay), and benefited greatly from associating with a geologist, two botanists, two entomologists, an ornithologist, and a jack-of-all-trades originally trained as a freshwater ichthyologist. The diversity of interests and backgrounds was, I believe, good for all of us — although one of the botanists (Tom Zanoni, who I knew from his days at the Botanical Garden in Santo Domingo) never ceased in his attempts to convince me that biology was the study of plants and their parasites (versus my contention that the discipline dealt with animals and their food).

Additional attractions included the spectacular light shows put on by massive thunderstorms that seemed to develop nightly to our north. The isolation was also appealing, as was the knowledge that much of what we were seeing had been experienced by very few scientists. Comic relief was provided by the crash landings of omnipresent White-crowned Pigeons into the fruiting Poisonwood Trees.

Foremost on the downside of the relationship was the weather. Not that conditions were unusual for the tropics, it's just that we couldn't escape the sun and heat. Although this provided some heartfelt

FIGURE 20. Our camp at the base of the Navassa lighthouse; notice the congregation on the shady side of our very large "sundial."

FIGURE 21. The *Metopium* overhanging the excavated trenches that served as our latrine.

appreciation for the trail-blazers of tropical biology, we longed for a break. We had to settle for nightly showers in solar-heated water (providentially provided by the welcome fact that the cistern under the old lighthouse keeper's home still held water) and following the shade around the lighthouse as the sun journeyed across the sky each day (Figure 20). We became so adept at the latter that we could have easily marked the hours of the day in sundial fashion around the base of the lighthouse and used the position of anyone in camp as our hour-hand.

A close second to the heat in the hate survey was the unavoidable ubiquity of Poisonwood Trees (*Metopium brownei*). Although I proved resistant to its charms (a good thing since a large percentage of the sphaeros and anguids I caught were in *Metopium* duff), even minimal precautions necessitated detours whenever one passed from the open habitats through the Poisonwood-lined edges and into the forests. The detours weren't a problem in terms of time (we had plenty), but contributed to spatial confusion once in the forest and out-of-sight of the lighthouse (we never got lost, but found ourselves temporarily uncertain of our location for several hours at a time on more than a few occasions). The *Metopium* overhanging the excavated trenches that served as a latrine also contributed to nightmares about rashes in unthinkable places (Figure 21).

The third entry in the hate parade actually began as a pleasant surprise. Early in the trip, we had commented on the near absence of mosquitoes, but evidence of a hard rainfall on the night before our arrival and nine days to allow completion of the all-too-rapid life cycle changed everything. Unless one had reason to venture out after dusk, one retreated into one's tent immediately after the dinner dishes were done, only to bathe in a pool of sweat until the temperature finally descended into a barely tolerable range during the wee small hours of the morning. Was I miserable while on Navassa? Sure. Would I ever go back? In a second.

Surrender

To summarize my capitulation to the allure of the islands, I recall an evening along the Río San Rafael. The cool mountain-fed waters create a microclimate in the river's valley that allows the rainforest to reach the very edge of the sea (Figure 22). After an exhausting, hot and dirty, albeit successful, day of collecting data in a dry scrub habitat, I found myself sitting chest-deep in a cool, shaded freshwater pool, sipping a cold one, listening to Jimmy Buffett sing about margaritas, looking over the incredibly gorgeous gravel beach, watching the ever-changing ebb-and-flow of the waves, and realized that this is "as good as it gets" (Figure 23).

FIGURE 23. This is "as good as it gets" (photograph by David M. Nieves).

Acknowledgments

In addition to the persons mentioned above, I thank the many students whose edification provided the excuse I needed to visit and become enamored with the islands. I thank Albert Schwartz, whose reputation as a crusty old coot didn't keep him from treating me with respect, even before I had earned it. José Ottenwalder has selflessly facilitated many of my trips and has become a friend in the process. Ellen Censky taught me the finer points of noosing *Ameiva* and introduced me to Anguilla. Blair Hedges has frequently provided invaluable advice and encouragement. I once blamed him for extirpating a population of twig anoles that we were unable to find; he admitted to the crime, but indicated that it had all been Richard Thomas's idea (I have yet to hear Richard's side of the story). Andreas Schubert is another new friend who has facilitated my exploits to the extent of opening his home to me. Rick Passaro, Kevel Lindsay, and Karim Hodge made my first visits to St. Maarten, Antigua, and Anguilla most enjoyable. Finally, although I hadn't planned on mentioning anyone who was featured in the body of the essay, I thank Gloria Santana, who has issued collecting permits, but, more importantly, left me with an indelible vision of hope for the future when she stood up to those machete-wielding campesinos.

FIGURE 22. The coastline south of Barahona where mountains meet the sea.

Suggested Reading

Gorst, M. 2001. *Measuring Eternity: The Search for the Beginning of Time*. Broadway Books, New York.

Larson, E. J. 2001. *Evolution's Workshop: God and Science on the Galápagos Islands*. Basic Books, New York.

Losos, J. B. and K. de Queiroz. 1997. Evolutionary consequences of ecological release in Caribbean *Anolis* lizards. *Biol. Jour. Linn. Soc.* 61:459–483.

Sifers, S. M., M. L. Yeska, Y. M. Ramos, R. Powell, and J. S. Parmerlee, Jr. 2001. *Anolis* lizards restricted to altered edge habitats in a Hispaniolan cloud forest. *Carib. Jour. Sci.* 37: 55–62.

Waugh, A. 1930. *Hot Countries*. Farrar & Rinehart, New York.

Williams, E. E. 1983. Ecomorphs, faunas, island size, and diverse end points in island radiations of *Anolis*, p. 326–370. *In* R. B. Huey, E. R. Pianka, and T. W. Schoener (eds.), *Lizard Ecology: Studies of a Model Organism*. Harvard University Press, Cambridge, Massachusetts.

Powell, R. 2003. As good as it gets, p. 169–183. *In* R. W. Henderson and R. Powell (eds.), Islands and the Sea: Essays on Herpetological Exploration in the West Indies. Society for the Study of Amphibians and Reptiles, Ithaca (New York). Contributions to Herpetology, volume 20.

The Shape of Things

RICHARD A. SAJDAK[1]

Feeling less like General MacArthur than probably anyone in history, I jumped off the deck of the boat into warm knee-deep water. After beaching the boat and helping my wife jump down, I grabbed duffle bags and other paraphernalia handed down by the crew. As a few villagers and kids gathered to witness the event, we piled the gear on the white sand beach.

Colorful Shacks and Long-nosed Snakes

It was 1980, but I was feeling the tiniest bit like Stanley on the shores of the Nile when I asked for Señor Henderson in my very best phrase-book Spanish. Directed mostly by gestures, I followed a smiling young boy up the beach into the shade of a line of Coconut Palms and along a row of turquoise, pink, and yellow shacks that made up the village of Mano Juan (Figure 1). As I followed, sweating in the unfamiliar heat, the boy led me to a weathered shack decorated with flakes of yellow paint, where Bob Henderson sat on the stoop writing up field notes (Figure 2).

So this was Isla Saona, paradise in the Caribbean, goal of a trip which had begun in Milwaukee several days before. After landing in Santo Domingo, the oldest European city in the New World, and a bus ride through sugarcane fields to La Romana on the southeastern coast of Hispaniola, we had met with people from Gulf + Western, who had arranged a trip on a gleaming white 24-foot sportfisherman to the island on the following day.

As we slowly motored through the coral reef into the protected waters off Isla Saona, the white beach, blue water, palm trees, and pastel houses did look dazzling. As I followed the boy along the path in front of the village past a pig scratching its butt on the doorstep of one of the pastel huts, I had a few other thoughts. This was only my second venture into the tropics, the first being a month-long trip to Costa Rica a few years earlier to participate in a radiolocation study of Neotropical snakes with co-workers from the Milwaukee Public Museum.

FIGURE 1. Mano Juan, Isla Saona, Dominican Republic.

Bob had learned of Hispaniola and its snakes from the grand old man of West Indian herpetology, Al Schwartz. While pursuing morphological studies of long-nosed tree snakes in different parts of the world, he had been looking for live specimens of Hispaniolan vine snakes (*Uromacer*) to use in behavioral observations, and had contacted Al for advice. Bob came back from a short visit with specimens and stories of tree

[1] 4 Callingham Road, Pittsford, New York 14534, USA.

FIGURE 2. Bob Henderson tagging specimens and updating field notes.

snakes as common as garter snakes back in Wisconsin. For someone used to the rarity of rainforest snakes in Central and South America, this was a revelation. While in the Dominican Republic, he had met Sixto Incháustegui and José Ottenwalder, who suggested Isla Saona, an island off the Dominican Republic's southeastern coast, as a vine snake hotbed. Bob resolved to begin field studies of the snakes — and here we were.

Vine snakes are a diverse assemblage of tree snakes occurring throughout the tropics. *Ahaetulla* in Asia, *Thelotornis* in Africa, *Oxybelis* in Central and South America, and *Uromacer* on Hispaniola, while of unrelated ancestry, share similar physiognomies: elongated snouts, very thin bodies, and a predilection for eating lizards. Their elongated snouts and large eyes allow for good vision and, rare for a snake, binocular depth perception. These are requirements for successfully hunting alert prey in the complex 3-D world of trees. Skinny bodies allow them to move easily through the trees without shaking the branches, and their eyes allow them to pick up movements as tiny as the shift of a lizard's eye. Binocular vision gauges the distance, and a quick strike grabs the prey, whose struggles are quickly ended by venom ejected from fangs at the rear of the upper jaws. Vine snakes employ a moving ambush strategy, moving slowly, imperceptibly from spot to spot, waiting at each location for movement to betray a lizard's presence.

Good Science

Bob and I were interested in the ecology of these snakes, among the most specialized and yet abundant arboreal species in the world. How and what do they hunt? How do they interact with other tree snakes? How do their strategies vary between species, locations, and communities?

Several problems confronted us. First, throughout most of the tropics, snakes tend to be very scarce. For example, during a 10-month field study of Brown Vine Snakes (*Oxybelis aeneus*) in Belize, working almost every day, Bob had seen or captured 38 vine snakes a total of 89 times. Second, in most areas, vine snakes are part of a very large community of tree snakes, complicating the interpretation of observations. For example, how much of the Brown Vine Snake's morphology and behavior is a result of ancestry and how much is a response to competition with the distantly related Parrot Snakes (*Leptophis* spp.) and other arboreal species? A paucity of observations in a complex system render such questions almost impossible to answer.

Hispaniola seemed to avoid both of these problems. As Bob found on his first visit, vine snakes occur there in numbers more like temperate than tropical ecosystems. Also, among Hispaniola's 19 species of snakes, only five were arboreal, three vine snakes and two boas (genus *Epicrates*). Even better, the three species of vine snakes were almost never found together, but either alone or in pairs. As researchers working on lizard and frog biology had long known, we were to find that studying abundant species in a simplified community made for good science. Likewise, Hispaniola lacks the immense diversity of prey species which complicate mainland ecosystems. Isla Saona, for instance, has only three species of *Anolis* lizards, vine snakes' primary prey, in contrast to the dozen or more found at many Central American sites.

FIGURE 3. The long snout and large eyes of *Uromacer oxyrhynchus* allow for good vision (photograph by Suzanne L. Collins).

FIGURE 4. Like other vine snakes, *Uromacer oxyrhynchus* employs a moving ambush strategy (photograph by Robert Powell).

Of the Hispaniolan tree snakes, the boas were active by night. Two smaller species, the Hispaniolan Vine Boa, *Epicrates gracilis*, and Ford's Boa, *E. fordii*, feed largely on *Anolis* lizards. A third, semiarboreal species, the Hispaniolan Boa, *E. striatus*, is a much larger snake. Adult Hispaniolan Boas are largely mammal and bird eaters, although juveniles commonly take anoles and other lizards.

The only other arboreal species are the three species of vine snakes. These are extremely thin-bodied snakes that occasionally exceed a meter in total length. Although their venom is highly effective on lizards and other prey, in the rare instances when they envenomate a human, only minor pain and local swelling result.

These three vine snakes differ greatly in appearance. The Hispaniolan Green Vine Snake, or Culebra Verde, *Uromacer oxyrhynchus* (Figures 3 and 4), is a pencil-thin, long-nosed species, lime-green in color with a white stripe low along its sides. The Hispaniolan Brown Vine Snake, or Culebra Pinta, *U. frenatus*, is not quite as thin or long-nosed as *U. oxyrhynchus* and is brown or grey in color, with a green back. By contrast, the Hispaniolan Short-nosed Vine Snake, *U. catesbyi* (Figure 5), although still an attenuated arboreal snake, is stouter, with a shorter, wider head. Covered in green scales outlined in black, the Short-nosed Vine Snake looks like a length of cord plaited from grass.

Creature Comforts

As we set up housekeeping, I looked over our rented digs — a two-room shack, about 15 by 30 feet on a concrete floor. Constructed of logs and boards (many of which seemed to be driftwood or at least salvaged lumber) with a sheet metal roof, it boasted a front and back door, three shuttered but unscreened windows, but no plumbing (we had brought 20 or so gallon jugs of potable water). In both rooms, bare light bulbs hung from a pair of insulated wires running the length of the hut. The current for these lights came from an outpost of the Dominican Republic's Navy at the far end of the village. They generously shared the output of their generator (run from mid-afternoon to evening) with the village via the pair of wires, which stretched from one end of the village to the other, running through each shack in turn. Cooking was over an open fire and bathing was in the surf.

We discovered we had a choice of keeping the doors and shutters closed, stifling any breeze that might have mitigated the heat, or opening them and sharing our dwelling with chickens, pigs, and an occasional wandering dog. The villagers were fascinated by our weird behaviors, and preserving the previous night's catch meant windows and doorways jammed with kids and adults watching the show. By night the cucarachas had the run of the place, and the arrival of flies with the dawn eliminated any need for an alarm clock.

On the Prowl

That first night we readied our headlamps and equipment by the lightbulb's yellow glare and headed out. Night hunting is romantic, an "anything can

FIGURE 5. The broader head of *Uromacer catesbyi* enables these snakes to consume larger prey (photograph by Robert Powell).

FIGURE 6. Coconut grove on Isla Saona.

happen" feeling, which becomes addictive. On this night, we sensed unfamiliarity, apprehension, strange terrain, strange sounds, and a fear of losing one's way — all of which led to tension. This night such feelings were mitigated by a clear, starry night, the presence of friends, and most unusually, the not-so-distant sound of surf on the beach.

FIGURE 7. The author foraging for vine snakes.

Columbus described Hispaniola as having trees that touched the sky. Throughout the Caribbean, such tropical forests are long gone, victims of building, agriculture, grazing, fuel gathering, and other human activities (Figure 6). The inland area within walking distance of Mano Juan was inhabited by free-ranging pigs and goats. Between them and firewood gatherers, the habitat that remained was one of thorny open shrubs and small trees on a rocky limestone substrate (Figure 7). Topsoil was limited to occasional hollows and depressions. Infrequent trees stood as much as 12 m high, towering over the 5–6-m acacia scrub. This habitat, although degraded, supported healthy populations of anoles and vine snakes, and was an easy place to work.

As we separated, each of us quickly became isolated in the black night, head moving up and down, side to side, playing the headlamp beam down to check footing, then back and forth, painting the foliage with light, searching for a hint of a snake. Tiny twinkling spider eye shines punctuated the dark, with the less frequent anoles sleeping on a branch tip or a hulking tarantula on the ground to spice up the night. Only a few moments had passed before the first shout. "Got one!" echoed through the dark.

Curiously, hunting tree snakes, even diurnal species, is primarily a nocturnal pursuit. By day, they are so incredibly well-camouflaged as to be impossible (at least for neophyte gringos) to find. During our stay, I never saw a vine snake by day, even in areas where they were common at night. The downfall to their camouflage, and the reason for our nighttime success, lies in their light-colored bellies. By day, the white undersides serve as countershading to minimize silhouetting from below. At night, the snakes sleep draped in a loose coil among the leaves on the ends of branches. When a headlamp beam shines on them, the belly reflects the light, betraying their presence. Capture is a simple matter of grabbing the slumbering snake before the light awakens it (Figure 8). For the most part, they chose convenient perches from waist height to basketball rebounding level. The occasional high-perching iconoclast necessitated improvised strategies ranging from knocking the snake off the perch with a scrounged branch to piggyback rides.

The Final Score?

We left Isla Saona after 16 days. Our observations plus those from a subsequent trip showed that, although both Green and Short-nosed Vine Snakes ate anoles, they differed greatly in kinds of prey and hunting strategies. The more specialized Green Vine Snake fed solely on anoles, whereas almost three-quarters of the Short-nosed Vine Snakes' diet consisted of frogs. Several times, we caught a Short-nosed Vine Snake with an immense bulge revealing the presence of yesterday's meal — a 10 cm-long female Treefrog (*Osteopilus dominicensis*) that weighed more than half as much as the snake. The shorter nose and wider jaws are necessary to handle such big and bulky prey. Also, the long-nosed Green Vine Snake was more sedentary, utilizing lower and thinner perches than the short-nosed species.

We had been quite puzzled by reports of ameivas and other ground-dwelling lizards in the diets of these seemingly arboreal snakes, especially the Brown Vine Snake. Was their skinny physique fooling us? Did they really forage on the ground? A brief visit by Bob Henderson to Ile de la Gonâve, Haiti, a few years later showed that Brown Vine Snakes use a deer hunter's strategy, hunting ground-dwellers from tree stands, draping themselves on branches hanging low to the ground, striking downwards to ambush unsuspecting lizards scurrying beneath (Figure 9).

FIGURE 8. Bob Henderson with part of one night's catch.

FIGURE 9. *Uromacer frenatus* frequently suspends itself from low-hanging branches to ambush terrestrial lizard prey (photograph by Robert Powell).

These three closely related and similar appearing species actually divided the available food resources rather finely: an anole ambusher, a forager on frogs, and an arboreal hunter of ground-dwelling lizards. Head shape and body weight are closely related to hunting strategy and diet. Such insights can be useful in investigating the ways arboreal snakes partition resources throughout the world.

Suggested Reading

Henderson, R. W. 1982. Trophic relationships and foraging strategies of some New World tree snakes (*Leptophis*, *Oxybelis*, *Uromacer*). *Amphibia-Reptilia* 3:71–80.

Henderson, R. W. and M. H. Binder. 1980. The ecology and behavior of vine snakes (*Ahaetulla*, *Oxybelis*, *Thelotornis*, *Uromacer*): A review. *Milwaukee Publ. Mus. Contrib. Biol. Geol.* 37:1–38.

Henderson, R. W., M. H. Binder, and R. A. Sajdak. 1981. Ecological relationships of the tree snakes *Uromacer catesbyi* and *U. oxyrhynchus* (Colubridae) on Isla Saona, República Dominicana. *Amphibia-Reptilia* 2:153–163.

Henderson, R. W., A. Schwartz, and T. A. Noeske-Hallin. 1987. Food habits of three colubrid tree snakes (genus *Uromacer*) on Hispaniola. *Herpetologica* 43:241–248.

Sajdak, R. A. 2003. The shape of things, p. 185–189. *In* R. W. Henderson and R. Powell (eds.), Islands and the Sea: Essays on Herpetological Exploration in the West Indies. Society for the Study of Amphibians and Reptiles, Ithaca (New York). Contributions to Herpetology, volume 20.

Isla Cabritos: Paradise Lost?

Sixto J. Incháustegui[1]

It was May 1968. The excitement of the trip and the anxiety to get there early had precluded any breakfast. We left the jeep, which had been driven as close to the water's edge as the soft mud permitted, and waded to the fishermen's rowboat. The boat was slowly pushed into deeper water, and the five of us climbed aboard. The water's surface was calm at 6:00 o'clock in the morning.

Initially, Juan and his two sons were rowing, but as the temperature rose, the wind picked up, the water became choppy, and rowing became increasingly difficult. Eventually, we all had to take turns. Terns frequently flew overhead, and an occasional solitary flamingo, with its long legs in perfect alignment with its equally long neck, passed by (Figure 1). The landscape was unique, and watching it change as we left the land farther and farther behind diminished the stress of the trip. Still, rowing in choppy waters, stifling temperatures, no breakfast, and having to constantly bail the boat was anything but relaxing. Aggravating the situation even more was the fact that water was sometimes coming in faster than we could bail it out with our two rusty tomato paste cans. We were below sea level, as incredible as it might seem, and, as the morning progressed, the surrounding mountains trapped the rising heat, which rapidly became oppressive. We had been rowing and bailing for four hours. Finally, exhausted, we reached the shore.

I was spent — but the sight on land was even more magnificent than the landscape we had seen from the boat. It alone replenished my energies. In fact, we had traveled from the southern shore of Lago Enriquillo on the mainland of Hispaniola to an exotic enclave called Isla Cabritos.

Lago Enriquillo

Hispaniola is the only West Indian island with paired species of several endemic, very characteristic faunal groups: two freshwater turtles (*Trachemys*), two Rock Iguanas (*Cyclura*), and two Todies (an endemic family of small, brilliant green West Indian birds). Hispaniola, although second to Cuba in size, is the most complex of the West Indian islands, both in its geologic history and present physiography and ecology. Lago Enriquillo is considered to be the largest remnant of what used to be an old marine channel that formerly separated present Hispaniola into two islands, often referred to as the northern and southern paleoislands. Researchers interested in the terrestrial animals of the West Indies and in West Indian biogeography are familiar with these terms and their implications for the biogeography of Hispaniola.

Lago Enriquillo covers over 250 km^2 (Figure 2). It is the largest and possibly the most peculiar of all insular Caribbean lakes. It is situated in the southwestern Dominican Republic, close to the border with neighboring Haiti, and its surface is 40 m below sea level. Cabritos, the largest of its islands, stretches from east to west for about 12 km, and the island's highest point is still about 4 m below sea level. Traveling by boat below sea level is unusual enough,

Figure 1. A Greater Flamingo (*Phoenicopterus ruber*) flying over Lago Enriquillo (photograph by Robert Powell).

[1] Grupo Jaragua, El Vergel 33, El Vergel, Santo Domingo, República Dominicana.

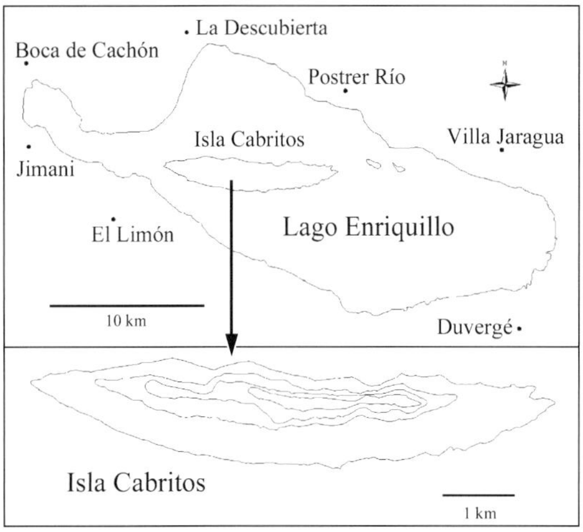

FIGURE 2. Lago Enriquillo (top) and Isla Cabritos. Contour interval = 10 m.

but, provided you have a good pilot, you can also fly below sea level. Besides its elevation (or lack thereof), Enriquillo is hypersaline. The salinity is much higher than that of sea water, and may nearly double the salinity of the sea during droughts.

The lake was named after Enriquillo, a Taino chieftain who was raised as a Christian by the Spaniards. However, after severe mistreatment, he became the first American rebel. He took to the mountains around the lake, successfully resisting the Spanish efforts to curb the rebellion until the region became the last stronghold of the Tainos before they disappeared completely. They knew the island as Guarizaca. Later, the first Europeans to come to this part of Hispaniola used it to raise goats; from this came its current name, Cabritos ("little goats" or "kids").

Cabritos is part of the relatively recent fossil reef that existed when the area was still part of the old marine channel. It is about 12 kilometers long, and about two in width. A central ridge extends for most of its length. Thus, a trip across the island takes you from a low sandy beach to the top of the ridge and eventually back down the other side to another sandy beach. Ecologically speaking, Cabritos is very interesting. It is small, but has a little of everything. Starting at the water's edge, sandy beaches harbor the Hispaniolan Curlytail Lizard (*Leiocephalus schreibersii*). Their namesake curly tails are usually brightly colored and appear to be used to communicate with other curlytails. On Cabritos, these lizards live primarily in the harsh beach habitats. These are very salty, very hot, and support only sparse vegetation. Lizards are found only at low population densities, and the few individuals must take advantage of every opportunity to exploit the very limited shade and find the equally limited food.

Not only the beaches, but the entire island is very dry, salty, and hot. Colonizing plants must be well-adapted to this harsh environment (Figure 3). Other than grasses and small halophytes, the most common plants on Cabritos are Acacia Trees (*Prosopis juliflora*), which were originally introduced from Africa. Those near the shore eventually are threatened by water-level fluctuations. Due to the prevailing high temperatures and limited rainfall, the lake loses water for years, mostly through evaporation — and, as the water level declines, the island gains land and the Acacia forest stretches toward the shore. However, every 20 or so years, when a hurricane passes by, the water level rises, and the Acacias and any other trees that are too close to the water are covered by the lake and die. Their trunks, denuded of vegetation and covered by the salts of the lake, harden under the burning sun and last for many years, providing a special accent to the shoreline scenery.

A Medieval Formula

"Iguanas are all over the island. At this time of the year, the mango season, they are easy to catch. Iguanas love mangoes, and there are none on Cabritos. Take a can full of mangoes with you and a couple of bottles of rum. Peel the mangoes, soak

FIGURE 3. Ermógenes, Cabritos National Park Administrator, in front of "typical" Isla Cabritos vegetation.

FIGURE 4. The author (right) and Michael Carey, back in Santo Domingo after the 1968 expedition to Lago Enriquillo, with a Ricord's Iguana (*Cyclura ricordii*).

them in rum for a while, and then offer them to the iguanas. They will come like ants to eat the mangoes. Then they'll become dizzy, and all you have to do is to grab them." As I remembered that conversation on the day before we rowed to Cabritos, it seems to me to be as fantastic as some of the medieval formulas for spontaneous generation — and, just like the formula for producing mice from old clothing and garbage, the plan for catching iguanas didn't work.

Background

In 1968, Michael Carey (Figure 4) was a young American graduate student. He had spent several months studying *Cyclura pinguis* on Anegada Island. Albert Schwartz, a renowned specialist on West Indian herpetology, had asked him to take advantage of the opportunity to pass by the Dominican Republic before leaving the West Indies. The purpose of his stop was to collect some specimens and gather some information needed to clarify the relationships among these very interesting endemic West Indian reptiles. At that time, I was just beginning college in Santo Domingo. Due to my interest in herpetology, I had stumbled across some papers published by Al. One of my college professors, Idelissa Bonnelly, encouraged me to write him. I did — and that was the beginning of a long and productive professional and friendly relationship.

However, in 1968, I had not yet met Al Schwartz personally, although I had been corresponding with him. When he asked me if I was willing to participate in a field trip to the southwestern Dominican Republic with Michael Carey and/or provide logistical support for his trip, I quickly agreed to do both. So, the purpose of that first trip to Isla Cabritos and to other parts of the southwest was to look for both Rhinoceros Iguanas (*C. cornuta*) and Ricord's Iguanas (*C. ricordii*).

Back to the Adventure

Michael Carey came as planned and stayed at my house before and after the field trip. A cousin lent us his jeep and offered a driver (Figure 5). Although the distance from the mainland to Cabritos is shorter from the northern side of the lake, we went to Duvergé, on the south side. The reason for this was that an old published paper dealing with Hispaniolan Rock Iguanas had described a local guide who was very able and knowledgeable. We asked around, but trying to find an old man with a very common Spanish name was like looking for a needle in a haystack. We never did find the right person, and instead hired another old man, who claimed also to be very knowledgeable. He insisted that he needed his two grown sons to help in the enterprise. Arrangements were made. We already had food for us, but we purchased some green plantains and canned sardines for them. Everything was ready, and we eagerly awakened on the morning we were going to the "land of the iguanas."

As soon as we landed on Cabritos, we pulled the boat as far inland as we could, tied it to one of the nearby Acacia Trees, took the scarce belongings out of the boat, and placed them on the sandy beach — which was just as peculiar as almost everything else on Cabritos and around the lake in general. As a relict of its time as an old sea channel, most of the sand is composed of millions of dead, small, approximately one-centimeter-long mollusk shells (no extant mollusks occur in the area).

Juan, the senior guide, started to make a fire, which generated a confrontation with Michael. Mike opposed making a fire, which he considered too dangerous in the very dry ecosystem. Juan replied that the habitat on the island was very uniform and that it was easy to get lost. The smoke from the

FIGURE 5. The author (left) and the jeep driver for the 1968 expedition to Lago Enriquillo.

FIGURE 6. Typically xerophilic vegetation on Isla Cabritos.

small fire was the only sign that could show us how to get back to the boat. After a few minutes of Juan's energetic opposition to leaving the boat unless there was a fire, the argument was over. The fire produced a thin column of gray smoke that rose perpendicularly into the sky — and the five of us started to explore the island.

I was fascinated by every detail. As I walked from the shore toward the center of the island, the soil started to change from sandy beaches made of millions of dead shells to large slabs of slate, several square meters in size. Many looked like the remnants of old cement foundations. In fact, evaporation under high temperatures and the mixture of sand, shells, and salt were responsible for their existence. Although hard, they broke easily when we walked on them.

Cacti and Acacia Trees were the predominant plants (Figure 6). Mourning Doves were heard here and there. Pigeons were nesting, and an occasional individual would fly away as I inadvertently walked too close to a nest. I continued to the center of the island, where I encountered a herd of goats. They were obviously not very domestic, thundering away as soon as they saw me. I enjoyed the sight of a male Vervain Hummingbird (*Mellisuga minima*) "singing" from the tip of a dead branch, nearly 3 m above the ground. Vervain Hummingbirds are the second smallest bird in the world (*M. helenae*, from Cuba, is the smallest). I began to think that I was really in paradise — and today, after 33 years and many trips, I still consider Cabritos a paradisiacal island.

That first day was very clear and sunny, exactly what you would expect most of the year — but this was May, the rainy season. Although part of Cabritos is in the driest life zone on Hispaniola, thunderstorms and heavy showers are common occurrences at this time of the year.

On summer days like this, as I later learned, midday temperatures regularly exceed 40°C. I was certainly feeling the heat, but suddenly, I came upon another magnificent sight. Without realizing it, I had crossed the top of Cabritos and was near the northern side of the island. The sky was a vibrant blue, with scattered white clouds. In the distance beyond the lake, I could see the Sierra de Neiba, which harbors the remains of a massive coral reef and petroglyphs of pre-Columbian inhabitants. After contemplating the scene for a while, I suddenly realized that I had left my companions behind. Also, I still had not seen a single iguana. It was well after noon, and the day had already been a long one. It was time to return and look for my companions and our "camping" site. At the time, I remember thinking that I would probably see iguanas on my way back.

My exhaustion was cumulative. We had been staying at a "family hotel" in Duvergé, the best we could find at the time. It was built of Royal Palm wood, painted dark green and orange, and had a tin roof. Whenever we were to have a meal at the "hotel," we had to give a day's notice or no food was available. Temperatures were so high at night that I had to soak two towels in water to cover myself in a vain effort to sleep. Besides, that morning we had awakened before five and had started our trip before six. I had helped row and bail, and had walked across Cabritos in the heat. I had eaten only a small pack of saltine crackers when we first arrived on the island. I was very thirsty — and our drinking water had been left at the "campsite." I had just about run out of gas.

I turned around to start my trip back. Now, more than marveling at every sight, I was merely trying to find my colleagues or the boat. I thought that they must be nearby, but that wasn't the case. I had started the trek on the right side of the group, and had walked away from the others. So, there I was, by myself, with no food or water, in the middle of an uninhabited dry and extremely hot island.

I had just crossed the ridge when I remembered Juan's words: "Everyplace on Cabritos looks pretty much like every other." Where was I? Where was the boat? Where was that thin column of smoke climbing to the sky? Where were my fellow colleagues? I could not see any sign of anything.

I decided to head westward along the shore. As I later learned, this decision was taking me away from the boat, but it also took me to one of the most spectacular wildlife encounters I have ever experienced.

An Antediluvian Spectacular

I was walking fast, trying to find either the boat or my companions. Then, I slowed my pace for a bit and once again contemplated my stark, but beautiful surroundings. As I approached the water, I suddenly saw, only a few meters in front of me, a 4-m long male American Crocodile (*Crocodylus acutus*; Figure 7). He had caught a big Tilapia (a fish introduced from Africa) and, at that moment, was in the typical posture of crocodilians dealing with food. The water was very shallow. All of the croc's feet were braced on the sandy bottom and his head and tail were arched well above the water's surface. I stood quietly while he manipulated and eventually swallowed the fish. He, in turn, stared at me, but didn't interrupt his meal. I was so close that I could clearly see every detail of his body, individual scales and patches of brown and black color. This was the first wild crocodile I had ever seen — but, very soon, I was to learn that it was not the only one. After the magnificent male finished his dinner and lowered his head and tail to a "normal" position, I glanced around and was surprised by the sight of a second, a third, a fourth ... and eventually ended my count at 60. Sixty adult crocodiles were in front of me, many of them feeding. There I was, all by myself, in the middle of nowhere, looking at 60 or so crocodiles. I was so impressed that, if I could paint, I still could document the scene in every detail.

American Crocodiles in Lago Enriquillo mate from December to January, lay their eggs in February and March, and the peak of the hatching season is in May and June. Thus, most adult females are very active in May, either guarding their nests or actually retrieving the hatchlings for transport to less saline areas where freshwater streams enter the lake. Through many subsequent years of fieldwork around the lake, I never again observed group feeding during this particular time of year. Apparently, that year had been one with many good rains, the water level was high, salinity was low, and Tilapia were abundant.

FIGURE 7. Female American Crocodile (*Crocodylus acutus*) in Lago Enriquillo.

Tilapia have been introduced throughout the tropics for use as a staple food or as one of several species of aquatic animals that are relatively easy to cultivate in aquaculture facilities. They were introduced to the Dominican Republic's main bodies of water during the late 1950s. They were not introduced into Lago Enriquillo, however, because it is hypersaline and Tilapia are freshwater fish — but, as so often happens, nature doesn't bow to human assumptions. After a few years, local fishermen began finding these fish in the lake, and they rapidly became more and more common. We subsequently learned that Tilapia had reached the lake when the Río Yaque del Sur had flooded. As it turns out, Tilapia are very resistant to salinity (a population in Cuba was even established in a coral reef close to shore). In Enriquillo, however, these fish are at their limits of tolerance. When the salinity increases during dry years, many die. When it decreases after hurricanes, Tilapia thrive. Apparently, such a time had coincided with my first trip to Cabritos.

Back to Camp

After spending half an hour contemplating the crocodiles, my priorities reverted to finding the boat. After a while, I began to have serious doubts that I was going in the right direction — and it was getting late. I decided to return to the center of the island. When I got there, I climbed the tallest tree I could find, and finally saw in the distance a thin column of smoke.

The sun had almost set when I finally got to "camp." As I approached, I could see Juan, one of his sons, and Mike. Immediately, Juan shouted to me: "Where is my son?" I didn't know what to think. They were supposed to be the "expert" guides. I had been wandering around all day long, and he was asking me where his son was. As it turned out, when we started out, we had unknowingly split into three groups. Juan, one of his sons, and Mike had stayed together. The other son and I had separated from the group and, inadvertently, from each other — and they thought the two of us were together. Obviously, Juan's son was also lost. Fortunately, he happened to be very nearby and we were all united a few minutes later.

The day had been a long one. Juan and his sons boiled the green plantains for dinner, and we had pre-cooked canned spaghetti. Neither the food nor our water was particularly splendid, so we decided to leave early the next day. We spent a long night accompanied by mosquitoes and sand flies and at sunrise headed, as planned, back to the mainland and eventually, back to our dark green and orange Royal Palm-wood "hotel."

Where Were the Iguanas?

We did not see a single iguana. Evidence of their presence on Cabritos was abundant. We found lots of feces in the form of loose pellets containing compacted, undigested remains of leaves and wild fruits — but no iguanas.

Both species of iguanas are present on Cabritos, with very healthy populations. Michael Carey's choice of Duvergé as the point of entry, based on written documentation of previous collecting trips, proved to be the worst possible choice. Besides being the greatest distance between the mainland and Cabritos' shore, it takes one to the side of the island with the lowest population of iguanas (due to less suitable habitats).

Although we were disappointed to have endured all of those hardships only to see no iguanas, the trip was nevertheless worthwhile — and we did eventually see both species of iguanas at many localities around the lake (Figure 8). A few days later, we drove back to Santo Domingo, and Michael flew home to Florida. I consider that trip to Lago Enriquillo and Isla Cabritos my first professional field trip.

FIGURE 8. The author with a Ricord's Iguana (*Cyclura ricordii*) back in Santo Domingo after the 1968 expedition to Lago Enriquillo.

The Dominican Perspective

Many of my colleagues who are contributing essays to this book are not native to the Caribbean islands. Undoubtedly, they explained why they were attracted to the West Indies and why they keep coming back. In this respect, I had no choice. I am a native, specifically of the Dominican Republic. When I say that I had no choice, I do not mean to imply that I am uncomfortable about being a native of the Caribbean or my country. On the contrary, I have always been very proud of my heritage. What I mean to express is that I do not have to explain why I became interested in this region of the world. What I can explain, however, is how I became interested in herpetology.

The Making of a Dominican Herpetologist

I grew up travelling throughout much of the Dominican Republic with my parents and sisters, often to wild or isolated places. That was a good beginning. I also have liked animals, plants, and nature in general in all of its manifestations. Choosing to focus on the herpetofauna, and particularly on reptiles, was probably a matter of convenience. Reptiles were by far the most frequently encountered animals as I grew up. I remember finding a hatchling Hispaniolan Slider in some sand stored inside my father's dark room when I was about four years old. Some wild-caught adult females had been kept there for a while for food. Also, almost every country boy (and many city boys) are exposed to the many ubiquitous populations of anoline lizards. In contrast, native mammals are very scarce, and birds can fly. So, the most prominent members of the local fauna are also the easiest to see and catch. Undoubtedly, it was the frequency of close encounters that biased me in favor of reptiles.

Return to Cabritos

When the Dominican Republic was approving its first law relative to protected areas, establishing the Dirección Nacional de Parques in 1974, I had the opportunity to return to Cabritos. By that time, I was working in the National Museum of Natural History. An assessment was necessary to decide whether or not to include Cabritos as a National Park in the National Protected Areas System.

Fortunately, the decision was positive. As the years went by, I also had opportunities to spend many hours in the field on Cabritos, mainly studying the breeding biology of the American Crocodile. I roughly estimate that, in all, I must have spent over 130 days doing fieldwork on the island. Undoubtedly, it is the wild place that I know best.

FIGURE 9. Endangered Rhinoceros Iguana (*Cyclura cornuta*) on Isla Cabritos.

A Premium Nature Show

In the late 1970s, Cabritos' crocodiles and iguanas became renowned, at least in the national and international conservation communities. F. Wayne King and other international specialists who came and visited the island considered the crocodile population in Lago Enriquillo to be in near pristine condition. Gatherings of over 125 adults were still seen. The average number of nests per season (crocodiles lay eggs only once a year) was around 150.

Crocodiles on Cabritos are in a very peculiar habitat. The lake is highly saline and the islands lack fresh water (two other islands have areas <1 km^2). The shorelines, both on the mainland and the islands, are mostly denuded of vegetation. However, sandy beaches are abundant, and crocodiles lay their eggs on them. The American Crocodile is known to build nests either in mounds of dead vegetation, much like the American Alligator (*Alligator mississippiensis*) and other crocodilians living in swamps, or to dig nests in sand or soil. Cabritos' crocodiles always dig their nests in sand. Nesting occurs both on the islands and on the shores of the mainland, although most are on the islands. Adult crocodiles can handle excess salts, but hatchlings seem to need freshwater for approximately the first year. Baby crocodiles hatched on Cabritos and nearby islands have to be transported for several kilometers by their mothers across the lake to sources of freshwater on the mainland.

During the late 1970s and early 1980s, a couple of days on the nesting beaches during the peak hatching season was a "premium nature show." Crocodiles use the same communal nesting beaches year after year, unless environmental changes force them to move (usually major changes in water level). With some 50 nests on a single nesting beach, hundreds of baby crocodiles could often be seen wandering around.

Iguanas

Hispaniola, as mentioned before, is the only Caribbean Island with two different species of Rock Iguanas. The Rhinoceros Iguana (*Cyclura cornuta*; Figure 9) gets its common name from the set of enlarged scales on the top of its snout. It can grow to nearly two meters, and is the more common of the two species. Evidence suggests that the species once had a broad distribution throughout Hispaniola, and populations are still scattered over many of the island's drier areas. In contrast, Ricord's Iguana (*C. ricordii*; Figure 10) has a very restricted distribution, at present limited to the extreme southwest of the Dominican Republic.

On my second trip to Cabritos, to assess its natural resources prior to its consideration as a national park, I approached the island from the north shore, near the town of La Descubierta. The distance to the island is much shorter and the logistics are far easier. Over the years, this route has become the traditional means of traveling to the island. Close to the landing site, I found a healthy population of Ricord's Iguanas. These

FIGURE 10. Critically endangered Ricord's Iguana (*Cyclura ricordii*) on Isla Cabritos.

lizards seem to be more social than the Rhinoceros Iguanas. They build burrows in the sand, and use them as diurnal retreats. Rhinoceros Iguanas also build burrows or merely excavate pre-existing crevices in rocky cliffs, but, when the population density is high, the burrows of Ricord's Iguanas are much closer together.

In 1985, I estimated a population density of eight adult Ricord's Iguanas per hectare. When examining the historical data on densities of Rock Iguanas (which is rather limited), three categories are evident: populations with very low densities, probably reflecting declining populations; populations with extremely high numbers, mostly reported from small isolated islands and cays where the populations are naturally well protected; and a middle group, with densities of 5–11 iguanas per hectare. *C. ricordii* on Cabritos, at least at one site in 1985, seemed to fall in this middle group.

The Ravages of Time

As I write this essay, my memories go back to the late 1960s, the 1970s, and the 1980s. We've since entered a new century and a new millennium — unfortunately, my lost paradise has suffered much over the years.

Teams of researchers from the National Museum of Natural History no longer maintained a permanent presence on the lake after the mid-1980s. Their absence was crucial, particularly during nesting and breeding seasons when the population is most vulnerable. Crocodiles are hunted locally for their penis, on the belief that it has aphrodisiac properties; for its fat, used in local traditional medicine; and, to a lesser degree, for its meat. Mass killings are unknown, but systematic harvesting of adults during the late 1980s significantly reduced the lake's population.

In 1991, the Wildlife Department initiated a new program at the lake. Surveys dating to that period showed that numbers of nests per year had dropped to fewer than 50. This program reactivated efforts to protect Enriquillo's crocodiles, and subtle signs of recovery have been noted. Young hatched in the early 1990s are now being recruited into the adult population. Nonetheless, in 2001, the number of nests registered was only 33. Much remains to be done.

The situation appears to be equally grim for iguanas. Gloria Santana, of the Wildlife Department, studied Rock Iguana population densities on Cabritos in 1997–1998 at the same study site where I recorded my 1985 data. Her results indicated a population density of only 1.33 *C. ricordii* per hectare.

Cabritos is one of the few known places where two species of *Cyclura* live under natural conditions sympatrically and syntopically. Few data document the ecological relationships of both species. Also, I have not seen any data on population densities of species living together. The only data of which I am aware are my data from Cabritos and Pedernales Province, both from 1985. At that time, not a single *C. cornuta* was recorded from the Cabritos *C. ricordii* study site. During the same year, we recorded 0.03 *C. ricordii* per hectare at a site 4 km east of Pedernales, and the density obtained for *C. cornuta* at the same site was identical. At least in this sample, the data suggested that both species were present in equal numbers.

Santana's recent data (1997–1998) gave figures of 0.70 *C. cornuta* per hectare, and 1.33 *C. ricordii*. A substantial decline of Ricord's Iguanas has apparently occurred since 1985, and the biomass of iguanas (both species considered together) per hectare appears also to have declined. A comparison of data sets also shows that at the Pedernales site, both species occurred in equal densities, whereas on Cabritos, *C. ricordii* still might have a higher population density. The main point I wish to raise here is the scarcity of data regarding these very interesting and endangered lizards and to note the apparent decline of the Ricord's Iguana population on Isla Cabritos.

A Haven for Wildlife

Lago Enriquillo and Isla Cabritos are merely samples of the magnificent wildlife found on Hispaniola and in the Dominican Republic in particular. The southwestern part of the country, which includes the Sierra de Neiba, the Enriquillo Basin, the Sierra de Baoruco, and the dry flatlands and coastal and marine habitats of Jaragua National Park, are especially blessed. A very high percentage of the native Hispaniolan flora and fauna is represented. Both endemic mammals, the Solenodon (*Solenodon paradoxus*) and the Hutia (*Plagiodontia hylaeum*) are present. Also, all endemic Hispaniolan birds, as well as many resident and migratory aquatic species, frequent the region. Most of the endemic reptiles also are common. Marine environments are important reservoirs for Manatees and Leatherback, Green, and Hawksbill seaturtles. The

Jaragua-Cabo Rojo area has the highest globally recorded concentration of juvenile Hawksbills (*Eretmochelys imbricata*). Yet, the southwestern Dominican Republic is but a microcosm of Caribbean habitats.

A Hotspot for Biodiversity

Those of us who are familiar with the wildlife of the Caribbean were not surprised when a recent article in *Nature* magazine listed the Caribbean region among the eight globally significant biodiversity hotspots. In addition to considering very high levels of endemism, the main factor taken into account when identifying hotspots is the degree of threat. Sadly, Hispaniolan biodiversity, both in Haiti and the Dominican Republic, is severely threatened.

Although designated a globally significant hotspot, Caribbean biodiversity is far from being completely known — much research is needed. A principal reason for the dearth of knowledge is a deficiency of human resources. Qualified researchers are not easy to find (all too few people heed the call to become a specialist in field biology and its several disciplines). Also, partnerships are very important, and support for developing ongoing collaborations among colleagues from countries in the region and from the generally more developed northern countries is lacking.

A Gloomy Picture and a Glimmer of Hope

The Dominican Republic occupies position number 86 on the Human Development Index of the United Nations Development Program, placing it among the countries with medium levels of human development. Haiti is in position number 134, which places it among the least developed countries in the world. Both nations have high human population densities. The current population estimate for both nations is approximately 15 million, and this is expected to double by 2030. To support this rapidly growing human population on only 77,800 km^2 of land surface is a great challenge — and one that must be met if we are to preserve any significant samples of our unique and valuable wildlife.

In the Dominican Republic, wildlife is protected by national regulations, and particularly by a system of protected areas. As in many other countries, official efforts are not enough. Since the late 1980s a number of non-governmental organizations (NGOs) have been established, and several of these support ongoing conservation programs. NGOs have been particularly active in efforts designed to integrate local people into the co-management of protected areas and into programs supporting the sustainable use of biodiversity. In addition, ecotourism has started to flourish, encouraging a non-exploitive use of the resources.

A Personal Look to the Future

Over the years, the threat to our biodiversity has taken me from university professor and government employee (at the National Museum of Natural History) to non-governmental organizations. I have helped found and have worked for a series of NGOs (particularly the Grupo Jaragua, a national conservation NGO). At the present time, I work for the United Nations Development Program in the Dominican Republic Office. Dominican conservationists of my generation are few in number, but we are beginning to see the recruitment of a new generation. Young conservationist leaders are emerging and are tackling with some effectiveness at least a few of the nation's growing problems. Esteban Garrido and Héctor Andujar, from Oviedo in the Jaragua area, and Hermógenes Méndez, from La Descubierta near Lago Enriquillo, are making their presence felt, at least on the local level — and that, of course, is where effective conservation efforts must begin.

When I see the kids from communities surrounding Jaragua National Park eagerly participating in the conservation summer camps offered by Grupo Jaragua, when I sit down to discuss conservation problems with Esteban or Héctor, or when I listen carefully to Hermógenes present his data on the season's nests or population surveys, I feel optimistic about the future. Yet, the sight of a big Rhinoceros Iguana sunning on the sandy beaches of Cabritos or of the scores of females fighting for a nesting site on Isla Beata seems at first glance to be incompatible with the needs of people living in the region. How long can you be a conservationist on an empty stomach? We will be able to conserve our biodiversity only if local people are effectively integrated into the management of sustainable development initiatives from which they can profit.

The challenge is great, but being an incurable optimist, I end this essay by declaring that Isla Cabritos, my lost paradise, a flagship for the conservation of Hispaniolan biodiversity, shall not die.

Epilogue

Rock Iguanas of the genus *Cyclura* are considered as a group to be the most endangered lizards in the world. Special efforts are being made by the Iguana Specialist Group of the World Conservation Union (IUCN) to preserve these very important and impressive animals. The Rhinoceros Iguana, with a much broader distribution on Hispaniola, is considered to be "vulnerable" (to extinction), whereas Ricord's Iguana, with a very restricted distribution, is "critically endangered." Both species are included in Appendix I of the Convention on International Trade of Endangered Species (CITES), designating them as species considered to be too endangered for purposes of international trade.

The American Crocodile, with a broad but fragmented distribution over much of the Western Hemisphere, is considered "vulnerable." It also is included in CITES Appendix I. Efforts to conserve this and other species of crocodilians are coordinated by the Crocodile Specialist Group of IUCN.

In the Dominican Republic, the National Zoo (ZooDom) has maintained a successful captive propagation program for Rhinoceros Iguanas for more than a decade. At present, the zoo is initiating a similar program for Ricord's Iguanas. It also has successfully bred the American Crocodile. The Department of Protected Areas of the Ministry of Environment administers important sites within the present distribution of both species of Rock Iguanas and of the American Crocodile. Grupo Jaragua continues to work for the conservation of these important species.

Suggested Reading

Alberts, A. C. (comp. & ed.). 1999. *West Indian Iguanas: Status Survey and Conservation Action Plan*. IUCN/SSC West Indian Iguanas Specialist Group. IUCN, Gland, Switzerland.

Carey, W. M. 1975. The Rock Iguana, *Cyclura pinguis*, on Anegada, British Virgin Islands, with notes on *Cyclura ricordi* and *Cyclura cornuta* of Hispaniola. *Bull. Florida State Mus. Biol. Sci.* 19:189–233.

Incháustegui, S. J., M. Mota, and P. P. Merejo. 1985. Densidad de población en *Cyclura cornuta cornuta* y *Cyclura ricordi* (Sauria: Iguanidae). I Congreso de Zoología "Dr. Rogelio Lamarche Soto." 2,3,4 Diciembre de 1985. Museo Nacional de Historia Natural de Santo Domingo, República Dominicana.

Incháustegui, S. J., J. A. Ottenwalder, C. Sanlley, and D. G. Robinson. 1980. Reproductive biology of *Crocodylus acutus* in the Enriquillo Lake, Dominican Republic (abstract). Abstracts, Joint Meeting SSAR/HL, Milwaukee, Wisconsin.

Myers, N., R. A. Mittermeier, C. G. Mittermeier, G. A. B. da Fonseca, and J. Kent. 2000. Biodiversity hotspots for conservation priorities. *Nature* 403:853–858.

Santana, G. 2001. Introducción al Estudio de las poblaciones de *Cyclura* en la Isla Cabritos, Lago Enriquillo, República Dominicana. Tesis Licenciatura en Biología. Departamento de Biología. Universidad Autónoma de Santo Domingo, República Dominicana.

Secretaría de Estado de Agricultura, Departamento de Vida Silvestre (SEA/DVS). 1993. Estudio y protección del cocodrilo americano (*Crocodylus acutus*) en la República Dominicana. Secretaría de Estado de Agricultura, Santo Domingo, República Dominicana.

Webpages

IUCN/SSG Crocodile Specialist Group
www.flmnh.ufl.edu/natsci/herpetology/crocs.htm#Top

IUCN/SSG Iguana Specialist Group
www.scz.org/iguana/

Grupo Jaragua
www.grupojaragua.org/

Incháustegui, S. J. 2003. Isla Cabritos: Paradise lost? p. 191–200. *In* R. W. Henderson and R. Powell (eds.), Islands and the Sea: Essays on Herpetological Exploration in the West Indies. Society for the Study of Amphibians and Reptiles, Ithaca (New York). Contributions to Herpetology, volume 20.

Puerto Rico

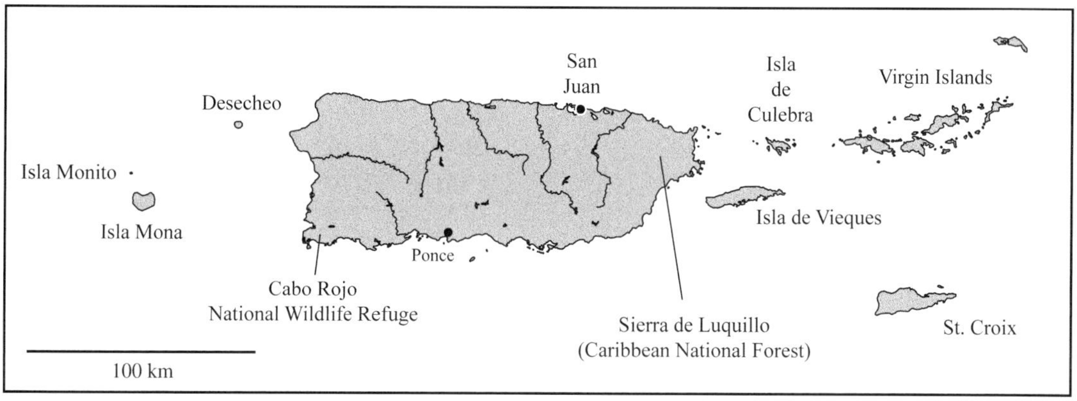
San Juan
Isla de Culebra
Virgin Islands
Desecheo
Isla Monito
Isla Mona
Isla de Vieques
Ponce
Cabo Rojo National Wildlife Refuge
Sierra de Luquillo (Caribbean National Forest)
St. Croix
100 km

An Island Enchanted by Frogs

RAFAEL L. JOGLAR[1]

I was born on an island where everybody is crazy about frogs. For that reason and others that I'm about to explain, I believe Puerto Rico is enchanted by frogs. As a biologist, I have been searching for frogs, lizards, and snakes on this enchanted Caribbean island for about 30 years. I know the island very well; I know its people, history, culture, topography, and geography, and, through the years, I have learned a little about its amphibians and reptiles.

La Isla del Encanto (The Island of Enchantment)

Puerto Rico is known as *La Isla del Encanto* (The Island of Enchantment). Just what this phrase means and whether any native Puerto Ricans can come up with a complete list of our island's enchantments are different stories. I will try to explain this enchantment, but the reader must understand four things: (1) this is no simple task, (2) someday I will get in trouble for such impertinence, (3) the enchantments of Puerto Rico apply also to other islands in the Caribbean, and (4) the list might change from person to person, depending on their experiences and cultural, ethnic, and educational backgrounds. Being a biologist, I must admit that my personal view could be biased toward biodiversity — but this is not a world without bias (and mine is as good as the next person's). So, here is an admittedly partial list of enchantments: (1) rainforests (e.g., El Yunque, Guavate, Toro Negro, Guilarte, Maricao); (2) waterfalls and river pools; (3) white sandy beaches and crystal clear waters (some with palm trees and some without, your choice); (4) old Spanish forts and 500-year-old cities; (5) *comida criolla* (local food); (6) rhythms and melodies that will make you jump and dance (even if you have two left feet like me!); (7) drinks (did you know that the *piña colada* was invented in Old San Juan?); (8) a phenomenal variety of fruits; (9) ecosystem diversity and biodiversity (e.g., rainforests, *mogotes*, dry forests, cloud forests, mangrove forests, swamp forests [*Pterocarpus*], bioluminescent bays, karst forests, endemics, coral reefs, and, sadly, many endangered species); (10) amazing cays and bigger islands (Mona, Desecheo, Caja de Muertos, Vieques, and Culebra); (11) a blend of histories and cultures (Taino, Spanish, African, American, Caribbean, and others); and (12) weather — you can choose your own: hot on the coast and in the lowlands or 5°C cooler in the mountains, wet and rainy in the north or dry in the south — but nowhere does the temperature drop below 15°C and only rarely does it exceed 32°C.

In addition to all of these enchantments, I personally would add one more. The frogs of the genus *Eleutherodactylus,* locally known as coquis, are one of the island's most precious commodities, but, at the same time, the island itself is enchanted by these frogs.

Everybody Loves Frogs in Puerto Rico

A walk through Old San Juan provides ample evidence that everybody in Puerto Rico loves frogs. Undoubtedly, you will discover different kinds of frogs in the stores: ceramic frogs, plastic frogs (Figure 1), and even wooden ones. In addition to numerous kinds of crafts that at least try to look like frogs, T-shirts, ties, necklaces, earrings, key chains, ceramics, and many, many others feature my favorite creatures. A visit to any other big city, small town, or even the stores at the airport will result in the same revelation. In addition, frogs are depicted in all kinds of art forms and artistic representations: poems, songs, paintings, musical compositions, posters, etc. In essence, frogs are everywhere! If you visit a house on any part of the island, you will

[1] Department of Biology, University of Puerto Rico–Río Piedras, P.O. Box 23360, San Juan, Puerto Rico 00931–3360, USA.

FIGURE 1. Frogs in Puerto Rican stores; made in Japan and green in color, they have little in common with real Puerto Rican species.

probably find a decorative frog, most likely in the living room. The San Juan phone book lists at least 17 businesses with the name "El Coquí" or "Coquí," but the number undoubtedly exceeds 500 across the entire island. Gas stations, ice plants, car rentals, liquor stores, computer services, and garbage collection services, just to mention a few, take their names from these beloved little frogs.

I know of no other country in the world that loves frogs so much. To put things in the appropriate perspective, the coqui is the national emblem of Puerto Rico. Need I say more? Judging from the petroglyphs and pictographs on our island (Figure 2), even our ancestors, the Taino Indians, were fascinated by frogs. I sometimes wonder how much of this fascination influenced me to pursue a career in herpetology and to study these interesting creatures for three decades.

More Species of Frogs Than Any Other Island!

The number of native amphibians in Puerto Rico is smaller than that on the other Greater Antilles (Cuba, 52; Hispaniola, 63; Jamaica, 22; Puerto Rico, 18), suggesting that the anuran (or amphibian) diversity in Puerto Rico is lower than that on the other islands. However if you consider that Cuba, Hispaniola, and Jamaica, respectively, are 12.3, 8.5, and 1.2 times bigger than Puerto Rico, the story changes. The number of species of amphibians per unit area is greater in Puerto Rico than on any of the other Greater Antilles (Table 1). On that basis, W. E. Duellman demonstrated that amphibian diversity per unit of area is higher in Puerto Rico than on many other islands of the world — including Borneo, Sumatra, Java, Sulawesi, Hainan, Taiwan, Madagascar, and New Guinea. Furthermore, we not only have more species of native frogs but also more species of introduced frogs (exotics) than any other West Indian island. Since the early 1900s, at least six species of amphibians have been introduced to and are established in Puerto Rico.

More Frogs Than Anywhere in the World!

When I noted that frogs can be found everywhere in Puerto Rico, I was not just referring to decorative frogs in the airport, stores, and residences. Real frogs truly are everywhere — and in big numbers! Densities for coquis in the mountains of Puerto

FIGURE 2. *Coquí Taino* petroglyph in Piedra Escrita, Jayuya, Puerto Rico.

Table 1. Amphibian diversity and endemism in the West Indies. Data presented are based on information in Hedges (1999) and Duellman (1999).

	Native Species	Endemic Species	Percent Endemism	Exotic Species	Total Species	Area (km^2)	Species/ 10^6 km^2
Cuba	52	50	96	2	54	110,922	468.7
Hispaniola	63	63	100	2	65	76,470	823.8
Jamaica	22	22	100	4	26	10,990	2,001.8
Puerto Rico Bank	20	20	100	6	26	9,511	2,102.8
Puerto Rico	**18**	**18**	**100**	**6**	**24**	**8,990**	**2,002.2**
Lesser Antilles	10	9	90	4	14	5,840	1,712.3
Bahamas	2	0	0	4	6	11,296	177.0

FIGURE 3. The Coquí (*Eleutherodactylus coqui*) is the most abundant of the 16 species in this genus found in Puerto Rico.

Rico are the highest reported so far for frogs anywhere in the world, reaching nearly 25,000 individuals per hectare. Although no obvious explanation exists for such high densities of frogs, the absence of large herbivores and predators has been suggested. Even though frog diversity and density is higher at high elevations in Puerto Rico, low elevations and urban areas too are full of frogs. In any backyard or school garden, regardless of elevation, you can hear at least four or five species of frogs calling simultaneously, and their densities are often enormously high.

More Ways to Make More Frogs

Frogs and toads have exploited various reproductive strategies, with the greatest variety occurring in the tropics, and some of the most spectacular and extraordinary occurring in Puerto Rico. Practically all anurans utilize external fertilization. Internal fertilization is known for only a very few species, two of which are found in Puerto Rico: the Coquí (*Eleutherodactylus coqui*) (Figure 3) and the Coquí Dorado (*E. jasperi*) — and other Puerto Rican species, such as the Coquí Guajón (*E. cooki*), also may use internal fertilization. To fertilize eggs, anurans engage in amplexus; nearly always involving a male embracing a female. This embrace may be inguinal (around thc hips), axillary (in thc armpits), ccphalic (around the head), straddled, glued (yeah, that's right, literally pasted to one another), or independent — but not in Puerto Rico! Our Coquí (*E. coqui*) does the "hanky-panky" in a very different way. The female embraces the male by using her hind legs in what is called a reverse hindleg clasp. Although this sounds like something one would encounter in professional wrestling, no one can say that our females are not liberated!

Among anurans, three major categories of reproductive modes are defined by egg deposition site and/or development: (1) aquatic eggs, (2) terrestrial or arboreal eggs, or (3) eggs retained in the oviducts. Puerto Rico is one of the few places in the world where you can observe all three modes among native amphibians. Most anurans are oviparous, as are 17 of 18 (94%) of Puerto Rico's native species. The Sapo Concho (*Bufo*, formerly *Peltophryne*, *lemur*; Figure 4) lays its eggs in water and the Ranita de Labio Blanco (*Leptodactylus albilabris*) lays its eggs in foam nests. All but one species of Puerto Rican coquis lay terrestrial or arboreal eggs. Females of only five of 4204 anuran species in the world (0.1%) retain eggs in their oviducts. One of these is the Puerto Rican Coquí Dorado (*E. jasperi*) (the other four species are African).

Although most anurans have a free-living tadpole (or larval) stage, most of the native frogs in Puerto Rico (16 of 18, or 88.9%) are direct developers (Figure 5). Their embryos develop directly into terrestrial frogs, resembling miniature adults. Another unique feature of Puerto Rican frogs is that several

FIGURE 4. The endangered Sapo Concho (*Bufo*, formerly *Peltophryne*, *lemur*) is the only native toad in Puerto Rico.

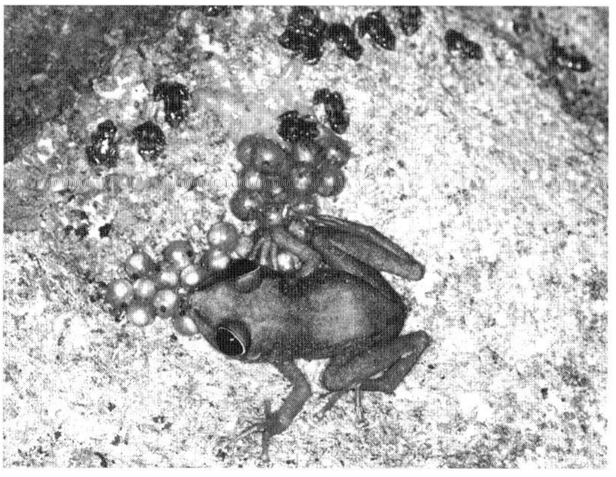

FIGURE 5. Male Coquí Guajón (*Eleutherodactylus cooki*) guarding a multiple clutch and recent hatchlings.

FIGURE 6. Coquí Duende (*Eleutherodactylus unicolor*), the smallest of Puerto Rican frogs, is restricted to cloud forest in the Sierra de Luquillo, eastern Puerto Rico.

species of *Eleutherodactylus* exhibit parental care (*E. coqui, E. portoricensis, E. cooki, E. cochranae, E. hedricki,* and maybe others). Males of these species brood or guard egg clutches to prevent desiccation and predation.

From Cloud Forest to the Coast in 55 minutes!

One of the greatest attractions for doing field work in Puerto Rico (or on any other West Indian island) is the proximity of different ecosystems and ecological associations. On the mainland, traveling from a cloud forest to a dry forest could take hours or even several days. In Puerto Rico, you can do research in the cloud forest of El Yunque and, in less than one hour, you can be in the coastal dry forest of Fajardo, or skin diving on a coral reef at Luquillo (both in the northeastern part of the island). The same is true for cloud forests in the central cordillera (Toro Negro, Adjuntas, Maricao) and dry forests or marine ecosystems in southern Puerto Rico. To give you an example of the amount of contrast between these two ecological life zones, annual rainfall in Puerto Rican dry forests is 600–1,000 mm, whereas average annual precipitation in the cloud forest ranges from 3,800–4,533 mm. I usually take my students to Luquillo to see marine biodiversity in the morning, move up to observe lower rainforest biota at El Yunque after lunch, and finish the day after dark in the cloud forest near Mt. Britton listening to the Coquí Duende (*Eleutherodactylus unicolor*; Figure 6) and observing other cloud forest species. Students are amazed by the biodiversity and ecological contrast in a distance of less than 20 km. It doesn't get any better than that! Because Puerto Rico is such a small island, all of its phenomenally diverse ecosystems and ecological associations are compressed into 8,990 km^2.

Biodiversity on Both Sides of the Shoreline

On my last trip to Isla Mona, I was snorkeling in the Sardinera area. In just one hour, I encountered an almost overwhelming diversity of marine invertebrates and vertebrates. Then, while approaching the shore, but still in the water, I saw a majestic silhouette of a large dragon-like creature (Figure 7), the Iguana de Mona (*Cyclura stejnegeri*). As I watched what is arguably the most spectacular species of the Puerto Rican fauna, I was in a state of sensory overload, overwhelmed by so much biodiversity on both sides of the shoreline: marine and terrestrial, so different but at the same time so close, separated by only a few meters. While exploring the marine world, I was looking into the terrestrial world through a window — and both were equally fascinating.

I recall a similar experience in the Virgin Islands. One night, while skin diving in very shallow water in Cinnamon Bay, I observed such an enormous diversity of marine life that, for a moment, I thought I had lost contact with reality. Diving at night is similar to exploring the rainforest after dark: your senses are exposed to a completely different world! What you see, smell, and touch at night is very different than what you experience if you visit the same area by day. Then, while passing close to shore, I was under the impression that somebody was staring at me. I looked up into the trees along the shore and saw a Yellow-crowned Night Heron (*Nyctanassa violacea*) just a couple meters away. As I observed the curious bird, I could hear several species of coquis calling in the background. Once again I hit that state of sensory overload and realized that the Virgin Island experience was similar to that on Mona. In both places, I was overwhelmed by the incredible biodiversity and its spectacular beauty.

FIGURE 7. Iguana de Mona (*Cyclura stejnegeri*), one of the largest terrestrial reptiles native to Puerto Rico and undoubtedly the most spectacular element of the Isla Mona fauna.

¡Que gente más buena! (What nice people!)

One of Puerto Rico's most obvious and enticing enchantments is its people. The same is true for other Caribbean islands where I have worked (Cuba, Dominican Republic, and San Andrés) and probably for all of the others. My wife, Patricia A. Burrowes, was born in the United States, raised in Cali, Colombia, and has lived in Puerto Rico since 1987. She says that Puerto Ricans are the nicest people in the world. I take this as a compliment, because she has traveled a lot and has worked and lived in many countries. Every time she says that, I smile and I tell her that, in Puerto Rico, as in any other country of the world, there are all kinds of people.

Since I was born and raised in Puerto Rico, I feel a little uncomfortable bragging about how nice Puerto Ricans and other Caribbean islanders are. Instead, I'll tell two stories that make the point.

Several years ago, close to Three Kings Day (6 January), I was working with two other biologists, José Enrique García and Fernando Bird, in the Sierra Pandura in eastern Puerto Rico. We were searching for a cave-dwelling frog locally known as the Coquí Guajón (*Eleutherodactylus cooki*). Because this species occurs exclusively on privately owned land, we had to ask for permission from the land owners to work on their property.

On this occasion, we had seen from a distance a place that seemed ideal for our search. It was covered with *guajonales* (giant boulders that form the caves in which these frogs live). When we reached a nearby house, the owner came out to greet us. I was planning to introduce myself, explain what we were doing, and request permission to work on his land. But before I could properly identify myself, he interrupted me and said with a gentle voice (in Spanish, of course!): "Please come in. My family and I are about to eat dinner and we invite you to eat with us. After you and your friends have eaten we can talk about whatever you want." Because of time constraints, we couldn't accept their invitation, but we were amazed by their hospitality. They obviously were not trying to impress us, since they were modest people. They were just being hospitable — and I've since learned to expect that in folks from that region.

Since 1991, Patricia and I, together with our students, have been working in the Sierra Pandura and the people there have always been extremely friendly. When we explain that we are studying coquis, they are happy to let us work on their farms — and, to my surprise, they don't seem to mind that we are in search of frogs, lizards, and snakes. From 1995 to 1997, Patricia worked intensively in this area, doing field work as often as three or four days a week. During that time, she met many people who were friendly and helpful. After a full day of field work, she would return home to San Juan (a 90-minute drive) covered with mud from head to toes, and tell me stories of people she had met and how helpful they had been. Often she would bring home fruits and vegetables that people had given her. We both regard Don Pablo, Doña Marcela, Don Victor, and many others in the Sierra Pandura region as the nicest people in the world.

In 1989, Patricia and I traveled to the Dominican Republic to do field work. My cousin Consuelo Joglar and her husband, both natives of the Dominican Republic, joined us. We worked the first night near Jarabacoa (an upland area on the edge of the Cordillera Central), but realizing how dry it was (making frogs hard to find), we decided instead to go to the dry part of the island and search for reptiles. When we arrived in Barahona (in the southwestern part of the country), we needed to find a cheap hotel for the night, but had no luck. Barahona was celebrating some kind of festival and we were without a place to sleep. We entered a local drugstore and fell into conversation with the pharmacist, who was fascinated by our research. When we told him that we didn't have a place to stay, he said: "I have a little house outside of town. I will write a note to the keeper so that he will let you in and you can spend the night there. I think you will like the place because it is out in the country and there are plenty of good places to look for frogs, lizards, and snakes." I could not believe what I was hearing! He gave us the note and we left in search of the place. When we found it, we were amazed by the beauty of the site and surprised by the comforts of the house, or better yet, the *hacienda*. The keeper read the note, greeted us, and opened two rooms for our use. That night, we worked until 3:00 AM — and then had a great place to crash for the rest of the night. We will be forever grateful to this stranger who saw our need and did not hesitate to offer us a degree of hospitality unheard of in many parts of the world. To me, these are the nicest people in the world!

Amoná, The Jewel of the Caribbean

At first sight, Amoná (Figure 8) looks pretty much like any other Caribbean island: crystal clear wa-

FIGURE 8. Isla Mona is considered the "Jewel of the Caribbean" because of its remoteness, spectacular beauty, and unique biota.

ters, white sand beaches, attractive (even seductive) vegetation, a unique fauna, and fascinating caves — but, actually, this island offers much more. No hotels of any kind clutter the landscape, so visitors must camp. For many of us, this is by itself an important attraction. Another fascinating feature is that, when you set up camp, you will inevitably get unexpected company: enormous iguanas. These are the Iguanas de Mona (*Cyclura stejnegeri*; Figure 9) — and yes, I am describing Isla Mona (Amoná is the Caribbean Indian name given to the island).

Politically Mona Island is part of Puerto Rico, but it is distinct biogeographically. Sitting in the Mona Passage 67 km west of Puerto Rico and 59 east of the Dominican Republic, it is not part of the Puerto Rico Island Bank (which otherwise includes Puerto Rico proper and nearly all of what are known as the Virgin Islands). Mona is the most isolated of the Puerto Rican islands. Because of this remoteness and its spectacular beauty and unique biota, Mona is considered by many to be the "Galápagos of the Caribbean." Others have called it *La Isla del Tesoro* (Treasure Island). I personally refer to it as *La Joya del Caribe* (The Jewel of the Caribbean).

Mona Island is a 57-km^2 kidney-shaped island. It is smaller than Vieques (138 km^2) and bigger than Culebra (27 km^2). It is 12 km long and 5 km wide and has 32 km of shoreline. Monito Island is 5 km northwest of Mona. This rectangular island is 0.17 km^2 with 0.6 km of shoreline.

Mona is a large block of eroded limestone with many cracks and caves. The bulk of the island is a flat-topped, raised platform or *meseta*. This meseta is surrounded on all sides by vertical cliffs. Along the northern coast the cliff drops vertically from an elevation of 80 m to a depth of 30 m beneath the sea. Along the southern and western coasts, a narrow 3–6 m high coastal shelf widens to a maximum of less than one km at Piedra Carabinero on the southern coast.

A Visit to Mona will Change your Life

Describing Mona in mere words is difficult, so I won't even try. For now, I will merely say three things: (1) Isla Mona is paradise! (2) When God created Eden, He probably thought of Mona, and

FIGURE 9. Juvenile Iguana de Mona (*Cyclura stejnegeri*).

(3) If you go to Mona — it will change your life! Since my first trip in 1974, I have taken many groups of students to the island as parts of courses I teach at the University of Puerto Rico. I have noticed that, after "the Mona experience," most of them become more sensitive to environmental issues and some of them become more devoted to conservation. For a long time, I wondered if a place like Mona has the power to change people — but now, I have no doubts.

Susana Rivera, a student in my "Ecology of Puerto Rico" class, went to Mona on a class field trip. She now works full-time on a successful conservation program dedicated to restoring the San Juan estuary. Changes also occur in people who are no longer students. Abel Vale, a successful businessman in San Juan, claims that his whole life changed when he visited Mona for the first time. He abandoned the business world and now directs a not-for-profit organization called *Ciudadanos del Karso* (Citizens of the Karst Region), which is devoted to conserving the karst region and its biodiversity in northern Puerto Rico. Many people who work or have worked on Mona have commented that, after their first visit, they knew that they had to come back.

My own first trip occurred while I was working on my bachelor's degree in biology. Once I saw the spectacular beauty and rich biodiversity, my life changed forever. I knew that I would be devoted to biology for the rest of my life, and I also knew that, like so many others, that I would go back some day. The first thing I did after finishing my degree was to apply for a job in the Department of Natural Resources of Puerto Rico, mainly so that I could work on Mona. Even though I didn't get that job, I still go to Mona every year, taking students and nurturing the possibility that Mona could change their lives in the same way it changed mine.

An Island Replete with History

Mona is one of those places that, no matter when and where you go, traces of history are just waiting to be discovered. The island was originally "discovered" in September 1493 by Christopher Columbus on his second voyage to the New World. Apparently, the admiral was not feeling very well after his departure from Cabo San Rafael on Hispaniola, and he had to stop on Mona before continuing his journey. While he recovered, his crew gathered provisions from the island, including *casabe* (yucca bread) prepared by local Taino Indians and fresh water. If you explore some of the caves on Mona even today, you will find evidence of its pre-Hispanic inhabitants. Taino (and maybe Pretaino) Indians produced the pictographs in Cueva de Espinar (and several others). These are not only fascinating, but testimony to the fact that these people had something in common with the contributors to this volume, namely an interest in biodiversity. In their pictographs and other artistic or religious expressions, they depicted bats, other mammals (perhaps monkeys), birds, lizards (Figure 10), and frogs. I wonder if this could have been their way of expressing the West Indian experience.

An Island of Endemics

According to Holdridge's ecological life zones, Mona is a dry forest similar to those in southwestern Puerto Rico (from Guayama to Cabo Rojo), Caja de Muerto, Desecheo, Fajardo, Ceiba, Culebra, and parts of Vieques. The dry forests of Puerto Rico are extremely rich in biodiversity and exhibit high levels of endemism. These two characteristics are accentuated on Mona, particularly in regard to the herpetofauna. Most terrestrial reptiles and the only amphibian on the island are endemics, and their common names have the phrase *de Mona* attached — indicating that they are exclusive to this island: (1) Coquí de Mona (*Eleutherodactylus monensis*;

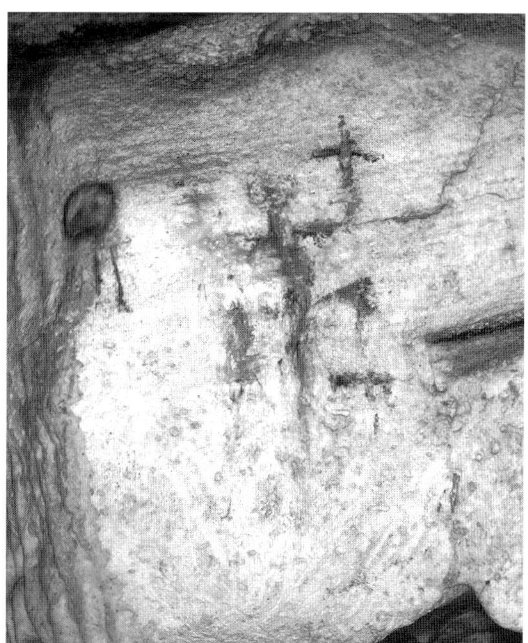

FIGURE 10. Lizard pictograph in Cueva Negra, Mona Island.

FIGURE 11. Coquí de Mona (*Eleutherodactylus monensis*).

Figure 11), (2) Salamanquita de Monito *(Sphaerodactylus micropithecus*), (3) Salamanquita de Mona (*Sphaerodactylus monensis*), (4) Siguana de Mona *(Ameiva alboguttata*), (5) Lagartijo de Mona (*Anolis monensis*), (6) Víbora de Mona (*Typhlops monensis*), (7) Iguana de Mona (*Cyclura stejnegeri*), and (8) the extinct Tortuga Terrestre de Mona (*Geochelone* [*Monachelys*] *monensis*). In addition, two endemic subspecies, the Culebrón de Mona (*Epicrates monensis monensis*) and the Culebra Corredora (*Alsophis portoricensis variegatus*) occur. Non-endemic reptiles are the Salamanca (*Hemidactylus mabouia*, of African or South American origin), the Lucía or Santa Lucía (in the *Mabuya sloanii* species complex, native to the West Indies), and the marine turtles: the Carey (*Eretmochelys imbricata*), the Tinglado or Tinglar (*Dermochelys coriacea*), and the Peje Blanco (*Chelonia mydas*). Overall, 15 species of amphibians and reptiles occur on Mona, and >80% of terrestrial species are endemic. So, now you know the main reason why I call Mona the "Jewel of the Caribbean."

Are We Losing the Enchantment?

Only by finding the balance between development and conservation can we guarantee that our enchantments will last forever. This is undoubtedly the greatest challenge facing the people and government of Puerto Rico — and all of the other Caribbean Islands.

Unfortunately, Puerto Rico, like all of the other islands in the Caribbean, has already lost some of its enchantment. In the last 25 years, three endemic species of Puerto Rican frogs have probably gone extinct: Coquí Palmeado (*Eleutherodactylus karlschmidti*), in 1976; Coquí Dorado (*E. jasperi*), in 1981; and Coquí de Eneida (*E. eneidae*), in 1990. The other anuran species are not doing very well either. A little more than two-thirds (77%) of Puerto Rican frogs are in trouble: 16.6% are probably extinct; 16.6% are confronting serious survival problems and are endangered; 16.6% are threatened; and 28% currently are declining.

The principal problem is habitat destruction, most of it related to development for tourism or for housing a rapidly growing population — but other issues may prove equally catastrophic. Patricia and I have been studying long-term climatic patterns. Our research identifies a warming trend from 1970 to 2001, particularly during the months when coquis are most active. Warmer and presumably drier conditions could easily inhibit reproduction in terrestrial nesters.

In collaboration with David E. Green, a veterinary pathologist at the United States Geological Survey, we have documented the presence of the lethal chytrid fungus, *Batrachochytrium dendrobatidis*, in skin infections of at least two species of coquis as early as 1976. We also have studied the effects of introduced species of frogs, not only as possible predators and competitors for our native frogs, but also as vectors of deadly pathogens and disease. The implications are obvious.

A Positive Note

To finish this essay on a more optimistic note, I am glad to report that, in 1987, Patricia and I founded Proyecto Coquí, a not-for-profit organization with the mission of conserving Puerto Rico's biodiversity, with an emphasis on its amphibians. To guarantee that Puerto Rico doesn't lose its enchantments, we have directed our efforts to scientific research, emphasizing field work, habitat protection with a focus on land acquisition, and education oriented to all segments of the general public and government. In 2001, Proyecto Coquí presented a proposal to create two wildlife reserves for the conservation of coquis in the sierras Pandura and Cayey. We dream and hope that we will be able to create the first wildlife reserves in the Caribbean (and in the entire Western Hemisphere) that are designated especially for the conservation of frogs.

Dedication

This essay is dedicated to two native Caribbeans, my parents, Doña Olga Jusino de Joglar and Don Rafael Joglar Ardura, who, by their example, taught me two important lessons in life: hard work and honesty.

Acknowledgments

Over the years, my work in the Caribbean has been supported by the University of Puerto Rico–Río Piedras (Department of Biology, School of Natural Sciences, School of Graduate Studies and Research, Resource Center for Science and Engineering, AMP Program, MRCE/CREST Program, Howard Hughs Program, and others), United States Fish and Wildlife Service, Toyota Foundation of Puerto Rico, Department of Natural and Environmental Resources of Puerto Rico, Comisión de Donativos Legislativos, and Proyecto Coquí.

Revisions to earlier versions of this manuscript by Patricia A. Burrowes and John Simmons resulted in significant improvement. I am also grateful to Robert Henderson and Robert Powell (*los dos Robertos*) for inviting me to submit an essay, and thereby providing me an opportunity to express my views. Patricia A. Burrowes and Gabriela Joglar provided constant support and advice, making this adventure and many others not only possible, but entertaining. May God grant us many more adventures together!

Suggested Reading

Cardona Bonet, W. A. 1985. *Islotes de Borinquén: Notas para Su Historia*. San Juan, Puerto Rico.

Dávila, O. 1991. *Arte Rupestre Indígena de la Isla de Mona. Revista Naturaleza. Edición Mona*. Sociedad de Historia Natural de Puerto Rico, San Juan.

Duellman, W. E. (ed.). 1999. *Patterns of Distributions of Amphibians. A Global Perspective*. Johns Hopkins Univ. Press, Baltimore, Maryland.

Duellman, W. E. and L. Trueb. 1994. *Biology of Amphibians*. Johns Hopkins Univ. Press, Baltimore, Maryland.

Frank, E. F., C. Wicks, J. Mylroie, J. Troester, E. C. Alexander, Jr., and J. Carew. 1998. Geology of Isla de Mona, Puerto Rico. *Jour. Cave Karst Stud*. 60:69–71.

Hedges, S. B. 1999. Distribution patterns in the West Indies, p. 211–254. *In* W. E. Duellman (ed.), *Patterns of Distributions of Amphibians. A Global Perspective*. Johns Hopkins Univ. Press, Baltimore, Maryland.

Joglar, R. L. 1998. *Los Coquíes de Puerto Rico: Su historia natural y conservación*. Edit. Univ. Puerto Rico. San Juan.

Joglar, R. L. 1999. *Que Cante El Coqui. Ensayos, Cartas y Otros Documentos sobre la Protección de la Biodiversidad en Puerto Rico (1987–1999)*. Proyecto Coquí, San Juan, Puerto Rico.

Joglar, R. L. and P. Burrowes. 1996. Declining amphibian populations in Puerto Rico, p. 371–380. *In* R. Powell and R. W. Henderson (eds.), *Contributions to West Indian Herpetology: A Tribute to Albert Schwartz*. Society for the Study of Amphibians and Reptiles, Contributions to Herpetology, vol. 12. Ithaca, New York.

Reagan, D. P., G. R. Camilo, and R. B. Waide. 1996. The community food web: Major properties and patterns of organization, p. 461–510. *In* P. Reagan and R. B. Waide (eds.), *The Food Web of a Tropical Rain Forest*. Univ. Chicago Press, Chicago, Illinois.

Rios, N. and R. L. Joglar. 1999. *Rana grylio* (Pig Frog) in Puerto Rico. *Herpetol. Rev*. 30:231–232.

Rivero, J. A. 1998. *Los Anfibios y Reptiles de Puerto Rico*. Edit. Univ. Puerto Rico, San Juan, Puerto Rico.

Rivero, J. A. y D. Seguí Crespo. 1992. *Anfibios y Reptiles en Nuestro Folklore*. Imprenta San Rafael, Quebradillas, Puerto Rico.

Stewart, M. M. and L. L. Woolbright. 1996. Amphibians, p. 273–320. *In* P. Reagan and R. B. Waide (eds.), *The Food Web of a Tropical Rain Forest*. Univ. Chicago Press, Chicago, Illinois.

Townsend, D. S. and M. M. Stewart. 1986. Courtship and mating behavior of a Puerto Rican frog, *Eleutherodactylus coqui*. *Herpetologica* 42:165–170.

Joglar, R. L. 2003. An island enchanted by frogs, p. 203–211. *In* R. W. Henderson and R. Powell (eds.), Islands and the Sea: Essays on Herpetological Exploration in the West Indies. Society for the Study of Amphibians and Reptiles, Ithaca (New York). Contributions to Herpetology, volume 20.

Lesser Antilles

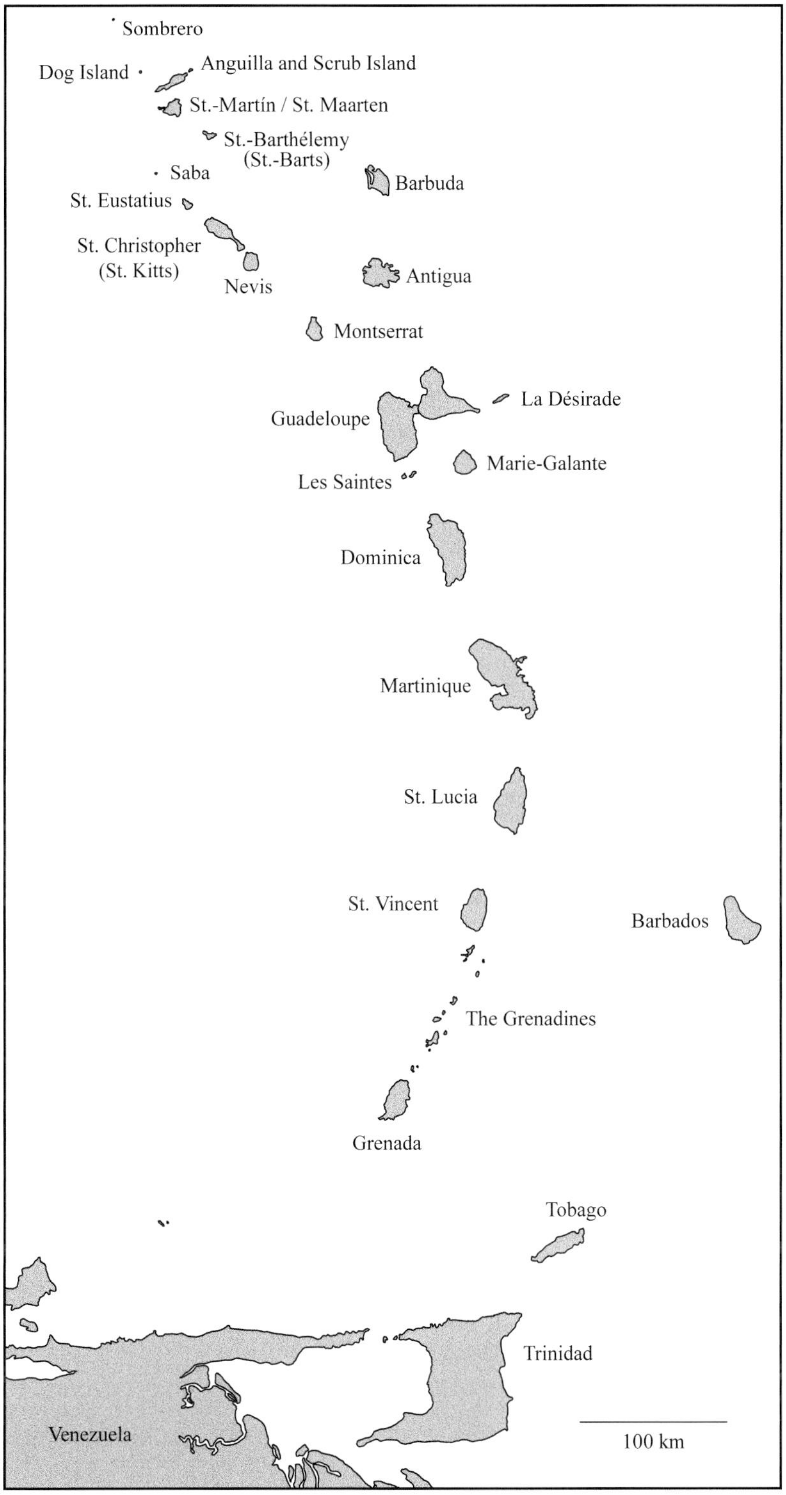

Introduction | Bahamas | Cuba | Jamaica | Hispaniola | Puerto Rico | **Lesser Antilles** | Indices

Looking for Lizards in All the Right Places

James Lazell[1]

On the day I graduated from high school, in June 1957, a group of five was assembled on the platform at North Philadelphia Station, Pennsylvania. My parents were there to see their only child, Skipper, and Mr. C. Reed Cary, distinguished elder of the Coulter Street Meeting of the Religious Society of Friends, was there to see his wife, Margaret, off on the train to New York City. We would be met by a hired car and taken to Idlewild Airport (none of us had even heard of John F. Kennedy) to board a flight to Jamaica and meet C. Bernard Lewis, Tom Farr, and George Proctor. Mrs. Cary and I were engaged in an animated discussion on the relative merits of the Port Henderson Hills versus the Cockpit Country as our priority collecting area.

Margaret Cary was a retired professor from Bryn Mawr College and a renowned entomologist specializing in sphinx moths. I grew up in Philadelphia and attended Germantown Friends School and, occasionally, the Coulter Street Meeting, where I met Mrs. Cary, who was a true mentor.

Alfred Wegener and G. G. Simpson

I was born, however, on Manhattan. Like many children of my generation, I was a contraceptive failure case. My parents never intended to have children; they had lived in New York City since the mid-twenties, right through the Great Depression. They loved New York. They lived in a walk-down apartment on Jane Street in Greenwich Village. Their friends were writers, artists, actors, socialists, and communists (who could discourse for hours on the difference). One of them, Christopher Coates, Curator of the New York Aquarium, then out on Coney Island, and his colleague, Jim Oliver, would be mentors, too.

My mother was what today would be called a social worker, employed by the YWCA in Hell's Kitchen. When I was inevitable, she made her first complete career change and began becoming an expert in early childhood education. My father, a United States Naval Academy graduate who had resigned from the Navy at its request in 1924 because "there would never be another war," worked for the Colgate Palmolive Peet Company as a packaging expert (you know, how to get the toothpaste into the tube). They reluctantly decided that New York City in World War II was no place to raise a child. Eventually my father took a job in packaging with a major grocery chain based in Philadelphia that was transitioning to self-service, and we moved to suburban Germantown. My mother's research had led her to believe that Germantown Friends School (GFS) offered the very best education available, if she could just get me properly prepared and into it.

I bother you with this history because it mattered; she was right. GFS had two excellent natural science teachers: Joseph Cadbury in the Lower School and Howard Platt in the Upper. I could skin and stuff a bird before I was out of first grade and, by eighth grade, I not only knew who Alfred Wegener was and what he had theorized, but why not to be deceived by the distribution of marsupials (there aren't any in Africa and apparently never were, but the Holarctic is bedded with their fossils). I knew about G. G. Simpson and southward displacement, too.

A Brief Acting Career

My mother always thought of me as an experimental organism: something on which to test her theories of education. My father never thought of me as more than expendable. "You don't have to buy him any presents," I once overheard my father say. "Just buy him a bus ticket to Fort Monmouth [over in the New

[1] The Conservation Agency, 6 Swinburne Street, Jamestown, Rhode Island 02835, USA.

Jersey Pine Barrens]; he'll go catch his own." When I was 13, I caught a Silver-haired Bat, a real rarity in Philadelphia. Joe Cadbury told Fred Ulmer, then Curator of Mammals at the Philadelphia Zoo, about it, and Mr. Ulmer came right over to photograph and measure it (it is in the MCZ [Museum of Comparative Zoology, Harvard University]). When the Zoo decided to go with a television program to run competition with Marlin Perkins's very popular "Zoo Parade," Mr. Ulmer thought of me as one of the youngsters who would ask the questions of the curators and other experts — and so I met Roger Conant. It was he who agreed to purchase crocodiles, boas, and iguanas for the zoo from the West Indies, if I could get them back alive.

Margaret Cary provided my plane fare and living expenses for the first two weeks in Jamaica, after which she departed and left me and Yale graduate student John Coutsis behind. The deal was that I would work as John's assistant on moth collection, but I could hunt reptiles whenever he did not need me. I would pay my own expenses from my savings (mostly generated by my brief TV acting career of several years before) and hope to catch enough big reptiles to repay myself preparatory to the next summer's travels. The zoo was ready to purchase specimens of large, exhibitable species, but Roger pointed out that "lizards of small size are of great interest to a taxonomist. . ." and "would indubitably be valuable. . . to some of the lads at Michigan" (in litt., 29 June 1957). Roger got me US$25 for my first of three crocodiles and sent the anoles off to Dick Etheridge.

Writing Letters

In those early days, I knew enough to write lengthy field notes on the actual specimens I collected. I was especially aware, through the tutelage of both Roger and Isabelle Hunt Conant, how important colors in life are, and that no preserved specimen can retain them. I did not write much, however, about habitat or behavior; all that would change rapidly. Fortunately, I did write letters, especially letters home. One of my mother's closest friends, the late Ann Matlack Bucher, took a special interest in my career. She edited a selection of my letters and published them. Some reappear here, along with many subsequent ones, and excerpts from my correspondence files. When I started to write of my long history in the West Indies, I realized just how much I have forgotten. Three file drawers of letters and 23 bound volumes of field notes saved me. They have not forgotten. Someday maybe I will write the whole story.

Love, Skip

The following letters that lack a salutation and a sign-off are letters home. They all began "Dear Folks" and ended "Love, Skip." For the rest, I left the identifications intact. I have strayed from chronological order when the story line demanded it.

Back in 1957, Mrs. Cary and I flew from Idlewild to Havana, Cuba, changed planes, and went on to Jamaica.

Kingston, Jamaica — 16 June 1957

I have just returned from two days in the heart of the Blue Mountains and a delightful visit with Dr. Coleman Goin, of the University of Florida, and his family. There he is working on the genetics of the tiny rain-forest frogs whose remarkable variability has puzzled naturalists since they first came to the West Indies. The magnificence of the rain forest stands out in bold contrast to the country of the south coast, which was to me the usual tropical "bush." Mighty tree ferns rise from the spongy forest floor like crowded Corinthian columns; banks of lace ferns overshadow the winding little trail, and high peaks rise up with startling acuteness. To stand on the edge of a mountain and peer down at the hamlet of Clydesdale, from which I had climbed, was truly awe-inspiring. I had certainly come very little distance from the old estate to reach the forest, if one measured it on a map, but it was all straight up!

Birds abound in these mountains: woodpeckers of brilliant coloration chase each other about in the canopy far above my head; somber ground doves and "mountain witches" waddle about on the precipitous ground beneath the ferns. Reptile collecting is absolutely effortless! Giant anoles peer down on me from their high perches with a look of callous indifference — that is, until a noose is slipped over their heads. (The noose is made of the fibrous center vein of the coconut palm leaf.) Under the stones are Dr. Goin's prey, lovely little russet and black frogs that whistle audibly even through the layers of rock and humus.

By contrast, the south coastal area is one of dense, nearly impenetrable thorn and cactus. It does not look much like what we think of as cactus country, for it is also jungle. The cactus tower overhead and the thorn trees are lashed soundly together with vines and strangler figs. The wildlife is rather different, too: anoles are still plentiful, though more of the small turquoise blue fellows with rich burnt umber throat fans are in evidence. Large saucy ground lizards provide the real sport, sitting just long enough to let a noose touch their noses before they streak off into the tangle and are lost.

Kingston, Jamaica — 26 June 1957
Mrs. Cary was a bit taken aback by my having a crate built yesterday for the crocodile I intended to catch today, but she now feels reassured since the little fellow is reposing in her bathtub until I can send him off to the Philadelphia Zoo. Catching a small croc in a large mud hole was tough enough, but was seriously complicated by the presence of a number of large ones. Personally I think the large crocs would be much easier prey, because they are more belligerent than the little ones, thus offensive rather than defensive, but John was not of this opinion. I'm afraid the whole experience has only reconfirmed his resolution to stick to moths and leave the reptiles to me. Moth collecting has been a bit disappointing, largely, I think, due to the weather. Nevertheless, Mrs. Cary and John have collected three exceptionally good species. One cannot expect the best results at the moth lights when twenty inches of rain falls in twenty-four hours! This happened last week at Port Antonio.

FIGURE 1. *Epicrates striatus* from near Savanette, Haiti. This species was welcomed at the Philadelphia Zoo.

John Coutsis was well acquainted in Haiti. His mother had served there with the United Nations. It was he who introduced a generation of researchers to the Pension Tourdot at Pont Morin. That would become focal for many of us, including Yale linguist Sid Mintz, and — soon — Ernest Williams.

Port-au-Prince, Haiti — 23 July 1957
Haiti is a truly fantastic country! John's old friend, Serge Brierre, whom I met in Jamaica, met us upon arrival and we were off immediately to do up the town. We got to a wedding and several nightclubs before time made it necessary to get back to the business of field biology.

Port-au-Prince at first impresses one with its sad combination of abject poverty and great wealth, but as one lives here day by day, one cannot help feeling the depth of spirit that is evident in the most humble man. There is an intangible atmosphere of freedom, and a Haitian seems able to survive merely on his pride of being a Haitian. It is wonderful to know and live with such people.

Cap Haitien, Haiti — 7 August 1957
This little city lacks the hustle and clamor of Port-au-Prince, the capital. It seems generally to be much more as all of Haiti must have been a half century ago. It is very late at night, and it has been a long day in the bush. I have enough of the common boa, *couleuvre endormie*, to keep the zoo stocked for many years, and am now trying to find a large pink snake called *couleuvre chovel*, the horse snake, which I hope will turn out to be one of the other species of Hispaniolan boas. So far, though, I have gotten no closer to one than to a fellow whose grandfather saw one once.

In Haiti the nights are noisy with the distant and indistinct pulse of drums. Not even dogs howl in the late evening in the Cap! The reverberations are picked up from one hill to the next, accelerated, retarded and superimposed on each other. Yet the basic pattern continues: the drumming at first seems to blend in with the pulse of one's body, then gradually takes over until it is impossible to escape its domination. The moon is dark, and only the light of stars silvers the hilltop above the Cap. This is the Haitian night at its best.

I caught some dandy boas (*Epicrates striatus*; Figure 1) in Haiti with the assistance of Serge Brierre and his family (Figure 2). I realized, however, that these would not be sufficient as zoo exhibits to fund me for long. In the mountains of Jamaica I had met a British veterinarian, Dr. Chappell, who told me there were enormous boas and gaudy iguanas in profusion on the remote island of Dominica. That became my destination for 1958. Dr. Chappell also told me that Barbuda was an utterly unknown and unexplored island. Mrs. Cary was intrigued and agreed to fund me to go there in search of sphinx moths. In the spring of 1958 (the dismal school year behind me), I first set off for the Lesser Antilles.

FIGURE 2. The Brierre family was crucial to my early successes in Haiti. I am second from the left. Flanking me are Max and Maxeau and to the far right is Serge Brierre (photograph by John Coutsis).

St. Johns, Antigua — 6 June 1958
On my way here from Porto Rico I met a man who has an estate on the northern end of the island of Dominica. He says that reptiles abound there, possibly because there is no mongoose, and I know that a number of little-known forms are found there. Consequently being completely foot-loose this year, I think I shall go on down with him. We leave on the steamer *Oluf Sven* tomorrow. I will be able to work on Great Bird and Long Islands, off the Antiguan coast, when I return from Dominica, since I must come back anyway in order to go on to Barbuda to get Mrs. Cary's moths.

The man I had met was George Bruney, who ran his family's estate at Moore Park on Morne Au Diable. The *Oluf Sven* deposited George at Portsmouth on the northern leeward coast of Dominica, close to home. I went on to the capital, Roseau, to officially enter the island. In those days an American could travel all over the Antilles on a driver's license or a birth certificate, but I had official letters of introduction from the Philadelphia Zoo, and I wanted to be known to local officials. The *Oluf Sven* off-loaded several passengers at Roseau and turned back north. That night it sank, short of Portsmouth, along what was then a roadless and virtually uninhabited coast. At least people said it did; but they said no one survived, so who was to know? I am glad I missed the trip. I did not inform my folks.

Roseau, Dominica — 11 June 1958
I have re-found Dr. Chappell, the veterinarian I met in Jamaica last year, and he is assisting me in getting around the country to locate my specimens. This is certainly the most beautiful island in the West Indies; it rises to a height of nearly five thousand feet, and is 40% rain forest. The population is small, and there is little of the ruthless sort of cultivation that has harmed so many of these islands. As for the water that you inquired about — it's magnificent! — whether you drink it, swim in it, or just watch it fall several hundred feet down a mountainside.

Roseau, Dominica — 17 June 1958
Since I wrote last, things have been happening fast. The boa here is a big fellow more closely related to the common South American *Boa constrictor* than to anything else. However, it is very different in body proportions and in coloration and pattern. This snake is indeed common — I have a whole closet full of them! Some are quite hefty fellows, nearly three times the weight of a Haitian Boa of the same length. They are fairly docile for large snakes, but their name *tet' chien* implies a certain quality of temperament that *couleuvre endormie* lacks. Besides reptiles, I have collected a great many friends, all of whom are eager to have me come and stay on their estates or in their homes, and collect all the snakes I want. I have not yet seen the iguana or a *menicou*, the opossum that Mr. Ulmer [Philadelphia Zoo] wants. But I am assured they are here, and in time I shall get plenty. So all goes very well.

High on that list of friends was George, expecting me to appear any day up at Moore Park. I took the regular coastal ferry up to Portsmouth and got a truck ride to (the rather vertical and long) driveway up to George's house. I had been collecting *Anolis* in the course of my wanderings. Those from up in the mountains east of Roseau — in good *tet' chien* country — were green and dark-fanned: very different from the brown, yellow-fanned Roseau lowlanders. Those at Portsmouth were utterly different from either; they had a bold, giraffe-like pattern, but in three colors. At Moore Park, anoles were most like the mountain form, but rather intermediate between that and the north coast type. I knew I was onto something.

However, I needed those gaudy iguanas. George said they hardly ever appeared very far from the coast, so he recommended I just hike down to the windward side, downhill all the way. Next morning, I did. This would be one of the most important days of my life, but apart from knowing it was in the latter half of June 1958, I did not record the date.

In those days, most people in the countryside of Dominica spoke Creole rather than English (this had largely changed by 1998). I had picked up a bit of Haitian Creole and was making the transition to the Lesser Antillean form.

"*Tini lezard ici?*," I asked as I walked south. "*Oui, tini ca, mais pas en grand pile*," yes, we have iguanas, but not in great heaps, was the standard reply. "*Si ou vli trouvee grand pile lezard la, allee a* Woodford Hill." If you want a great heap of iguanas, go to Woodford Hill. They pointed down the road, south, the way I was going. Frequently I would detour with someone who thought he might find an iguana, and, of course, I turned over rocks, logs, and junk as I went along. Still, the advice after failure was always the same: go to Woodford Hill, on down the road (20 miles from Moore Park, over headlands and down into beautiful little turquoise bays; up, up again over another ridge. . .).

In mid afternoon the very first motor vehicle of the day came along, occupied by a white couple. Yes, they were going to Portsmouth, back the way I had come; yes they would give me a ride, but, well, er . . . I would have to ride in the boot; I was that dirty and sweaty. Thanks anyway, I said, I'll go along to Woodford Hill. It was nearly sunset when I turned off onto the graveled driveway, bordered by Royal Palms, that I had been told to recognize. I stumbled along, dehydrated and famished, past rows of lush bananas, eventually to a large bungalow set in a beautifully tended garden. I had reached the estate house, Woodford Hill.

On the veranda sat a strikingly handsome ebony man, sipping a frosty drink. Charles Winston was manager of Woodford Hill Estate (which was owned by a British Corporation), formerly of the Imperial College of Tropical Agriculture, and O.B.E. (Order of the British Empire) — I did not find all this out that evening, only later. Next to him sat a beautiful *cafe-au-lait* woman, his wife, Mena, who certainly deserved accolades, even if not formally bestowed. The Winstons fixed me up with a rum punch and listened to my story.

"Yes," Charlie said, "plenty of iguanas here; we'll work on that in the morning." He had a truck going to Portsmouth next morning. He would have the driver stop in at Moore Park and tell George I was alive and well (not many people in Dominica had telephones in those days; George was not among the few). Meantime, my condition called for a shower, clean clothes, dinner, and a bed. If I wanted to get lots of iguanas, all the various lizards, boas, and specimens of the several kinds of small snakes, I had better plan on staying a few days, at least.

The next morning I was awakened by Charlie. He already had the first iguana — beautiful! It was apple-green suffused with sky-blue and flushed with pink on the head. It turned out to be the very first *Iguana delicatissima* scientifically recorded from Dominica, but it was only the first of many. We went snake hunting at the dump that afternoon — with a bulldozer. Wow! I never had it so good! The Winstons took me in as the prodigal son; I was eight years older than Josette, their oldest. They received the BBC, but had no other regular contact with the world beyond shipping schedules and banana prices. Mena and the kids were fascinated by my tales of life on other islands and back in the United States. Charlie took an immediate interest in natural science. He knew all about the geography and physiography of rainfall. He provided the data I needed to begin my understanding of the *Anolis* variation I was seeing — for here, on the windward coast, was yet a fourth sort of *z'andolie*.

Looking back on those days in their historical context is fascinating. This was just four years after the United States Supreme Court had ruled racial segregation in the public schools illegal (1954), and six more years would pass before that court abolished the poll tax and thereby initiated some sort of racial equity in American voting. I had been raised in a liberal society and had attended an integrated school — albeit very selectively and carefully integrated (of course, the white kids were ostensibly very carefully selected too, although with me in there some didn't think carefully enough). It was one thing for visiting white boys to make fast friends with the guys like Serge Brierre and George Bruney, but to come to feel part of a family was new and different. It was a wonderful thing at age 18 to discover that some aspects of life really are the way they are supposed to be, and need to be — even if today, in the twenty-first century, they still are not always so.

Eventually, I left Woodford Hill, went back to Moore Park, Portsmouth, and Roseau, packed up my animals, and sailed for Antigua. No air transport was available in or out of Dominica then, so I could not airfreight my animals back to the zoo until I reached Antigua. Then I went to Barbuda, and eventually back to Haiti.

Port-au-Prince, Haiti — 28 July 1958
Barbuda was a bitter disappointment from the standpoint of moths. I am afraid that a lantern just doesn't do for me what the ultra-violet lamp was doing for John. Nevertheless, that island is a fascinating place; low, flat, and sandy like the Bahamas, and with only a few people who survive very well selling lobsters to the Porto Ricans. Their menu consists of deer and turtles.

I have received a letter from Mr. Conant saying that the Dominican animals came through to the zoo very well. The iguanas are a new record for the island, and the boas are probably the first of their species seen alive in the United States. I think I will gamble the money from this catch in Haiti trying to locate the other two species of boas that have been recorded there.

I caught some good stuff, but neither of those elusive boas. I sent what I got back to the zoo and Conant distributed the specimens. I knew the zoo could not continue to purchase Antillean specimens for long; I had about flooded the market. Smaller specimens and an academic or museum connection would enable me to continue, to return to islands where I felt more at home than anywhere else I had ever been.

As it turned out, it was those anoles from Dominica that did it. Conant sent them off to Ernest Williams at the MCZ at Harvard. Williams wrote me in the autumn of 1958, summoning me to a meeting in Washington during Christmas vacation.

"We have known for some time," Williams told me, "that there are two species of *Anolis* on Dominica. You have specimens of four. How many species do you believe actually live there?"

"One," I said, "it's geographic variation."

"I will pay your expenses," said Williams, "if you will go back and prove that."

I did not have to wait long for good news.

MCZ — 19 February 1959
Dear Skip,
I received enough guarantees from the museum to promise funds for the full program we discussed in Washington. Actually I see no point in your making hard and fast commitments on time spent in each place. We might as well be opportunist and say that you keep going while and where the going is good and the money holds. . . .

At all events I think Dominica will repay attention to the smaller beasts — *Anolis* and *Sphaerodactylus* in particular. What I want from you is almost the reverse of what the zoo wanted — not large showy beasts — but mostly large series of small animals — to the extent possible. . . .

But in addition to specimens, I want notes taken at the time on how and where the animals were found — color notes — behavior notes. I offer the possibility of joint publication on the herps of Dominica if you can get me the material for it. . . .
Regards, Ernest Williams

Unfortunately, the trip was not smooth sailing. I had developed *vivax* malaria subsequent to my return to Philadelphia. It was easy enough to cure, once the doctors were sternly told by my father to pay attention to what I told them: "I have malaria, give me chloroquin." "But there hasn't been a case of malaria in Philadelphia in twenty years," they whined. "There has been in the Fer-a-Chevalle Valley of Haiti," I said. That was scary for my folks, but once again it is important to look at the historical context of the times.

Trujillo was the dictator of the Dominican Republic. He was pro-American and encouraged tourism and scientists, but there was plenty of gun play in the hills. "Papa Doc" Duvalier was the first really popularly elected president of Haiti. My friends, like the Brierres, had not voted for him, but things had not yet gone bad for them, although there were ominous signs. Castro was becoming famous and successful in Cuba. Many of us, who knew how bad Cuban dictator Batista was, were pro-Castro, but my parents were not. My parents really did not want me to return to the West Indies.

MCZ — 28 April 1959
Dear Mr. and Mrs. Lazell:
I can only accede to the wishes of you as Skip's parents in regard to my financing part of his West Indies trip. I have already sent him $300 for the Lesser Antillean part of the venture — half the sum I promised him. To this extent then I have taken an action from which it will be somewhat difficult to retreat.

I also feel morally committed to Skip in regard to the remainder of the trip. However, your wishes override that obligation, but the denial to Skip of the second portion of my funds must come as your choice and must be defended to him by your persuasion and your authority. I feel that I would be in a false and untenable position if I personally cancelled Skip's plans for Haiti on the grounds that the country is unsafe. . . . My judgment, which could very well be wrong, is that <u>this</u> summer will most probably be safe though I am not confident about any succeeding summer.

In political matters none of us can be sure that our judgment of the future is a correct one. I vividly recall that during part of the last Oriental crisis the family of one of the Chilean visitors here wanted to send him funds to fly back because the <u>United States</u> might be unsafe.

In the matter of the tropics too my judgment differs from yours. . . . The tropics are in reality very varied. There is a great deal of difference between the (possibly malarial) lowlands and the high mountains or the valley at 3000 feet in the Dominican Republic where Trujillo has his summer home, and where I plan, if things work out, to spend most of my time in Hispaniola.

There are dangers in the West Indies but they are less than on the mainland or the islands of the continental shelf; they are less than on a rattlesnake ledge in Pennsylvania.

Forgive me if I write too much about the West Indies. I shall probably be involved for the rest of my life with them. . . .
Sincerely, Ernest E. Williams

Taunton Lakes, New Jersey — 2 May 1959
Dear Skipper:
It may be that your parents are alarmed by some of the things they have read in the newspapers. Drew Pearson, for example, has had several columns on Cuba, Haiti, and Trujillo, and he has several times mentioned that the route for an invasion from Cuba to Santo Domingo lies through Haiti. . . .

Another point, and this is one that young people are very apt to overlook, is that your parents (like all mothers and dads) must realize that they are soon to lose you. It is natural for them to cling to the feeling that you are still their boy, and perhaps they want to have the pleasure, just once again before you take off for some distant part of the world, of your company for a few relaxed weeks during the summer. This sort of thing is most difficult for impatient young males to understand. . . .

So do be patient with them (at least as much as you can). I think they have been most lenient with you and that you, young man, have had some really marvelous opportunities to see the West Indies. How many other parents of an only child would have permitted their pride and joy to go off on his own at such a tender age?
Sincerely, Roger Conant

MCZ — 8 May 1959
Dear Skip:
I've written to your parents explaining as carefully as I knew how my own attitude toward the West Indies and the Tropics. I have also promised to bear part of the cost of travel and accident insurance for you.

One thing only worries me. I don't want you to stretch your money beyond the bounds of reason. There is no percentage in taking undue chances. You may after all want to go again, and you will be kinder to yourself (and me) if you make quite sure that nothing goes wrong. . . .

Apart from this, remember that I want information as much as specimens. A specimen is just one kind of information and lacks information on color in life, on just exactly where, under what conditions and in what association it lives. A little of this extra information can be enormously useful.

Specifically: On Dominica I want representative collections from several localities — as different as may be, with you the judge of that — and with notes on where the animal was taken (on the ground, on a tree, how high on the tree) and what it was doing. I doubt that we have many new species to find in Dominica, though that remains to be determined, but we do need to know much more about the animals as animals.

Outside Dominica I have nothing positive to suggest at all. Your itinerary may be mostly already determined by your tickets and traveling arrangements. The one area in the Lesser Antilles that we know least about is the Grande Terre section of Guadeloupe. Next in order of our ignorance are Les Saintes, St. Barts, and the Grenadines.
Good luck, Ernest Williams

In the end, I got to go. My father argued that for me to keep doing what I was good at was my only hope of making a living.

Maracas Valley, Trinidad — 7 June 1959
Trinidad is a fascinating island; it is completely South American in both plant and animal life, which means that while there are many more species of animals available, only a few of them are little known enough to be valuable. There are over thirty species of snakes ranging from the tiny blind burrowing *Typhlops* to *Eunectes*, the tremendous Anaconda. Variety is added too by the poisonous forms, both the Bushmaster and the *Fer-de-lance*, to say nothing of three species of coral snakes, one of which reaches six feet in length.

St. Georges, Grenada — 15 June 1959
In Trinidad I managed to get my *Fer-de-lance* for the zoo. I hope they will have better luck with this one than zoos have had in the past. Apparently this beast is much more delicate in requirements for survival than his disposition would indicate! It is a nice sized specimen, and the species is one of larger New World vipers — really quite a hand-ful. Before sending him off, I had to de-tick him so that the virus laboratory could study these small parasites.

In Tobago I got some lovely tree boas and a Rainbow Boa for the zoo. I broke into the film industry when Arthur Greenhall, the excellent herpetologist in Trinidad, asked me to help him in collecting some big *Constrictor* for Walt Disney to use in his forthcoming production of *Swiss Family Robinson*. Of course Mr. Disney wasn't there himself, but his art director was working up the collection for the filming, so perhaps someday some of my beasts will appear on the silver screen! Here in Grenada I have managed to get Mrs. Cary a rather nice collection of sphingids. I think she will be delighted.

I also discovered Dr. John Groome, down at Point Saline, Grenada. Groome was a medical doctor who had come out from England because his wife had a lung condition that needed a dry, tropical climate. Point Saline was perfect, and she did well. Groome took an immediate interest in local politics and natural history. He managed to get a law passed in Grenada protecting all snakes on the island, ostensibly because they are good ratters. So far as I know, this was the very first snake protection statute ever passed anywhere on earth, beating one in Arizona by several years. In 1959, though, I did not tarry; I island-hopped north.

Woodford Hill, Dominica — 29 July 1959
To date, my work for the Museum of Comparative Zoology has been the tedious task of getting to every possible locality and collecting large series of *Anolis*. The results have been very gratifying; the animal does seem to be one species, as I had suspected, but they break nicely into four distinct subspecies, each with its own geographic range and climatic preference. Reports from the museum say they are happy with my collections from the islands to the south, and especially with a series of rare and little-known tiny lizards that I found on Morne Tecuche in Trinidad.

Those tiny lizards were *Proctoporus shrevei*, named for MCZ's Benjamin Shreve and reported by naturalist-writer (novelist?) Ivan T. Sanderson to glow in the dark. None of mine did that!

A hand-written letter from Ernest Williams caught up with me.

Somewhere in Puerto Rico — 26 July 1959
Dear Skip:
A Dr. Seronde visited Dominica with a parrot-hunting expedition and turned up a remarkable new ground anole from near "The Lake" [Fresh Water Lake]. It has a rather short head, keeled ventral scales, and a double row of enlarged scales down the middle of the back. It is related to the ground anole of Antigua [presumably *Anolis wattsi*] and of St. Kitts and Nevis, but it is more different from these than they are from each other. . . .

What you are doing with the *Anolis* sounds perfect, but keep your eye out for any small ground herps even if they seem just young of the usual forms: first, we need growth series; second, I feel that passing up these is the easiest way to miss something. . . .

I leave it to your discretion whether to stay on Dominica or go to Guadeloupe. The latter I know to be a still more complicated island than Dominica. Your choice. . . .
Good luck, Ernest Williams

My response was immediate because the characters noted on this anole fit perfectly the montane form of *Anolis oculatus* I was at that time documenting. This form is often rather terrestrial in open country, like that around Fresh Water Lake. I knew where Williams was scheduled to be next and wrote him. He responded:

Port-au-Prince, Haiti — 8 August 1959
I am at the Pension Tourdot and it and the country quite live up to your recommendations. . . .

Do not disturb yourself about the mountain anole. I do not describe animals from one specimen unless I have to, and it would be foolish indeed with you actually in Dominica. Still I am doubtful that the mountain form intergrades with the common lowland type. On scale characters it appears to belong to a different group, and while surprises are always in order, this one would be a major surprise.

As regards frogs, not all *Eleutherodactylus* call. . . . There is . . . supposed to be one silent *Eleutherodactylus* in Puerto Rico. I am here telling you what is received opinion — standard doctrine — at the moment. There are people who doubt this. This much I can tell you with confidence: there are at the moment more frogs than voices in Puerto Rico; thus at least some voices are being confused with others or some frogs call rarely or not at all. . . .

Do not forget *Sphaerodactylus*. As regards these I wonder how much season has to do with finding them. In both Haiti and Puerto Rico I know there are plenty of sphaeros, but this trip I've seen very few in P.R. and none in Haiti.

I'd like you to spend a few days on Guadeloupe on your way back. Just a little information on this island would be so welcome.
Good luck, Ernest Williams

I was never able to make sense out of the amazingly variable Dominican eleuths and hope Hinrich Kaiser has got them right. Back then I also struck out on sphaeros, never found a single one. When my wife, entomologist Dr. Wenhua Lu, and I went back in January 1998 (almost 40 years later!) and traveled the island with Stephen Durand, the excellent naturalist with Forestry and Wildlife, we got lots of *Sphaerodactylus* of both species. We also got lots of *Gymnophthalmus*. They were abundant, conspicuous, and widespread. I never saw one on Dominica or adjacent islands in the fifties or sixties. The *Anolis*, however, were exactly as I described them all those years ago, and we saw lots of *Iguana delicatissima*.

The Winstons had moved to Castle Comfort, on the coast south of Roseau. They were fine, and delighted to meet my bride, having despaired that I would ever "settle down." Marriage — to a fellow field biologist — has not solved that problem yet.

Anyway, back to yesteryear (1959) and time to make plans for the next summer.

MCZ — 24 November 1959
Dear Skip:
Guadeloupe and all its satellite islands would, I think, be too big a job for one summer. The major island seems to me more complicated than Dominica, which is certainly complicated enough. . . .

Grande Terre is really unknown, and I don't know why. It is lowland as compared with mountainous Basse Terre; perhaps the mountains have been more attractive to collectors. But there must be anoles on Grand Terre somewhere.

I definitely won't be here December 29th but Benjamin Shreve will be here. He is my assistant, an older man but a pleasant if peculiar personality. He has his own ideas on Guadeloupe and he will not neglect to tell you about them at length. . . .
Best regards, Ernest Williams

Ben Shreve and I became good friends. He was an extraordinarily successful taxonomist, meticulous and accurate beyond any other I have known. I have not seen his story written before, and will here briefly recount it as he told it to me.

Ben was the only son of a very wealthy Boston family of jewelers. Because he was an avid bird-watcher throughout his youth, his parents decided he must become a scientist and sent him to the Massachusetts Institute of Technology (MIT). He hated it; the science offered at MIT did not interest him and he flunked out after his freshman year. So, his parents sent him to Harvard. Things were a little better there. He discovered the MCZ and its bird department. Alas, however, that could not provide all that was needed for the baccalaureate. He hung on for a year or so, then flunked out.

His parents were incensed. No more coddling, he would have to get a job. He went back to the bird department at MCZ and virtually begged for employment. Without a degree, that was not to be. In Harvard's arcane world of ornithology, he was unemployable.

Disconsolate, distraught, and heart-broken, Ben pushed the wrong button in the elevator. He missed the ground floor and plunged to the basement. Symbolic indeed, he thought. Well, he would walk out the back door. As he passed the Herpetology Department, its door ajar, a voice boomed out of the murky depths.

"Boy! Come in here. I need some help." A huge man hunkered over a dissecting microscope, a dead snake trailed across its stage.

"I've counted the ventral scales on this damn snake six times and only got the same number twice. Can you count? Count those for me."

Ben counted, breathed deeply, and issued his verdict.

"Hah! That's the same number I got twice," bellowed the great man. "I believe you're right." The man was Thomas Barbour. Ben Shreve got a job.

I never asked Ben about his personal finances, but the story around the MCZ was that he worked for a dollar a year. Obviously, his parents were concerned only with seeing him employed; salary was not an issue. Ben never lost his love of birds. He worked in the Herpetology Department from nine to five, punctually. However, he would arrive early in good weather, and take time then and during his lunch and afternoon coffee break, to scan the big trees in the quadrangle for warblers, orioles, and tanagers.

In December 1959, Ben and I conferred on Guadeloupéen anoles. He had an uncanny ability to spot good scale characters. I was primed for Guadeloupe, but it would be postponed. Prescient, Serge Brierre wrote to me.

Cap Haitien, Haiti — Christmas 1959
How are you doing with your studies now? Is your English course getting any easier now, or do the snakes prevent you thinking hard enough to realize how easy it is? Should I still think of you as a hunter or have you changed any at all?
We do hope to see you soon in Haiti. . . .
Most sincerely, Serge

Over the academic year I worked my way through English courses, and plenty of biology too. I frequently said I learned far more about *everything* in the three months of summer than in the nine months of the academic year — *except* for biology. I learned more biology in lab and lectures, but it was a close race. In the meantime, we developed a plan.

MCZ — 8 June 1960
Dear Skip:
From what Dr. Williams tells me, the funds for the Haitian trip will be forthcoming. We had better make at least some tentative plans at this time.

The purpose of the trip as outlined in the grant application, is to visit the mountain hardwood forest in the north of Haiti. This area has not been collected and there is reason to suspect it has a rather distinctive fauna. . . . I am particularly interested in collecting ecological and behavioral data on these animals as well as in collecting specimens.

If this does not pan out, a trip to the southwest peninsula looks like the next best bet.

We certainly will need to rent a car, preferably a jeep or a Land Rover. Can you drive. . .?

In my previous trips I have lived off the country for housing. Do you think this will be possible this trip?

I plan to arrive in Port-au-Prince July 11 or 12. Could you get there a week or so earlier and make the necessary arrangements about car, papers, etc. and case the situation?
Sincerely, Stanley Rand

Yes, yes, and yes again! I was off again.

Port-au-Prince, Haiti — 8 July 1960
During the first week here I am supposed to be getting arrangements all set up so that when Stanley Rand, from the museum, arrives, we can start immediately for the trackless northwest. This will be a long tough trip, and requires quantities of paper and good friends. The former I am getting and the latter are fortunately always on hand in this country. In addition, I have been collecting iguanas for the zoo, and at long last I have collected one of the

FIGURE 3. Traveling through Haiti in the late 1950s was always an adventure. This breakdown occurred on the road to Savanette near Colombier.

FIGURE 4. Much of Haiti was still lush in 1960. Stan Rand at Morne Cabrit near Port-au-Prince.

other two species of boas that eluded me so long. This particular snake turns out to be very common, right under my unsuspecting nose: while I had been busily scaling mountain tops in search of it, it was literally abounding all over the low, dry country around Port-au-Prince! Such is the life of a herpetologist.

Port-au-Prince, Haiti — 25 July 1960
Well, we are back from the Northwest. It was indeed an expedition! With Père La Roche on one end in Bombardopolis, and Jean-Baptiste, the agronomist of Port de Paix, at the other, we were in good hands. These two men are the sole source of supplies and transport in this part of the country. There is only one track, which is only possible for a jeep in some places [Figure 3], and so most of our traveling was on foot, with a couple of burros to carry the gear. The results were fine: among the vast quantities of lizards and snakes is at least one new species. We are off for the southwestern peninsula the day after tomorrow, and we must go by air since all the roads are flooded out.

That trip yielded *Sphaerodactylus shrevei* and my second publication. The next would produce *Anolis koopmani*, which was described by Stan Rand.

Cayes, Haiti — 5 August 1960
I have spent the afternoon sitting on a huge pile of moulding coconut husks and systematically rearranging them. The result is a long series of rare and beautifully colored little geckos, with which Stan is delighted, and a great mass of tiny snails that are twitched out of the fibrous husks with another bit of fiber. The snails are so small that it takes thousands to fill a matchbox, and I was forced either to give up or go blind after the first few hundred. Following a final assault on the Massif de la Hotte tomorrow, we will be finished for this trip, and it looks as if we have a second new species of lizard. Stan amuses himself by poking lizards attached to long poles at other lizards hopping around in the bushes, and then writing down their reactions to each other [Figure 4]. This is how one supplements the actual specimens with data on their habits, and, as Stan put it, "How we make a living." Besides, this is what the museum needs.

In the spring of 1961 I graduated from university. However, to continue on in graduate school, I needed several more courses, especially in chemistry. I returned to Philadelphia and enrolled at the University of Pennsylvania. That pretty much ate my summer, but I did finish up in time to get to Guadeloupe and make extensive collections of *Anolis marmoratus* there. By now I was fully committed to the Lesser Antilles (Figure 5) and a doctoral thesis on the iguanid lizards there. In the summer of 1962, I headed right down the islands. I worked Martinique

FIGURE 5. I quickly became enamored with the Lesser Antilles: Rocher du Diamant, Martinique. This is where I "rediscovered" *Liophis cursor*, which supposedly was extinct.

and St. Lucia, and "rediscovered" the "extinct" vipers on those islands. They were a side issue, but I was fascinated that the myth had been perpetrated that they were introduced, the same as South American *Bothrops atrox*, and now extinct. I also found out how different what was then called *Constrictor orophias*, of St. Lucia, was from the Dominican *tet' chien*. I deviated briefly into the world of snakes (Figure 6) and considered lizards other than anoles and iguanas, too.

MCZ — 14 August 1962
Dear Skip:
On the *Ameiva* from Maria Island, a description of this new species is already almost finished, but I have had only three specimens received via Underwood, two in really vile condition, and two better specimens from Pere Pinchon. I'm glad to hear that you have more and notes on colors in life. Schwartz has collected specimens of this form too and I have been trying to borrow them.

I surely would like to have more of the Guadeloupe *Ameiva* and I am willing to pay for the privilege. Incidentally I found a new melanistic *Ameiva* in the collection of the BM [British Museum]. It is from Little Scrub Island, NE of Anguilla, and labelled "*corvina*" which it clearly is not.

I've always been inclined to regard [*Anolis*] *luciae* as a species. There is a combination of scale characters by which it can always be distinguished. I hope your judgment that it is closest to *richardi* is not based on size only. Your proposition that it must be closer to one of the two species on St. Vincent is not too safe. Logically it might be equidistant from both.

In the matter of intergrades I agree with you that intergrades are intergrades and no time should be wasted assigning them more to one side than to another. I have watched Ben Shreve spend three days making up his mind whether to call a few frogs (which he agreed were intermediate) one thing or another. . . .
Good luck, Ernest

Royal Victoria Museum, Trinidad — 2 November 1962
Dear Skip:
I have done some checking for you on *Bothrops* and can find nothing here which would suggest that *Bothrops* or Fer-de-lance or for that matter any highly venomous snake occurs on Tobago or ever has been recorded from Tobago. Tobagonians make no reference to any poisonous snakes on the island. The snake names on Tobago are different from those used in Trinidad. *Bothrops* and *Lachesis* are as you know called "Mapepire" in Trinidad along with some harmless types which people think are venomous such as *Leptodeira*, *Imantodes*, etc. The name "Mapepire" is unknown in Tobago.

Could the confusion have been caused by workers confusing Tobago with the island of Toboga, off the coast of Panama — which I believe does have *Bothrops* on it? The other thought that occurs to me is that I have always been under the impression that *Bothrops* was introduced into Martinique and St. Lucia. Have you checked back to find out where the original snakes came from? If these introduced species are different in taxonomy, ecology, and behavior from their original stock (presumably from either Central or South America) you would have a rather interesting problem in a rather quick type of evolution. Maybe I am playing this out in left field. Let me know what you and your excellent faculty advisor have to say about this. Say hello to him. . . .
Sincerely, Arthur [Greenhall]

That faculty advisor was Hobart Smith, then (like me) at the University of Illinois. I finished a master's degree there and moved on to Harvard in 1963. I continued my snake projects, and fell under the spell of George Simpson, who had promulgated the first truly sensible species concept that could be used

FIGURE 6. In addition to "rediscovering" the "extinct" vipers on Martinique and St. Lucia, I also "rediscovered" supposedly extinct snakes on Great Bird Island, Antigua, and Rocher du Diamant, Martinique. This is the MCZ paratype of *Alsophis antiguae sajdaki* (photograph by Allen C. Smith).

on disjunct samples, isolated in either space or time — the real, original evolutionary species concept, not to be confused with the irrelevant concept of cladists who have misappropriated Simpson's name. My zeal for snake systematics seemed to get me into a bit of hot water. Williams received a letter he showed to me. I wrote back.

MCZ — 15 November 1964
Dear Dr. Dowling:
Dr. Williams has turned over to me your letter of 9 November. I certainly regret that this situation has arisen also, especially as your interest in the problem predates my own. When we discussed these forms at the AAAS [American Association for the Advancement of Science] meeting in Philadelphia in 1962 you seemed to feel that you would not be working on these forms taxonomically — at least for a very long time. I wrote them up that spring (1963), and Dr. Williams, prior to letting the paper go to press, wrote you twice that I know of — knowing that you were at least interested, and wanting to clear the situation before letting me proceed. The paper finally went to press last spring (1964), and should be out within the month (page proofs returned two weeks ago).

I transparently used *Bothrops* in the Lesser Antilles as a contrast to *Constrictor* for the purpose of demonstrating two different sorts of differentiation in oceanic island forms: *Constrictor* populations forming a stepped-cline sequence (for me indicating conspecificity with *C. constrictor* [Figure 7]); the *Bothrops* failing to do this with respect to any of the mainland forms, and therefore, for me, indicating speciation [Figure 8]. Most of the questions you mention in your letter were not treated in my paper. I separated the island *Bothrops* from all the mainland species and each other on scale counts and color characters, adding a good deal of ecological information to indicate that they are really very different sorts of animals — not geographic representatives. . . .

So far as I know (though I neither brought up the question nor really investigated the situation), *B. lanceolatus* must be the type of the genus; *atrox* was considered the type species only as long as *lanceolatus* was its junior synonym.

B. caribbaeus (I assume *Bothrops* to be masculine, which, according to classical language scholars here, is perfectly permissible) is closer to *atrox* than to *jararaca* in a number of morphological characters, but *jararaca* is closer to *atrox* in general than *caribbaeus* is. I would postulate a sequence of relationship something like this: *caribbaeus* - - *atrox* - *jararaca* though I can't guess in which direction the sequence goes or what a dendrogram would look like, since I have no information other than morphological similarity at this time. . . .

Actually, *caribbaeus* is probably "most similar" to *B. insularis*; in the paper, I suggested that this was convergence. . . . As you may guess, I'm not much impressed

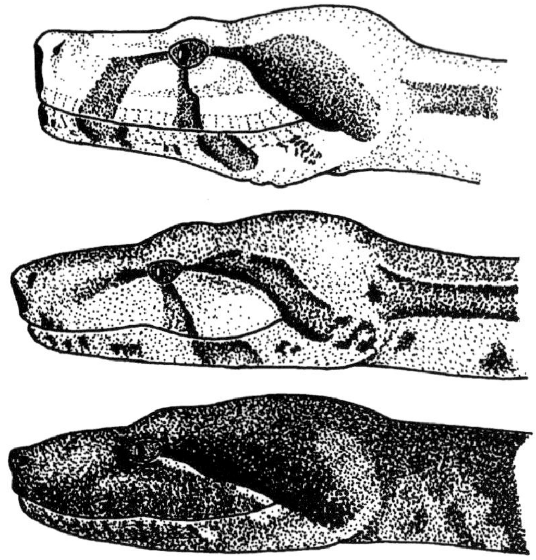

FIGURE 7. From top: *Boa constrictor constrictor* from Trinidad, *B. c. orophias* from St. Lucia, and *B. c. nebulosa* from Dominica (from Lazell 1964).

by "total similarity" or "degree of difference" at the species level. That is to say that I haven't any information on the adaptive values of the various species level characters, so merely assume that consistent ones are "good" and inconsistent ones are "not so good," sometimes downright "bad." I don't know and can't imagine what the significance of that temporal stripe is in *Bothrops*, but it certainly separates *caribbaeus* from *lanceolatus* every time, so I take for granted that it is at least the external expression of something of adaptive value.

Certainly a much more sophisticated analysis of the relationships of various *atrox*-group (or I suppose we should say now "*lanceolatus*-group") *Bothrops* should be undertaken by a snake specialist, and that is not me.

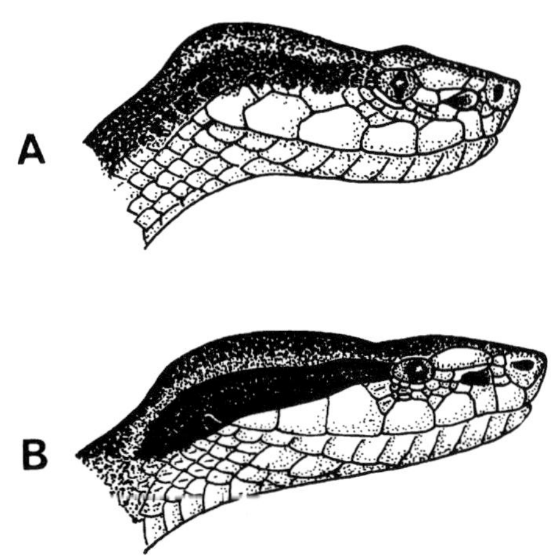

FIGURE 8. *Bothrops caribbaeus* (A) and *B. lanceolatus* (B) (from Lazell 1964).

In any case, I don't think that in any way I have exhausted the possibilities for work on these forms, and would certainly think that for you to examine our specimens would be most useful. I don't know what arrangements Dr. Williams could make for getting them to you, for there is a very large quantity in terms of poundage; *lanceolatus* is a larger species than *caribbaeus* by quite a bit, and possibly the largest in the genus.

I appreciate your offer of a bunk in New York, and may be able to take you up on it someday; for the moment, however, I'm swamped with academic work. Incidentally, I sent a number of young *lanceolatus* to Roger Conant back in 1962. Though some lived for more than a year, they have all now died — none ever reaching exhibition quality. I am sure that if you have devised a procedure for keeping them alive up here, he would be delighted to hear of it. My work in the islands may well be endless, and I can always get lots more of either species, though unless they will survive it seems a waste.
Yours most sincerely, J.D. Lazell

I got a rapid response.

The Zoological Park, Bronx, N.Y. — 18 November 1964
Dear Mr. Lazell:
I am glad to have your letter telling me what you are doing with the West Indian *Bothrops*. I am also very sorry that we had a mix-up on this and hope that it does not interfere with your continuing work in this area.

I cannot say what luck we will have in maintaining these forms, but more than 20 of the youngsters have fed at least once. I will look forward, with pleasure, to the appearance of your paper. My invitation to visit the Zoo will stand and I hope that you will take advantage of it to come down and see the animals at some future date.
Yours very sincerely, Herndon G. Dowling

Still more snake activity demanded some follow up.

MCZ — 24 April 1965
Dear Dr. [John R.] Groome,
Please accept my apologies for not answering your letter sooner. The ulcer sounds wretched, and its side effects (no booze) sound no more pleasant. I think I shall not return to Grenada until you have recovered, since the evenings with rum in hand are among my fondest memories. (I hope the thought of keeping me away from the place will not slow down your recovery though.)

I'm glad you're teaching; isn't it hell? I do it to survive, . . . it keeps me on my toes. . . .

Your snake is proceeding along. It turned out that *Pseudoboa* was the older generic name, but Mr. Greer feels *Cloelia* should be a separate genus. He feels further that the Grenada form should be a subspecies of the mainland form, despite the difference. So, the snake I would have called *Pseudoboa groomei* will probably be published as *Cloelia cloelia groomei*. Since the arguments all revolve around such intangibles as where to draw generic lines, and whether two beasts which will never meet in nature are reproductively isolated *potentially*, the whole bit becomes academic. Both Greer and I agree it will be Something-or-other *groomei*.
All the best, Skip

The summer of 1963 started slowly, but sped up nicely.

Codrington, Barbuda — 4 May 1963
Barbuda is the same sleepy little place it was before — Mr. Beazer fixed up the Guest House for me, and Mrs. Drinkwater waddled over to say she'd come prepare my evening meals. Quite untouched by change, it seems. They remember me, too.

Sandy Ground, Anguilla — 22 May 1963
This is an incredible little bit of rock. Low, dry, covered with salt ponds and mangrove swamps, it has no source of fresh meat, offers the world's smallest fish in amazingly small numbers, can raise no vegetables, and boasts the thickest mosquito population in the tropical world. But people — and very friendly, hospitable ones — live here in great quantity. So, of course, do lizards. . . . The object of the game, however, is not Anguilla itself, but the myriad islets around it — and Sombrero — far away from this or any other land.

I'll be until 1 June collecting on small islands here; at that time the boat goes across to tend the lighthouse on Sombrero, and I go with it. . . . Sombrero is less than 2 miles long, and a few hundred yards wide. It is said to be all rock. During Hurricane Donna the sea went right over its 40 ft. heights — but there *are* lizards there — and no zoologist has ever seen the place.

A French geologist, Julien, spent four *years* there in the 1800's. He got a few black *Ameiva* — the only presently known lizard and a very distinct species. Naturally, we want more of these, and expect *Anolis* and *Sphaerodactylus* too. If these two do not occur on Sombrero, it is the largest land area in the West Indies that lacks them. Only a few wave-washed, vegetation-less rocks lack anoles, and Sombrero is too big to lack at least some vegetation. . . .

There is obviously no mail service to or from Sombrero, so you won't hear from me until I get back. . . .

I had initially believed I would have to spend two weeks on Sombrero, but as luck would have it Commander T. A. Pack-Beresford scheduled his annual inspection tour for early June and would spend a whole day there. I reckoned that was plenty of time (if the boat had just gone and returned, as it usually did, I would scarcely have had 15 minutes ashore). However, the commander really did not want to take me. My trip was pure futility, he said:

°there were no lizards there at all. Hurricanes — like the recent Donna — overwashed the entire islet, surely drowning any lizard.

"Have you ever tried to drown a lizard?" I asked my hostess at the Windale Guest House. "It's a hard job." I did not say that to the commander.

The commander tried again. He had been to Sombrero on inspection many times, he said, and he never saw a lizard there. . . . But I persisted and he relented; all right, I could go.

Access to Sombrero is (or was then) by means of an iron ladder imbedded in the sheer cliff face. The commander went up first, and I was graciously allowed to go second. As I pulled myself up on the last rung, I surveyed the near-alabaster white rockscape while the commander was striding toward the lighthouse. Everywhere I looked were big, black lizards, *Ameiva corvina*. I did not say a word. I just caught my lizards — and not just the *Ameiva*. I had seen lots of movies featuring flogging and keelhauling in the British Navy, and reckoned one smart remark out of me like "gee, Commander, whatcha' call those black things, crocodiles?" would earn me, if not one of those punishments, at least being left on Sombrero without a bottle of rum. Since no one back home expected to hear from me for several weeks, I took a break from letter writing.

Philipsburg, St. Martin — 16 June 1963
Sombrero was a complete success: far from conforming to the rule that it was supposed to be the prime example of — *i.e.*, that the smaller the island the fewer the species it could support (Sombrero was supposed to have only the big, black, endemic Ground Lizard), Sombrero has the full complement of expected (by me) forms: an *Anolis* and a *Sphaerodactylus*. The *Anolis* appears to be *gingivinus*, the species occurring on Anguilla, here, and St. Barts, but the *Sphaerodactylus* is an obvious new species. Thus one rule of zoogeography has been crushed before another: no matter what else it may have in addition, every island has at least one species of *Anolis* and one species of *Sphaerodactylus*. That is Lazell's Law of the Lesser Antilles. . . .

I passed my Sombrero data on to Ed Wilson. He and Robert MacArthur thus culled Sombrero from Phil Darlington's data that they used to typify the species-area curve from the "theory" of island biogeography. Data culling worked perfectly for them (for over three decades now), but they missed the point: small islands are not like big islands in the Antilles. I would modify, codify, and split my "Law" into two "Rules," the "Rule of Three," which states that all Antillean islands have at least an anole, a gecko (maybe not a *Sphaerodactylus*), and some third species — and the "Rule of the Irreducible Anole," which notes that every rock or spit with at least herb-state vegetation will support an *Anolis*. This latter rule has failed at least three times: French Cap (United States Virgin Islands), and Watson and Carval rocks (British Virgin Islands) lack an anole — but support a *Sphaerodactylus*. Apparently the Bahamas do not fit these rules, but the Florida Keys do.

The *Sphaerodactylus* of Sombrero has never been described. The late Ben Shreve examined my two specimens; his comments are in the bottle: he wanted an adult. Dr. Jenny Daltry has collected three more that I know of, in 1999 (and more anoles, too). I repeatedly bugged her to do something (if, for example, one concludes they do not represent a new form, then publish a refutation of my 1964 opinion). Nothing happens. My specimens, in MCZ, remain unborrowed and I have to presume unexamined.

MCZ — 10 June 1963
Dear Skip:
Congratulations on the Sombrero success and the maneuvers which shortened the trip there. I had a faint clinical interest in the psychosis that might result from 15 days on that island, but I'm just as glad it worked out otherwise. . . .

Your comments on the zoogeography of the *Ameivas* are quite to the point. *Ameiva* certainly is usually a dry lowland animal — the variations between species being the degree of dryness they prefer. The only possible exception is *taeniura* in Hispaniola and this I think is again a relative thing. Even *taeniura* is rare in the uplands and its occurrence at Furcy may be a recent phenomenon after the extreme deforestation.

In Jamaica the extinction of *Ameiva* by the mongoose has always seemed to me a bit of lunacy. It was said to be holding out on the Liguanea Plain, and when I was there it certainly was, but it was holding on also in every other suitable dry sandy area — as I suspected, wherever it had ever been.

Regarding *Ameiva* and breakdown on islands, Schwartz told me that there were a number of races to be described on Anguilla — apparently color variants, and he believes this to be true of a number of islands. . . .

Incidentally Schwartz has turned over the Anguilla *Ameiva* (including Little Scrub) to a protégé of his in Miami. Your *corvina* may go to this man. . . .
All the best, Ernest

I got some of the weird and wonderful Little Scrub black *Ameiva*, but it would be decades before Ellen

Censky and Dennis Paulson finally got them named (*A. corax*). I wrote home a bit of history, not learned in school.

Oranjestad, St. Eustatius — 29 June 1963
St. Eustatius is a drowsy little rock that seems not to have changed in the nearly 200 years since Admiral Rodney shelled it in 1776 for being the first place on earth to recognize the American flag. He wiped out most of the population, and the town was reduced to ruins. Aside from a few minor repairs, nothing has been done here since. The anoles are not so dull though!

New windows in herpetology opened up. Over the next couple of years, George Gorman and I would develop a cordial friendship that, I believe, provided a great deal of mutually beneficial discussion and information exchange. Often, our independently derived opinions would comfortably coincide.

St. Augustine, Trinidad — 10 March 1964
Dear Mr. Lazell:
You will perhaps recall our meeting — before I left for T'dad in September. . . .

I would like to know, more or less precisely, what you are presently working on re *Anolis*. It would be foolish for me to duplicate some of your work, and we are covering a similar part of the world. . . .

Basically, I am working in behaviour — trying to elucidate relationships in the *roquet* group — this has taken me as far north as Martinique, and I'm sure I have followed, in places, in your footsteps. For example, in Bequia, no one was surprised that I was collecting lizards, for they remembered a chap with a beard doing the same thing a couple of years back.

As I told Williams in a letter written today, I'm contemplating a small paper on *A. trinitatis*. I feel that *A. t. trinitatis* and *A. t. vincentii* should be lumped; that the two St. Lucia subspecies should be united; and, as a working hypothesis, that the St. Lucia form is a good species (I am gathering behavioral evidence for this).
Sincerely yours, George C. Gorman

St. Augustine, Trinidad — 19 March 1964
Dear Skip,
I'm glad to see you place *luciae* next to *richardi* in your scheme. Based on my uncritical observations — they sure look damn close. Also, whenever I'm working with *luciae* and Victor Quesnel or Jack Price drops into the lab they ask, "What's that, a *richardi*?" Or else they derisively comment, "You mean that's a *trinitatis*!"

As I told Williams, I am fairly convinced that *trinitatis* is introduced here. In St. Vincent it is quite widespread, and is found in natural situations. Here its distribution is very patchy, and it only seems to be in major cities.

Because of a complete absence (perhaps not quite complete) of character displacement in behavior, it seems obvious to me that *trin.-aeneus* have not been in contact with one another for a long period of time.

Aeneus is known from one or two "natural" areas, but its distribution also bespeaks human intervention. I have yet to find what keeps the populations apart and therefore suspect it may be an artifact. Temperature data and perch preference, which Rand and Ruibal found important in explaining sympatric relationships of anoles, seem to be close to identical.

While the range of variation in Trinidad *trinitatis* seems to be encompassed by the St. Vincent animals, I am not sure you could say this for Trinidad *aeneus* and Grenadine *gentilis*. I *think* I can tell them apart — take a good look at live ones this summer. . . .

In closely related allopatric forms, we naturally have great difficulty distinguishing species from subspecies. . . . I have recently begun series of species choice experiments — in which a free *trinitatis* (or *aeneus*) can settle down with a member of the opposite sex of one or the other species. Preliminary results indicate that females make the right choice, *i.e.* settle down with males of their species. If this proves to be a good experiment, it may help clear up the *cinereus-roquet-gentilis-extremus-aeneus* mess.

Quite apart from behavior, I've gotten interested in the Dessauer-Fox techniques of protein electrophoresis for elucidation of relationships. In my last letter to Williams, I mentioned this and also mentioned that Fox suggested I spend at least a semester at LSU. . . .
Sincerely, George Gorman

I spent almost three weeks of that winter, until 10 January 1964, in the mountains of the Dominican Republic. It poured most of the time; I was usually wet and miserable, but I got lots of the sorts of ecological and behavioral data needed — including details of color pattern and color change — for such then little-known species as *Anolis aliniger*, *A. christophei*, *A. etheridgei*, and *A. shrevei*.

In the spring I headed for the southern Lesser Antilles. On St. Vincent, Dr. Earle Kirby worked things out so I could go south through the Grenadines on the government fishing sloop *Flamingo*, but it was a couple of years later before I reached Aves Island, the real Far Tortuga. It is, however, worth an aside. I went on the turtling schooner *Frankmark* out of Dominica. As with Sombrero, I was assured my trip was futile: there were no lizards there. The captain and most of the crew had been all over this atoll and never saw a lizard. Only the first mate, named Tarzan, held out hope. He recognized the stochasticity of natural processes. His reasoning was exquisite: "Look mon," he said, "every year hundreds and hundreds of turtles come up there to lay their eggs. Every turtle lays hundreds and hundreds of eggs. That's *thousands* and *thousands* of eggs. Surely not *all* those eggs hatch out baby turtles." If I had found a lizard on Aves, it would be named for Tarzan.

Back in 1964, I reached Point Saline, Grenada on 18 June. By that time I had seen the Grenada Bank small anole from dozens of islands and reluctantly concluded I could not sort out any diagnosable subspecies. I continued to amass anoles from Grenada's peaks, points, and coastal cays. I had come down the islands from St. Vincent and was confident I could tell *Anolis trinitatis* from *A. aeneus*, even if I could not make out subspecies within either. I went on to Trinidad and was unprepared for what I found.

On 9 July 1964, in Port-of-Spain, the type locality of *Anolis trinitatis*, I collected 20 anoles that ran the gamut from *A. trinitatis* as seen in southwestern St. Vincent to *A. aeneus* as seen in southwestern Grenada. I meticulously tabled ten color and pattern characters for each individual. These were not two different kinds of anoles with an occasional intermediate or hybrid. Only one kind of highly variable anole was here; it was a vast mass of total introgressive hybridization. I found similarly mixed and intermediate populations elsewhere on Trinidad. I fixed the testes of four males, wanting to follow up on Gorman's view that (the rare, I had thought) hybrids might be sterile. I wrote to MCZ.

MCZ — 20 July 1964
Dear Skip:
I never know how much to believe. Your enthusiasms and emphasis are always partly calculated — a version of one-upmanship — and partly sheer enjoyment of enthusiasm and emphasis — quite oblivious of reality.

The situation in Trinidad may be as you describe it. I know nothing of Queens Park Savanna from personal observation. It is, of course, the area from which Quesnel — whom you don't mention — reported hybrids long ago. However, for me the situation increases in interest if what you say is true. It is comparable to, but more enigmatic than, the Jamaican green and brown *lineatopus*. It is subtler because ecological difference is less evident — as you claim, absent.

Your reaction of sole concerns whether populations are nameable or not is unintelligible to me. This is a mania of yours on which — as on other points — I have long given you up. . . .
Sincerely, Ernest Williams

Ah, so — the cold, hard slap in the face. Shortly thereafter, Williams formally terminated his role as my faculty advisor and sponsor and officially banned me from the Herpetology Department. The result was that I could not count the scales on the lizards I had caught.

I got some great and generous offers to come to other universities, but I had that enormous practical problem of specimen access. From afar, Roger Conant took up my cause. He wrote Victor Hutchison, then at the University of Rhode Island (URI). Hutchison had a colleague on board there as an adjunct professor, Herndon Dowling. Initially, I was a little apprehensive because of the *Bothrops* issue, but I need not have been. Dowling signed me on as his student at URI, came up to MCZ in his Volkswagen van, formally borrowed all the Lesser Antillean specimens and drove them to my digs nearby. Furthermore, he found financial support for one more much needed field season in the islands, February and March 1966.

Thereafter followed a 14-year hiatus in my herpetological work in the Antilles proper. I did finish my thesis at URI and published on other aspects of Antillean herpetology, notably *Anolis reconditus* of Jamaica. In 1971, Williams reinstated me at Harvard so my thesis appeared in two MCZ bulletins.

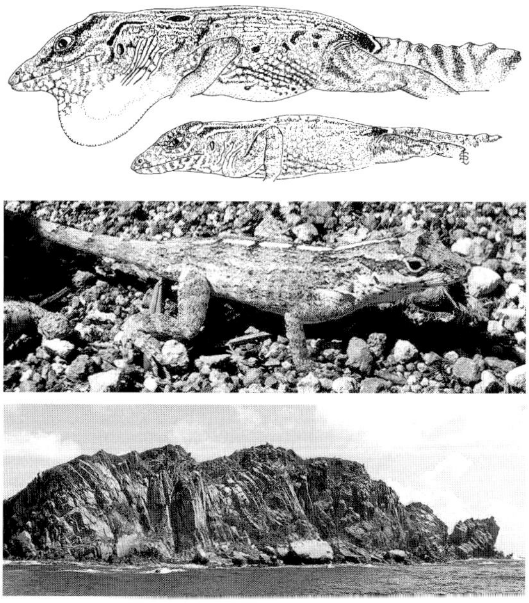

FIGURE 9. *Anolis ernestwilliamsi* (top, from Lazell 1983; middle, an adult female) from Carrot Rock (bottom), British Virgin Islands, was named in tribute to Williams's work in the West Indies (photographs by Joe Burgess).

FIGURE 10. I'm working on untangling a fruitbat from a mistnet on Guana Island, British Virgin Islands, after Lianna Jarecki gave up — it's all in how you hold your mouth (photograph by Tom Jarecki, 1984).

Williams also arranged that I be formally appointed an academic fellow of Harvard University in 1977.

In 1980, my erstwhile Harvard colleague Robert Jenkins arranged funding from The Nature Conservancy, enabling me to take a position with the Department of Natural Resources and the Environment, Government of the British Virgin Islands. I took up my duties (island inventories and recommendations for parks and refuges) in March of that year. That led to the discovery of what would become *Anolis ernestwilliamsi* (Figure 9) and the remarkable fauna and flora of Guana Island (Figure 10). This isle of 300 ha and with a maximum elevation of 206 m harbors as many reptiles as my beloved Dominica, which covers 79,000 ha and has a maximum elevation of 1620 m — 263 times the area and nearly eight times higher than Guana.

Before I complete my account, however, I must tell one more Ernest Williams anecdote. Way back, about 1960, several of us were gathered with Alfred Sherwood Romer in his office and library, before a wall map of the world. With a swirl of his left hand, Williams indicated the West Indies. "You understand," he said, "this is just a warm-up exercise." With his right hand he pointed east. "That's the real thing, over there." It would be two decades before I got there. Along with Williams's words, I remember those of the British historian George MacDonald Fraser: "The islands lie ahead of you in a great brilliant chain that runs thousands of miles from the South China Sea. That's the East — the Islands."

That is true, but I can never give up the West Indies. Both will always lie ahead of me.

Suggested Reading

Bucher, A. M. 1960. Herpetologist at work. *Germantown Friends School Alumni Record* 27 (1):1–2, 6.

Lazell, J. 1961. A new species of *Sphaerodactylus* from northern Haiti. *Breviora* 139:1–5.

Lazell, J. 1964. The Lesser Antillean representatives of *Bothrops* and *Constrictor*. *Bull. Mus. Comp. Zool., Harvard Univ.* 132:245–273.

Lazell, J. 1964. The reptiles of Sombrero, West Indies. *Copeia* 1964:716–718.

Lazell, J. 1966. Studies on *Anolis reconditus* Underwood and Williams. *Bull. Inst. Jamaica Sci. Ser.* 18:1–15.

Lazell, J. 1967. The ternery on Aves Island in March. *The Condor* 69:87–88.

Lazell, J. 1967. Wiederentdeckung von zwei angeblich augestorbenen Schlangenarten der westindischen Inseln. *Salamandra* 3:91–97.

Lazell, J. 1972. The anoles (Sauria: Iguanidae) of the Lesser Antilles. *Bull. Mus. Comp. Zool., Harvard Univ.* 143:1–115.

Lazell, J. 1973. The lizard genus *Iguana* in the Lesser Antilles. *Bull. Mus. Comp. Zool., Harvard Univ.* 145:1–28.

Lazell, J. 1981. Field and taxonomic studies of tropical American raccoons. *Natl. Geogr. Soc. Res. Repts.* 13:381–385.

Lazell, J. 1983. Biogeography of the herpetofauna of the British Virgin Islands, with description of a new anole (Sauria: Iguanidae), p. 99–117. *In* A. Rhodin and K. Miyata (eds.), *Advances in Herpetology and Evolutionary Biology*. Mus. Comp. Zool., Harvard Univ., Cambridge, Massachusetts.

Lazell, J. 2003. Looking for lizards in all the right places, p. 215–231. *In* R. W. Henderson and R. Powell (eds.), Islands and the Sea: Essays on Herpetological Exploration in the West Indies. Society for the Study of Amphibians and Reptiles, Ithaca (New York). Contributions to Herpetology, volume 20.

Dog Island Days

Ellen J. Censky[1]

I woke suddenly with my heart racing and my voice caught in my throat. A bright light was shining in my eyes and there were footsteps outside my tent, lots of footsteps — and someone was tapping on my tent. What was I doing here? The fisherman had warned me and the police had forbidden me to stay on this uninhabited island. They said that it was used as a drug drop. It was too far from the main island for them to patrol, and therefore they couldn't guarantee my safety. They didn't want a lone person, let alone a lone woman, camping on the island. So why was I here? The drug runners were at my tent, shining a light in my eyes, ordering me out. I was a goner, but I couldn't move because fear had paralyzed my limbs and my vocal cords. I had to think fast but my thoughts were still clouded with sleep — wake up, wake up! As sleep left me, it became more and more clear that the footsteps weren't made by humans. Instead, hundreds of hermit crabs (Figure 1) were rustling in the leaf litter around my tent, and the light that blinded me was not a beam from a drug runner's flashlight, but instead that from a very, very bright moon shining directly in the window of my tent. I had incorporated my fears and the surrounding sights and sounds of the night into a dream. This was the first of many nights spent on Dog Island, off the coast of Anguilla. If I was going to survive there long enough to do my research, I definitely would have to curb my imagination.

Captivated by Ground Lizards

Why was I in the Caribbean doing research on an uninhabited island? I had first been introduced to the islands in 1986 by Dave Watters, Curator of Anthropology at the Carnegie Museum of Natural History. He was conducting archaeological research on Anguilla, among the northernmost of the Lesser Antillean islands. He had asked if I would join him on one of his trips to make a collection of reptiles that he might use to compare with the material that he was uncovering in his dig sites. At the time, I was Collection Manager of Amphibians and Reptiles at the same museum. While my duties included field research and collection trips, these had to be approved by the curator, Jack McCoy. When asked for permission to do the trip, he responded: "What do we have from the island?" Shortly after we determined that the museum had no specimens from Anguilla, I was on my way. This was the beginning of my love affair with Anguilla, its satellite islands, and its residents — particularly the Ground Lizards, *Ameiva plei*. Although I was sent to survey all reptiles on the island and my research there has culminated in publications on many species including frogs (*Eleutherodactylus johnstonei*), turtles (*Geochelone carbonaria*), lizards (*Gymnophthalmus underwoodi*), snakes (*Ramphotyphlops braminus*), and an over-water

Figure 1. Incredible numbers of Dog Island hermit crabs led to nightmares.

[1] Sam Noble Oklahoma Museum of Natural History, University of Oklahoma, Norman, Oklahoma 73019, USA.

invasion of Green Iguanas (*Iguana iguana*), the animals that truly captivated me and captured my time were the ground lizards (*Ameiva*).

On that first trip in 1986, I noticed that the ground lizards on Dog Island, a satellite of Anguilla, were different than those on the other islands. Not only were the lizards of Dog Island smaller than those on Anguilla, males and females were less dimorphic in size. I was intrigued. What was causing these differences? This question led to my Ph.D. dissertation research and, between 1991 and 1993, I spent several months each year gathering data on the two islands.

Roughing It

Designing a research project is the easy part. Logistics, arranging for lodging, permits, permission to use study sites, etc., can all be far more difficult. However, logistics for Anguilla were relatively easy and arrangements plush. Until that time, fieldwork accommodations had always meant a tent and sleeping bag; luxury accommodations meant a tent, sleeping bag, toilets, and a shower. Camping was impossible on Anguilla, which has no campgrounds. So, over the years, I rented apartments with toilets, showers, and a refrigerator!

My study site on Anguilla was also ideal. It was in the Katouche Valley, a lush (by Anguillian standards) valley with tall trees, lots of undergrowth, and a salt pond in the middle. I could step out the door of my luxury apartment, walk about 15 m down a road, cross over a small beach, and I was in my site. Easy.

Dog Island Travails

Dog Island was a whole different picture. It is an uninhabited island that sits about 16 km to the west of Anguilla. Access to the island was by boat, of course. Several fishermen set fish traps in the vicinity, so transport back and forth would not be a problem. The problem was that I could not get to and from the island on the same day or day after day because the fishermen did not set and check their traps every day. The solution was being dropped off by the fishermen and camping for several days. After meeting with several fishermen, the decision was made that the main person responsible for my transport would be Ed Carty, a fisherman who has a large fishing operation with several boats. In reality, because of the uncertainties of fishing, boats under repair, and other problems, my actual transport fell to two other men who were both nonfishermen: Roland Hodge, the Director of Fisheries, and David Carty, a boat designer and builder and also the Anguillian Parliamentary Secretary.

FIGURE 2. My Dog Island campsite.

Permission to work on the island was another sticky problem. The police officially did not want me camping on the island alone because they could not guarantee my safety. The island was apparently an ideal spot for drug drops. A naturally flat, clear area on the island was big enough for small planes to land and take off. Boats could bring drugs from South America, drop them on the island, then planes could land, load up, and take off for the United States. The island was too far from the Anguillian coast for regular police patrols. Even though the police, aided by the United States Coast Guard, had dug deep trenches across this makeshift runway, some concerns about safety still existed. Once again, David Carty saved the day. He understood that this was the only island that had smaller lizards and therefore, to do my research, I needed to study this population. As a member of the Anguillian Government, he told me to go ahead and do the research. He outfitted me with a two-way radio, and I had instructions to call in to the Fisheries Department every morning at 9 AM and every afternoon at 3 PM. The idea was great, but the execution was less than desirable. In order to call Anguilla, I had to emerge from my camp or study site, walk down the beach, climb the rocks onto a high point on the island, or walk around the island to the other side and call out. The two-way radio was not powerful enough to carry from my side of the island. This was not a problem when nothing was happening, but you can imagine the problem this would pose should I really have needed help, for example, if a drug runner had come to the island. I would have had to emerge onto the beach making myself fully visible. . . . Fortunately, this never became an issue.

FIGURE 3. The chatter of Brown Boobies was sometimes deafening.

stand bathing in salt water. I decided to spend three to five days on Dog Island, and then return to Anguilla for a similar time span to gather data.

First Season

Dog Island is a beautiful little island of about 18 km^2. Like Anguilla, it is a low-lying and receives very little rain through the year. The vege-tation is a thorn-scrub. A few mangrove "forests" contain relatively tall trees, but the bulk of the island is covered by low bushes and several different kinds of cactus, many of which I came to know personally. The vegetation covering most of the interior is so dense that I could not penetrate it except along goat paths. The island was once a sheep station, although no sheep reside there today. However, wild goats occur in incredible numbers. The island also plays host to an array of seabirds and is a rookery for Brown Boobies (Figure 3). At times, on the east side of the island, the chatter of these birds is deafening.

My study site was on the west side of the island. It was similar to my site on Anguilla in that it was a forest of Manchineel trees and both sites had an abundance of *Ameiva* and *Anolis*. However, that was where the similarity ended. The Anguilla site had a small salt pond in the middle of it, the Dog Island site had none. The Anguilla site had many different kinds of fruit-bearing trees, an abundance of land snails, and very few hermit crabs. The Dog Island site had only one kind of fruit-bearing tree (Manchineel), no land snails, and an overabundance of hermit crabs. The Manchineel forest on Dog

So I was set. I had "permission" to camp on the island (Figure 2), I had transportation, and I had my two-way radio to communicate regularly. However, with no fresh water on Dog Island, my stay was limited by the amount of fresh water that I could carry in and by the amount of time that I could

FIGURE 4. Great Bay on Dog Island.

Island was unique in that the trees grew out of the ground almost parallel to the forest floor for about 2–3 m, and from that point each tree grew straight to the sky and tall. Apparently, many years ago, a hurricane had knocked down these trees, but they remained rooted and continued their upward growth after the hurricane. This peculiar growth pattern made walking through the study site, let alone catching lizards, difficult.

My first stint on Dog Island (Figure 4) was uneventful, aside from my dreamland adventure. Each day began at sunrise with a walk around the island in search of sea turtle tracks. I had told the Fisheries Department that I would monitor the beaches while I was on the island in order to supplement their sparse information on what species and how many individuals were nesting. By the time I returned to camp and had breakfast, the lizards were out and about.

Those first days were spent catching lizards, measuring and weighing them (Figure 5), determining their sex, and then marking them with a series of colored beads sewn onto the base of their tails, so that I could follow each individual and record interactions. After most of the lizards at the site were marked, I would spend my day observing them and recording behavior. The lizards were active from early morning to just after sunset, very different than the daily activity cycle of the lizards on Anguilla. There the lizards were slackers. They would arise later in the morning, around 9 AM, forage until about 1 PM, and then head underground. Some of the more energetic individuals would come back out at about 3 PM and hang out until about 5 PM, when they'd retreat to their burrows for the night.

FIGURE 5. The author weighing an *Ameiva plei* on Dog Island.

Once lizard activity on Dog Island died down, I would take a quick dip in the ocean, grab some dinner, look at the lights on Anguilla and St. Martin and wonder about the activities in which people on those islands were engaged. The loneliness one can feel on a deserted island is truly amazing, even though you can see evidence of human activity just a short boat ride away.

Each time I left Dog Island, I would break camp and stash the tent and all non-perishable food items in the bush. Since very few day-trippers came to Dog Island and since none of them ever made it off the beach, I felt confident that no one would break into my cache and make off with anything. Wrong. One time, I arrived on the island just as the lizards were emerging from their burrows. Rather than setting up camp, I elected to study the lizards and postpone camp resurrection until later in the day. Sometime around noon, I was feeling extremely hungry. Approaching the site where my camp supplies were hidden, I didn't immediately see anything out of place and headed straight to the peanut butter jar. Much to my amazement, the peanut butter was almost gone. Someone had broken the cap on the jar and consumed most of the breakfasts and lunches I had planned for this trip. Disappointed, I threw the jar on the ground and, within five minutes, the culprits appeared. I counted 232 hermit crabs in and around the jar. This incident made me rethink my research design — perhaps competition for food with hermit crabs was what was limiting the size of the Dog Island lizards!

Second Season

With new insight into the island life of these lizards, I returned for my second field season. I had modified the design of my research, which now entailed setting up enclosures, or rather exclosures, to keep hermit crabs out of certain areas. This, of course, would require more work. Luckily, I had some help.

Mary Chewning was an expatriate from the United States living in Anguilla with her husband, Taylor. She became one of my chief volunteers. She was a freelance editor for a North American publishing company and, therefore, was able to arrange her schedule so that she could join me. She was not a lizard catcher, but was willing to learn. She was a great help with the processing of lizards, but an even greater asset as a conversationalist. Apparently, I had acquired the habit of talking to lizards in my first year on Dog Island, and I evidently didn't

FIGURE 6. The eastern, windward coast of Dog Island.

confine my conversations with lizards to Dog Island. People on Anguilla were starting to look at me strangely. With Mary for company, I managed to break my habit and thereby rebuild my reputation as a scholarly herpetologist!

One evening, just as Mary and I were about to step out of our hidden campsite onto the beach to take a dip in the water, we noticed a very fancy cigarette boat speed into the bay, cruise around, and then zip back out. This was a very odd time for a boat to arrive because the sun was setting and navigating the reefs was difficult. In addition, the bay at Dog Island is not ideal for overnight mooring. A little later, a sailboat moved across the bay and was joined by the cigarette boat. Both moved into a little cove to the east of our bay. The sun set and, for the next two hours, lights signaled on the horizon. Then, suddenly, both boats moved out of the cove, the cigarette boat leaving under full speed and the sailboat more slowly, but under power. We suspected that we had just witnessed a drug drop and were somewhat relieved to see a United States Coast Guard ship patrolling the coast the next day.

I was happy to have the company and help of Mary on these trips, but she was not always available. So, many times I was again alone. However, the truth was that someone always seemed to be showing up. One time, I awakened to find a group of Frenchmen from St. Martin setting up camp down the beach. They arrived with tents, tables, chairs, and cases of wine — now that is luxury camping! For the next week, I was entertained with wonderful French cooking and French wines and stories of life in French St. Martin. Between the eating and drinking, I managed to gather more data on the lizards and, at the same time, educate these men about the sex and violence that characterized the lives of Dog Island lizards.

Third Season

This would be my final season of gathering data. I was close to understanding the environmental impact on these lizards' lives. I needed about one more month of data. My hope was that Jack McCoy, Curator of Amphibians and Reptiles at the Carnegie Museum, member of my dissertation committee, and my mentor, would come to the islands to help me — and to relax. He was all set to come but pulled out at the last minute. Dave Duvall, Professor of Biology at Arizona State University, took his place.

Dave spent nine days helping me gather data and process lizards. Having conducted his own field research, he was a great help, but when the waves started to build and the sounds of the crashing surf became too much, Dave's California beach bum identity emerged.

One day, a storm was brewing to the east, the wind was raw, and the waves were as high as I'd ever seen them on the beach near the campsite. That meant that they were really rolling in on the other side of the island where the wave action is always greatest (Figure 6). This area of the island is characterized by a narrow passage between rocks and a small beach at the end. That's where Dave was headed to do some body surfing. The weather was so blustery that the lizards weren't out, so I decided to tag along for no other reason than I wanted to know where to pick up the body when Dave came crashing onto the rocks. As we headed across the island the wind intensified, the sky grew dark, and a sheet of rain hit us like a rogue wave. The wind was so strong that the rain was moving horizontally. We found shelter in a large *Coccoloba* tree and waited out the storm. After about ten minutes the storm had passed, but it had been so violent that we decided we'd better check to see if we still had a camp.

We had pitched our tents on the beach. The stakes on mine had been coming out earlier in the week, so I had used really long sticks instead. They had held, but the straight metal A-frame skeleton of the tent had bent into a graceful arch. The now-useless tent lay flattened on the beach, and many of our supplies were strewn about haphazardly. We began cleaning up and had nearly finished when the wind picked up again. I looked at Dave's dome tent and it was flapping like a spastic seagull. Held down by only one stake, it was getting ready to go surfing. I called to Dave and jumped into it. He said he'd be there in a minute. I could hear him rummaging through things and wondered why he was trying to tidy up just as a big storm was threatening to flatten our only remaining tent. I needed his extra weight in the tent to keep it from taking off on a flight to Sombrero Island, the next stop west of us. After what seemed an eternity, Dave popped his head in the tent: "Okay, I'm ready." He had a light, a box of crackers, and a bottle of rum. We spent the rest of the afternoon with the wind howling around us. As the night drew on, the wind seemed to calm, or was that merely the effects of the rum?

The next day David Carty showed up to find out how we'd fared. He also mentioned that he had received a phone message from the assistant director at the Carnegie Museum. He had not returned the call yet, but would do so when he got back. He said he would return if it was anything urgent. He did not return for three days.

During those three days, though I was catching lizards and collecting data, my mind was preoccupied with the phone message from the assistant director. What could she have been calling about? Though I kept rationalizing with thoughts like: "Maybe she wants to vacation in Anguilla and needs information immediately," in my heart, I knew it was about Jack McCoy and I was pretty sure that it could not be good news. When David Carty returned, he told me that Jack had passed away the previous week. Jack's wife, Patsy, had told David to tell me to finish my research: "That's what Jack would have wanted." They would hold the memorial service when I returned to the states.

It seemed impossible that Jack was gone. He had been my mentor and my best friend. He had cultivated my interest in herpetology and nurtured my professional development. He had taught me the value of collections and the meaning of museum work. He was my toughest critic and my closest ally. He had been there to help me get started with my research program, and he was supposed to be there to see it to fruition.

As I finished my field season, my thoughts lingered on Jack and his many contributions to my life. One lizard on the island had eluded me for the whole trip. It was too nervous and never allowed me to get close enough to catch it. On that final day, I caught that lizard — and I cannot help but think that Jack McCoy was there, lending a hand.

A Limited Commodity

So, what was I doing on the island? Why were the lizards on Dog Island smaller and less dimorphic in size? The lizards on Anguilla are dimorphic, with males much larger than females, because the males fight with each other for the opportunity to mate with females. The females also choose to mate with the larger males. Large males spend much of their time during the mating season guarding females (Figure 7), sticking close to them and chasing off any other males. They, in fact, spend less time feeding when they are guarding females than when they are not. Guarding behavior is very effective; of the 34 marked adult males in my study site, the four largest males mated with 84% of the females.

Males on Dog Island also fight over females for the opportunity to mate, but Dog Island males do not guard females to the extent that Anguillian males do. The reason for their lesser effort is that food is a

FIGURE 7. Male *Ameiva plei* guarding a female on Anguilla.

limiting factor. On Dog Island, food is less abundant and the lizards must compete with hermit crabs for this limiting commodity. These lizards need to forego guarding females in order to feed. Without a large male guarding a female, a smaller male can and will mate with her. Thus, the restricted food supply limits both the size of lizards and extent of dimorphism in this population.

I look back at the days and nights spent on Dog Island fondly. I have returned to Anguilla numerous times — and Dog Island remains one of my favorite places in the entire Caribbean. Perhaps I feel this way solely because Dog Island is where I was initiated into the wondrous world of lizard life — or perhaps I feel this way because I grew steadier on my research legs while on Dog Island. As I think back to those early years in Anguilla, however, I've come to believe that the real reason Dog Island is so special is that I left a piece of me on the island, along with the spirit of Jack McCoy.

Suggested Reading

Busack, S. D. and E. J. Censky. 1995. Obituaries: In remembrance of C. J. McCoy. *Herpetol. Rev.* 26:65–66.

Censky, E. J. 1995. Mating strategy and reproductive success in the teiid lizard, *Ameiva plei*. *Behaviour* 132:529–557.

Censky, E. J. 1995. Reproduction in two Lesser Antillean populations of *Ameiva plei* (Teiidae). *Jour. Herpetol.* 29:553–560.

Censky, E. J. 1996. The evolution of sexual size dimorphism in the teiid lizard *Ameiva plei*: a test of alternative hypotheses, p. 277–289. *In* R. Powell and R. W. Henderson (eds.), *Contributions to West Indian Herpetology: A Tribute to Albert Schwartz*. Society for the Study of Amphibians and Reptiles, Contributions to Herpetology, Volume 12. Ithaca, New York.

Censky, E. J. 1997. Female mate choice in the non-territorial lizard *Ameiva plei* (Teiidae). *Behav. Ecol. Sociobiol.* 40:221–225.

Censky, E. J. 2003. Dog Island days, p. 233–239. *In* R. W. Henderson and R. Powell (eds.), Islands and the Sea: Essays on Herpetological Exploration in the West Indies. Society for the Study of Amphibians and Reptiles, Ithaca (New York). Contributions to Herpetology, volume 20.

Culture, Politics, Lizards, and Me

Karim V. D. Hodge[1]

Six hundred years ago, when the devil was still a boy, Anguilla reclined in complete innocence at the junction of the Caribbean Sea and Atlantic Ocean. Anguilla is a sleepy, little, low-lying limestone island in the eastern Caribbean shrouded in dry woodland scrub and surrounded by several offshore cays. Both the island itself and some of the cays are home to some very special reptiles.

A Little History

Today's Anguillians have been touched by various cultures throughout the island's history. At the time of its so-called "discovery" by Europeans, Tainos inhabited Anguilla (and many of its Caribbean neighbors). Tainos were part of the South American Arawak group who had moved north from island to island. They lived in harmony with nature, taking no more than necessary for their survival. This approach ensured a continual supply of resources and showed respect for all life forms — but conflict was already present. Another Arawak group, the Caribs, was less inclined to coexist with neighbors, and had already displaced the Tainos throughout most of the southern Lesser Antilles.

The conflict between the Caribs and Arawaks, however, paled in comparison with the struggle that led to the European domination of the entire West Indian region. On Anguilla, this clash of cultures ended with the complete obliteration of almost every facet of Amerindian culture with the exception of the Fountain Cavern and Big Spring, both archaeological sites believed to have been ceremonial centers of great significance. Christianity became the dominant religious influence, and island economies developed according to plans designed to enrich European coffers. Phosphate mining occurred on Anguilla, but the offshore island of Sombrero was nearly denuded centuries later for the same purpose. Most colonies became entirely dependent on agriculture — and this led in turn to the establishment of the trans-Atlantic slave trade. Although colonized by the British in 1650, Anguilla was spared the worst ravages of this practice. Several factors undoubtedly contributed, but perhaps the most significant was simply that the harsh, dry climate was not conducive to crop production (Figure 1).

A Forgotten Island

Without easily exploited riches, Anguilla was forgotten. Prolonged droughts and frequent famines persisted into the early 1900s. Even in the latter years, a lack of electricity and running water forced a reliance on intermittant water holes, often unreliable springs, and wells which were frequently contaminated with sea water. Dependent on oil lamps for light and farming land incapable of producing much food, Anguillians of all races struggled to exist. Forced to turn to the sea for sustenance and fuel, Anguilla became a nation of seafarers (Figure 2). The communal misery that characterized this forgotten period of history, however, united the nation, developed a strong character in its people, and may have been responsible for the near total absence of racial conflict.

Figure 1. Brimegen on the northern coast with characteristic Anguillian scrub habitat (photograph by Ellen J. Censky).

[1] The Anguilla National Trust, P. O. Box 1234, The Valley, Anguilla.

FIGURE 2. Middle Cay, one of Anguilla's many satellite islands (photograph by Ellen J. Censky).

Through much of the colonial period, Anguilla was part of a three-island state that included St. Christopher and Nevis. Due to its lack of wealth, Anguilla often was slighted by its politically powerful neighbors. In 1967, Anguillians rejected the leadership of St. Christopher, formed a secession movement, and declared its independence in a series of events now commemorated as the Anguilla Revolution.

An Awakening

Anguilla had been essentially unknown on the international front. Then, in 1969, at the instigation of the government on St. Christopher, the British invaded. Various news magazines documented the island's passive — and successful — resistance in stories entitled "The Lion That Meowed" or "The Mouse that Roared."

The arrival of electricity in the late 1970s launched massive coastal development, primarily for tourism, which rapidly became (and still is) the single most important element in the island's economy. Soon, large sections of both the northern and southern coastlines became the exclusive province of sun-seeking tourists. Previously pristine natural landscapes, seascapes, and ecosystems were altered to accommodate the new arrivals.

FIGURE 3. Anguilla's native population of Lesser Antillean Iguanas (*Iguana delicatissima*) is critically endangered (photograph by Ellen J. Censky).

The tourism-fueled boom resulted in a sudden influx of people from other islands, Europe, and the United States — none of whom shared Anguillians' unique "Weltanschauung." Tourism and the increased need for new housing led to a massive escalation of land values, particularly in coastal areas. The new emphasis on wealth and the multiethnic array of peoples representing a variety of cultures have combined to retard all efforts for a nationwide implementation of environmentally sustainable development.

An Environmental Crisis

Modernization has substantially reduced involvement in traditional activities. Agriculture and fishing have become effectively marginalized. As a consequence of the latter in particular, concern for coastal communities and the surrounding sea is minimal. Extensive coastal development has impacted mangrove forests and coral reef ecosystems. Unregulated exploitation has resulted in the widespread depletion of inshore reef fisheries and local populations of highly threatened species such as sea turtles. Damage, however, is not restricted to the coasts, even the xeric woodland scrub is retreating in the face of development.

"Politricks & Protected Areas"

Now for some "politricks," as we Anguillians say. Political will is essential for ensuring that conservation concerns are addressed when formulating investment programs and policy frameworks. Three key factors that affect the political will in Anguilla are a general lack of awareness of environmental issues and limited capacities of institutions and people to influence policy and land tenure on the island.

In Anguilla, as in most nations, political leaders and government institutions determine policies and the degree to which these policies are implemented or enforced. Furthermore, these same people and agencies control resource allocations and developmental initiatives. The level of political commitment to conservation and sustainable use of biological resources is expressed in the Country Strategic Programme; but, like those everywhere, Anguilla's political leaders face a range of competing choices for improving the well-being of their nation. Privileged groups often prevail by enticing politicians to make decisions for the short-term benefit of the chosen few.

FIGURE 4. In 1997, the Anguillian postal service issued a commemorative sheet featuring *Iguana delicatissima*.

Ultimately, Anguilla's political leaders and other interest groups need to be more aware of the economic benefits of biodiversity and the habitats that sustain it. As in many developing nations, external agencies often play a significant role in strengthening local conservation efforts — but the effectiveness of these interventions is limited without political support.

The new World Conservation Union (IUCN) promotes a positive relationship between people and protected areas, and recognizes that they can coexist productively. The situation in Anguilla is complicated by a land tenure system that historically placed nearly all (95%) of the nation's land in private ownership. Consequently, an enlightened government and an educated populace must work cooperatively to solve the problems of the future.

For example, protected areas in Anguilla can generate large amounts of revenue if managed effectively. Protected areas do not mean setting aside land as isolated areas without human interaction, but instead they can figure prominently in profitable ecotourism initiatives and can be used in educational programs designed to generate a sustainable relationship between people and nature — ensuring the survival of cultural traditions and the availability of resources for generations to come.

This is where the National Trust comes into the picture. Established in 1993 to act as custodian of Anguilla's heritage, the Trust has four major goals: (1) establish environmental education programs for all Anguillians (the educated populace), (2) promote and preserve the expression of Anguillian culture, (3) advise on matters relating to natural, cultural, and historic resources (an enlightened government), and (4) manage protected areas, heritage sites, and the national museum — and the Trust, in turn, leads to me.

My Humble Contributions

In early January of 1996, I was recruited from my job as a postal worker by the National Trust to assist Dr. Judy Dudley in her quest to learn more about Anguilla's most reclusive land animal, the Lesser Antillean Iguana, *Iguana delicatissima* (Figure 3). After acquiring the skills for catching and tagging these fabulous lizards, I was soon doing it full time. In 1997, Glenn Gerber (Fauna and Flora International) and I initiated the first conservation assessment of this critically endangered species on Anguilla. Spending up to eight hours at a time in the unbearably hot and humid bush was not the most fun I'd ever had — but I was becoming increasingly aware of the important role I could play in understanding and eventually preserving Anguilla's wildlife (Figure 4).

During that same period, I spent over a week in Antigua working with researchers from the University of Georgia on the sea turtles of Jumby Bay. I learned to mark turtles, and I also learned about the vagaries of fieldwork — namely that it often turns out to be less than fruitful. I also met John and Sarah Fuller who lived in Hodges Bay. They are wonderful people, deeply concerned about the island's environment, and willing to take action to conserve it. Interacting with such enthusiastic

FIGURE 5. A Green Iguana (*Iguana iguana*), emaciated upon its arrival in Anguilla.

FIGURE 6. The mass of floating debris on which Green Iguanas rafted to Anguilla was largely intact weeks after the event (photograph by Ellen J. Censky).

people certainly enhanced the working experience in a region where the conditions (hot, humid, or very dry and often isolated from any creature comforts) are frequently less than ideal.

I returned to Anguilla in time to face the ongoing issues surrounding the Lesser Antillean Iguana. The final report on our prior fieldwork painted a bittersweet picture. The species was maintaining a viable population (albeit barely), but was threatened not only by human activities (e.g., housing needs and development for tourism), but also by the arrival of the common Green Iguana (*Iguana iguana*; Figure 5). If unchecked, these newcomers might compete with the native species and perhaps even dilute its unique genetic makeup through hybridization.

Green "Frenchmen"

The Green Iguanas were called "Frenchmen," because they probably came from Guadeloupe, arriving on a floating island of logs and debris in the aftermath of Hurricane Luis in September 1995 (Figure 6). Because of the potential threat to the native iguanas, advisors recommended that we initiate an effort to extirpate the aliens (the new arrivals and a previously established population of escaped pets). Most herpetologists think catching lizards is fun, but finding and killing iguanas was grueling — in part because they tended to hang out in the most inhospitable places and in part because I was dependent on help from locals whose reaction to iguanas varied from fear to fascination, but inevitably entailed little desire to get too close to one.

FIGURE 7. A Cuban Treefrog (*Osteopilus septentrionalis*) from Anguilla. These frogs became established after arriving in Anguilla in containers of decorative plants shipped from Florida (photograph by Robert Powell).

Politically Astute

At about the same time, one of Anguilla's offshore cays was targeted for development by the Royal Caribbean Cruise Line. Because limited suitable iguana habitat was unavailable, this particular cay had been identified as a possible safe haven. For once, instead of working against conservation efforts, Anguilla's land tenure issues precluded the acquisition of the cay by the developer. Seeing an alternative perspective to what I had always perceived only as a problem caused me to become more politically astute.

Cuban Frogs from Florida

In the course of efforts to extirpate the Green Iguana, I discovered a large concentration of Cuban Treefrogs (Figure 7) around a coastal development on the western end of the island. Although the appropriate authorities and the manager of the development were alerted, no action was taken — and then came Hurricane Lenny in November 1999. The torrential rains provided ample opportunities for reproduction, and the frogs quickly spread across much of the island, invading cisterns and rendering them unfit for human use. Anguillians, unfamiliar with frogs, were not happy — and the frogs are now sufficiently well-established to defy any efforts to exterminate them. However, the Environmental Health Department has taken on the task of educating the public of what measures they should take to control the frogs getting into their water supply.

Sombrero Rockets

The year 1998 was an exciting time. A United States–owned aerospace company (Beal Aerospace Technologies Inc.) had initiated efforts to purchase and develop Sombrero Island (Figure 8) as a rocket facility for launching communication satellites. This 38-hectare rock sits on its own island bank, but is

FIGURE 8. The harsh landscape of Sombrero is the result of phosphate mining (photograph by Ellen J. Censky).

politically a part of Anguilla. The National Trust quickly pulled together a regional team of experts to critique the questionable findings of the developer's Environmental Impact Assessment. This team became known as the Caribbean Environmental Advisory Committee (CEAC), and it recommended that an alternative study be done and the Trust called for a rapid environmental assessment. Experts were brought from the United States, the United Kingdom, and the wider Caribbean. However, all efforts appeared futile, mainly because the government favored development.

Sombrero is home to hundreds of nesting sea birds, some of which are threatened regionally and internationally, to over 65 species of invertebrates, many new to science, and to the endemic Sombrero Ground Lizard (*Ameiva corvina*). In addition, the island's coastal waters provide food and habitat for Hawksbill and Green sea turtles. All of this had been ignored, until two conservation agencies from the United Kingdom, the Royal Society for the Protection of Birds and Fauna and Flora International, contributed their expertise to the rapid conservation assessment. The assessment confirmed numerous inconsistencies in the developer's environmental impact study, forcing the production of a supplementary document to address those issues. After approximately a year of turmoil, the aerospace company went out of business — but my political sophistication had grown by leaps and bounds.

Little Scrub Island

I subsequently left Anguilla for two years to complete a course of study in England. Upon my return, I found myself once again in the field. This time, however, I wasn't catching and killing iguanas. Instead, I was conducting my dissertation research on the Little Scrub Island Ground Lizard, *Ameiva corax* (Figure 9), for the purpose of completing requirements for a Master of Science degree in conservation biology.

Little Scrub Island is tiny (1.2 ha), harsh (no beaches and no fresh water), and basically inaccessible (Figure 10). In an article in *Fauna* magazine, Censky and Powell (2001) described the island experience as "exciting, wonderful, and loud" (primarily due to the abundance of sea birds). They went on to say:

"Humans rarely visit Little Scrub, which is accessible only on the calmest days. Even then, the visitor must balance precariously on the bow of the boat, waiting

FIGURE 9. *Ameiva corax*, the black ground lizard found only on Little Scrub Island (photograph by Robert Powell).

FIGURE 10. Little Scrub Island has no beaches and no fresh water (photograph by Ellen J. Censky).

patiently as the driver approaches the ominous rocks, timing with the utmost precision the exact moment when the boat and a tiny shelf of rock are at the same level, then jumping just as the driver reverses the thrust of the motor to avoid crashing on the jagged shore. Obviously, the adventure occurs in reverse when leaving the island. This is not an exercise for the timid, and even the most adventuresome must question their reasons for attempting the landing. Failure could easily be lethal, and the return for the risk is, at best, marginal — unless one happens to be a silly biologist enamored with the prospect of interacting with birds and lizards that, like those Darwin encountered in the Galápagos Islands so many years ago, show no fear of humans."

The vegetated portions of the island are dominated by "Pope's-head" cacti and the rocks are honeycombed with an infinite number of crevices and sharp edges just waiting to catch the biologist guilty of only a moment's inattention (Figure 11). Temperatures frequently exceed 40°C, and the lack of significant vegetation forces one to hide behind large boulders left by previous hurricanes or to try and snuggle into crevices or under ledges just to find a little shade — often at the cost of losing the benefits provided by the steady Atlantic breezes.

The lizards of Little Scrub, like those on Sombrero, are very dark. They are also very inquisitive, approaching closely and climbing on your shoes, hat, and arms in hopes of a meal. One once ran up the pant leg of a colleague, causing pure hysteria — followed by outrage with me for having one of the laughs of my life. Intent on the task at hand, I once sat on a cactus while trapping a lizard. For days, I felt those spines — and no position available to man alleviated the discomfort. Although the lizards appear to be thriving, with a range restricted to this tiny rock, they are vulnerable to natural disasters, especially hurricanes.

The Future

In the course of fieldwork with Ellen J. Censky and Robert Powell, we decided to write a book on the reptiles and amphibians of Anguilla (the second to be published by the Anguilla National Trust on various aspects of Anguilla's natural heritage). What quickly became apparent is the relative dearth of information on the nation's herpetofauna. With the exception of a few well-studied species, the possibilities for research are nearly endless — and even less is known about Anguilla's birds and invertebrate fauna. Anguilla could be a paradisiacal place to do research, with its sunny, breezy climate, friendly people, wonderful animals, and a plethora of unanswered questions, but the prospective researcher should bear in mind that that pesky land tenure issue might make an investigation a lot more "interesting" than a first glance would indicate.

Acknowledgments

I thank Robert Powell and Robert Henderson for providing me an opportunity to bring my writing skills out of the closet. I also thank Amanda Adams for her valuable comments on earlier drafts. Ijahnya Christian and Karlyn Liburd, of the Anguilla National Trust, read a later draft and made several valuable suggestions. Finally, I thank Dr. Floyd Homer for his inspiration and encouragement of my efforts to put words on paper.

FIGURE 11. Only a small portion of Little Scrub Island is vegetated. Scrub Island and the mountains of St. Martin are in the background (photograph by Robert Powell).

Suggested Reading

Censky, E. J., K. Hodge, and J. Dudley. 1998. Over-water dispersal of lizards due to hurricanes. *Nature* 395:556.

Censky, E. J. and R. Powell. 2001. Little black dragons of Little Scrub Island. *Fauna* 2(3):24–31.

Day, M. and K. V. D. Hodge. 1998). Sleepy dragons: "Uncover the mythical iguana." *Liat Islander* 43:22–27.

Gumbs, J. C. and K. V. D. Hodge. In press. An assessment of six mangrove swamps in Anguilla: a strategy for biodiversity conservation. Unpubl. report, Wildlife Trust and the Anguilla National Trust, The Valley, Anguilla.

Hodge, K. V. D. 1998. A conservation review of the Lesser Antillean Iguana (*Iguana delicatissima*) in Anguilla, B. W. I.

Unpubl. dissertation, Durrell Institute of Conservation and Ecology (DICE), The University of Kent at Canterbury (UKC), United Kingdom.

Hodge, K. V. D. 2000. Conservation assessment of the Little Scrub Island Ground Lizard (*Ameiva corax*). Unpubl. dissertation, Durrell Institute of Conservation and Ecology (DICE), The University of Kent at Canterbury (UKC), United Kingdom.

Hodge, K. V. D. 2000. Conservation assessment of the Little Scrub ground lizard: an endangered species. *Anguilla Life Mag.* (Winter 2000/2001) 13(3):23.

Hodge, K. V. D. 2001. Implications for biodiversity conservation in Anguilla. *The Anguillian*, January, The Valley, Anguilla.

Hodge, K. V. D. 2001. Political will to conserve and sustainably use biodiversity in Anguilla. *The Anguillian*, August, The Valley, Anguilla.

Hodge, K. V. D., E. J. Censky, and R. Powell. 2003. *The Reptiles and Amphibians of Anguilla, British West Indies*. The Anguilla National Trust, The Valley, Anguilla, British West Indies.

Hodge, K. V. D. and G. Gerber. 1997. Status of the Anguillian *Iguana delicatissima*. *Anguilla Life Mag.* (Winter 1997 10(3):29.

Hughes, J., K. Hodge, I. Christian, J. Stevenson, S. Holliday, I. Fisher, and M. Bryer. 2001. *Towards a bird conservation strategy in Anguilla*. Anguilla National Trust, The Valley.

Petersen, J. B. and G. J. Crock. 2001. Pre-Columbian Amerindians in Anguilla: a long, rich cultural heritage. Unpubl. rept. Fountain Natl. Park Corp., Department of Anthropology,University of Vermont, Burlington.

Petersen, J. B. and G. J. Crock. 2001. Fountain Cave is an internationally significant archaeological site. Unpubl. rept. Fountain Natl. Park Corp., Department of Anthropology, University of Vermont, Burlington.

Townsend, J. H., J. M. Eaton, R. Powell, J. S. Parmerlee, Jr., and R. W. Henderson. 2000. Cuban treefrogs (*Osteopilus septentrionalis*) in Anguilla, Lesser Antilles. *Carib. Jour. Sci.* 36:326–328.

Hodge, K. V. D. 2003. Culture, politics, lizards, and me, p. 241–247. *In* R. W. Henderson and R. Powell (eds.), Islands and the Sea: Essays on Herpetological Exploration in the West Indies. Society for the Study of Amphibians and Reptiles, Ithaca (New York). Contributions to Herpetology, volume 20.

Of Copper Eyes and Ample Nymphs

Hinrich Kaiser[1]

When I was growing up, a career chasing frogs in the Caribbean was really never on my mind — the Caribbean was, however. I caught the travel bug early and, starting at the tender and impressionable age of eight, used to get all kinds of exotic travel catalogs. I also liked nature, and I owned lots of nature-oriented books. So, presumably when I tried to determine how to make a life as a graduate student, thoughts pertaining to the Caribbean and nature collided. The result flung me into Caribbean herpetology. This turned out to be exceptionally rewarding, and my experiences in the Caribbean have provided me with my professional footing. For that, I love the Caribbean — particularly the Lesser Antilles.

My Caribbean experiences have been highly variable, and all make for great stories. I sat out Hurricane Luis on Montserrat while Galway's Soufrière Volcano was smoking. I backed a French "Rocket" (a very small rental car) into a ditch at the top of Soufrière Volcano on Guadeloupe (Figure 1). I was accosted at sunset by gun-toting drug runners at the top of St. Vincent's Soufrière Volcano. I nearly severed my Achilles tendon slipping in a ravine on Martinique while chasing after a frog (the nearly-torn tendon popped only a month later when I merely turned on one foot). I conquered Redonda, and camped there for three days among dusty ruins, small lizards, and goats. I walked across the devastation waged on Guanaja, Islas de la Bahía, by Hurricane Mitch. I chased after frogs in a US $1500-per-night resort on Barbuda (no, we did not stay there). I ate lobster. I body-surfed and lost my swimming trunks. I danced on Barbados during Crop Over (the local carnival) right after a dive counting corals next to the parade grounds. I talked to a local prime minister about declining frogs. I ate more lobster. All of these experiences make the Caribbean very special to me, a feeling reaching much deeper than science alone could explain. However, it all started with finding frogs on islands.

Copper Eyes

In the gentle climate of Martinique, my undergraduate field assistant, Heather Gray (now a Ph.D. herpetologist in her own right), and I had already been out for a couple of nights collecting frogs of the genus *Eleutherodactylus* for my research — but I was looking for something else as well. Another type of frog was supposedly there, but, unfortunately, I was unable to explain to Heather exactly what that other frog looked like. The frog I was looking for, you see, had not yet been described. I needed desperately to find it, and by now, our last day, this caused a little tension.

"It's right there on the leaf in front of you!" Heather said, irritated. Slowly turning my headlamp towards the leaf centered in the cone of Heather's flashlight, I focused on a tiny, dark-brown frog.

"Looks just like another juvenile *Eleutherodactylus* to me," I replied.

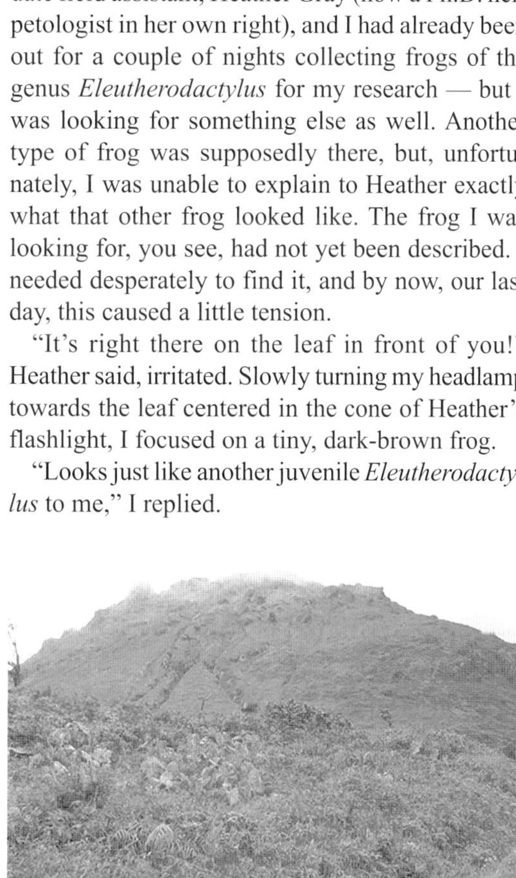

Figure 1. View of Soufrière Volcano on Guadeloupe.

[1] Department of Biology, La Sierra University, Riverside, California 92515, USA. Current address: Department of Biology, Victor Valley College, Victorville, California 92392, USA.

FIGURE 2. Morne Pelée habitat.

"Not *that* one," Heather countered, "the *other* one, with the squared snout." I swung my head a little farther to the left and finally saw what she was talking about. I was sure this was the animal we had been trying to find for nearly three nights. With visibly shaking hands, I carefully approached the leaf and its occupant. I had learned the frog-catching business in Canada's wilds a couple of years earlier, and I hoped my reflexes wouldn't fail me now. "I hope this isn't the only one of these around," I thought. "I mean, what if I miss, and lose that new species forever?"

With Heather frozen in her tracks, I pounced — and, in one rapid sweeping arm motion, I ripped off the entire leaf with my hand. Gotcha! That frog was not getting away now. Heather rushed to my aid with a plastic bag and, with my hand enclosed inside the bag, I released the collected greenery. A small, disoriented, saffron-bellied frog with copper-colored eyes sat at the bottom of the bag and stared into the light. Although this frog was small, the copper-red coloration of its iris and its raised posture gave it quite an illustrious appearance. It was soon to be joined by a companion with a yellowish gray belly and a black throat — and that was all we found on that trip.

That exciting night in 1991 was a stepping stone for a discovery that had begun several months earlier when, on one of my collecting trips, my cousin Helmut Schwarten had first heard a high-pitched "peep" on the slopes of Martinique's Pelée Volcano (Figure 2). While touring Martinique in our quest for *E. johnstonei* and *E. martinicensis* localities, we ended up waiting for sunset on the slopes of Montagne Pelée. That in itself is a perk of frog research — waiting for the sun to go down and watching the colors on the horizon change is not something for which we usually have time in our busy world (Figure 3). Distracted with finding *Eleutherodactylus*, I was sitting, staring at that colorful horizon, and not tuned into life on that slope in general. Suddenly, Helmut cocked his head and said:

"What's this?"

Predictably, I responded "What's what?"

"What's all the peeping over there?" It took me a moment to tune my *Eleutherodactylus*-ear to the higher-pitched frequency.

"That sounds like a whole bunch of other frogs," I ventured, "and I don't think those should be here. Let's go look."

With approaching darkness, we identified the general area where a single peeper among the group was hiding. Walking in on some frogs with flashlights and gear makes them stop calling, and that was the case here. With serious determination, Helmut began tearing apart the small, moss-based bush. We both intermittently glimpsed a tiny frog-like shape retreating deeper and deeper into the dark bush until, cornered and out of options, it launched itself into our lightbeams and away into the darkness. But, we had recorded the call and the thought "new species" had now entered my mind. The hunt had begun!

After sending the two specimens Heather and I had collected to Luis Coloma at the University of Kansas, we received news that we were, quite unexpectedly, dealing with a dendrobatid frog. Although Luis assessed the specimens carefully, they were not easily placed into any existing dendrobatid genus. I returned to the ravines on Martinique twice more, and found that these easily-heard frogs were exceedingly hard to catch — but that's what keeps frog-catching such a challenge! Although collection of some additional specimens allowed us to describe the species and its tadpole, this frog remains a taxonomic and biogeographic enigma. It is now known as *Colostethus chalcopis*, Martinique's Copper-eyed Rocket Frog (Figure 4), and it is the only dendrobatid frog not part of a continental herpetofauna.

FIGURE 3. Caribbean sunset.

Lesser Antillean Journeys

I began my work in the Caribbean in 1988 as a graduate student of David Green's at McGill University. I had visited Barbados for a summer as an undergraduate marine biology research assistant and had fallen for the balmy weather, the phenomenal biodiversity (and I was only thinking *ocean* then), and the incredible people. Those were pressing reasons for me to return to the Caribbean for research. When David asked me for my Master's research project, it was pretty straightforward. David's a frog guy and I like the Caribbean. . . . Does 2 + 2 = 4?

During my fact-finding about Caribbean frogs in the library, I was suitably impressed by how much work had been done — and began to realize how much could yet be done! With a simple population-level genetic comparison in mind, I set out on a four-island tour to collect *Eleutherodactylus johnstonei*, a species now common on nearly all Lesser Antillean islands. With cousin Helmut along as a research assistant, I visited Barbados, Grenada, St. Vincent, and St. Lucia for three weeks over New Year's 1988/89. This was merely a taste of what was to come, sitting with Helmut in a cane field on Barbados at midnight on 31 December, sipping champagne from urine-cup frog containers while surrounded by chorusing *E. johnstonei*.

Aside from the resolve to return to the Caribbean again (and again), I brought back containers with about 300 specimens of *E. johnstonei*, half of which flew to Canada on a jumpseat in the cockpit of a Wardair Boeing 747 — to the great amusement of the crew. Flying with live frogs is an experience I can recommend highly to anyone wishing lots of conversation on a business trip! As it turned out, some of these frogs were not *E. johnstonei*, but a subspecies of the Trinidadian *E. urichi* that looked quite different in life. After consulting with David, I resolved to write a grant and take on the systematics and biogeography of the entire Eastern Caribbean *Eleutherodactylus* fauna as a Ph.D. project. Once that grant had come in, several returns to the Caribbean were assured.

Although known as a playground for the rich and famous, the Caribbean offers so much more than sun, sand, hotels, and highballs. One's attitude when seeing the "Caribbean Way," however, determines whether one sees just a little or all of it. So many vacationers come only to be passively entertained, to sit on the sand, float in the pool, or test their mettle at one contrived water sport or another. They generally get what they expect, but it is merely the equivalent of a Las Vegas poolside in a fancier locality.

FIGURE 4. *Colostethus chalcopis* from near Morne Rouge, Martinique.

Frogs?

To really appreciate the Caribbean, you have to foray inland and stop and talk to those who live there. Among the many experiences of that nature, one instance stands out that helps me now understand just how my relationship with the Caribbean deepened.

By 1993, the leeward road on St. Vincent had suffered tremendously as a consequence of torrential rains and inadequate maintenance. A government can do only so much when the bottom falls out of the market on a monoculture crop, and local residents had no choice but to put up with these deteriorating conditions. Field assistant (now lawyer) Courtney George and I trucked along northwards feeling like the king and queen of the road in our nice, air-conditioned Mitsubishi Pajero. We were looking for a new, unexplored place to find frogs when we saw a woman and her daughter walking in the heat at the foot of a hill, loaded down with various bags and looking miserable. We looked at each other, and I hit the brakes. We took the two of them all the way up the road to a small house in a tiny fishing village and had a great conversation about island life. Our frog story was met with disbelief. What would such a fine-looking couple of folks do with *frogs*? No explanation seemed satisfactory, and the conversation switched to America instead. Then we were invited for dinner, for which we promised to return once we had scouted our frog locality.

As darkness fell, Courtney wondered out loud whether we should honor our dinner invitation, since we might eat something that our hosts would miss later. I submitted that we were people of our word and should not give all tourists a bad name by not showing up — and what a good decision that was! The unfolding evening was one of the nicest I have had in the Caribbean. By candlelight around a small table, we ate a Spam-based dish rich in onions and local spices that was placed between crackers — and there was lots of it! As we ate this simple meal, conversation flowed openly and a friendship was formed. Both Courtney and I came away with a feeling of belonging that evening. Subsequent similar occurrences on several islands have made the Caribbean a home away from home.

Different Countries, Different Customs

Becoming better acquainted with the customs of each island has been a great education, and it has become clear to me that "The Caribbean" or even the "Lesser Antilles" are not a "monophyletic group." Cultural influences are as diverse as those of the previous colonial powers, native Carib nations, and West African traditions. All appear in diverse nuances on each of the islands. I am often asked, "So, which is your favorite island?" — and my answer is always, "What exactly are you looking for?"

When I began work on these islands, my sense of discovery was acute, since I knew almost nothing about the fauna and flora of this region. I have become a better Caribbean naturalist after a decade of research in the region. Now, in addition to taking pleasure in discovery-seeking, I have added another dimension: showing others the *real* Caribbean. Interestingly, this becomes important not only to those who come with me to these lovely islands and begin their Caribbean experience. It also applies to many local residents, who finally get to see their forests and the diverse fauna and flora by night, often for the very first time. Regardless, the experience leaves everyone with something new and something different than they expected — some new appreciation, a new dimension of "Caribbean" is achieved.

Ample Nymphs

One instance, when the combination of discovery and "showing people" exceeded my wildest expectations, occurred while I was working on the frogs of Dominica. This island was seen by Christopher Columbus on a Sunday, and described by him as a "crumpled piece of paper God had flung into the blue Caribbean waters." The topography (Figure 5) indeed is quite dramatic, and one mountain or another is always in the background, or one precipice or another just ahead. My cousin Helmut and I had come to collect *Eleutherodactylus martinicensis* and *Leptodactylus fallax*. I needed the former to complete my data set for this species, whereas the latter was necessary to establish a conservation-oriented captive-breeding program at Metro Toronto Zoo.

FIGURE 5. Morne Macaque and associated forest habitat on Dominica.

When coming to a new island, I had always made it a point to select topographically diverse and geographically widespread localities. This involves a lot of driving around on each island to find appropriate localities (my record was 643 km = 402 mi in 48 h on St. Lucia, an island with dimensions of 42 x 20 km). Among the places we were exploring on Dominica was a new road, cut into the hillside of Morne Trois Pitons National Park. The new road led around Morne Macaque toward Freshwater Lake, which had been created to provide power for a hydroelectric development, and the road provided access for construction workers and vehicles to an area that was accessible only on foot just months earlier. Although this caused some environmental trauma, I was relieved to find, just a few years later, that the montane forest was reclaiming the road and the construction clear-cuts. With road access thusly provided, that evening Helmut and I planned to run a transect across an altitudinal gradient using chorus strength to determine the density of *E. martinicensis* along the road.

On our drive to our starting point that evening, we discussed ways of obtaining reliable data, and we decided to just drive and record whatever we could hear. Driving up our transect roadway from Trafalgar Valley, we were able to identify several areas where chorusing *E. martinicensis* were concentrated. Once we entered the four-wheel drive track to Freshwater Lake, pockets of *E. martinicensis* continued to be present, with at least a few individuals calling pretty much everywhere along our drive. At an altitude of around 700 m, just as we entered the windy and foggy windward slopes of Morne Macaque, we were struck by the fact that, all of a sudden, the calls all had clicks at the end. I do not have perfect pitch, but it also seemed that these calls had a lower frequency when compared with those we heard a few hundred meters back down the road. We backed up until the clicks disappeared, right around 600 m. We then realized that a "dead zone" existed, a quiet zone over an altitudinal difference of about 50 m along about 250 m of the steep road. When we reached Freshwater Lake, the "wheep-click" call was everywhere, on vegetation and among rock piles, and all of the calling males were quite orange, rather than the vivid *martinicensis*-brown.

'What a neat population difference,' I thought—and then I saw her. Looking downhill from the road, I was eye-to-eye with a giant orange *Eleutherodactylus* sitting on the branch of a small tree. The frog was so heavy, it was bending the branch considerably, and it moved slightly in the slope-side gusts and swirling fog. I leaned over and was just able to grab the bottom of the branch. I pulled the tree a little closer and snatched my orange prey from her perch.

"Helmut," I called out. "Check out this one. It's huge!"

"Oh," he answered, "I just caught a really big one, too."

FIGURE 6. *Eleutherodactylus amplinympha* calling, close-up, and egg clutch, all from near the type locality on Dominica.

We found several of these orange giants that night, all of them females. I was sure that we had just found a new species, probably somehow related to *E. martinicensis*, but quite distinct. This species has considerable sexual size dimorphism (females an average of 1.5 times the size of males), a unique call with a lower frequency than *E. martinicensis* and that noticeable click invariably incorporated, and a variety of morphological characteristics that are easily seen in a living specimen but tougher to score in long-preserved material. On an island where tales of forest spirits are an integral part of local lore, I thought that the name, *amplinympha*, was highly appropriate for a high-altitude species whose large females are suddenly encountered "forest-spirit-like" on fog-enveloped branches.

Reporting the find to our friends at the Forestry and Wildlife Division of the Ministry of Agriculture and the Environment and at the Springfield Center for Environmental Protection, Research and Education (Clemson University) caused quite a stir. We had people accompany us to check on these new frogs (Figure 6), the only endemic vertebrate on Dominica. As we began talking to people, we became aware that the distribution of these frogs was wider than we had initially thought, basically encompassing all higher elevation habitats on the island. With the help of some foresters, we were able to find additional areas where we scouted for the "wheep-click" calls. I felt that our discovery was fundamentally changing the way some of these people thought about their environment. They now had something uniquely Dominican they could relate to — notwithstanding the plethora of native plants and insects. A picture of the species even made the cover of the local wildlife magazine, the *New Forester*.

So Much More to Discover

So, this is what it has been like, working in the Caribbean for a decade. Although I love the thrill of new discoveries, I feel that introducing others to such thrills and to this wonderful and unique group of islands is equally rewarding. With over 300 years of herpetological research in the Caribbean behind us, much remains to be discovered in this concentration of tropical splendors — which reminds me, I am still trying to find an assistant to accompany me on my next trip. Any takers?

Acknowledgments

My research in the Caribbean has been funded by a variety of organizations and I have benefited from some tremendous frog-catching expertise along the way. For lending a helping hand and for companionship in the field, I am grateful to César L. Barrio-Amorós, Raul E. Díaz, Wolfgang Feichtinger, Courtney A. George, Heather M. Gray, Michael Schmid, Helmut Schwarten, Timothy F. Sharbel, Dominic von Stösser, Gudrun von Stösser, Jason D. Trujillo, and Axel Werres. For financial backing, I am very grateful to Boehringer Ingelheim Fonds, the College of Arts and Sciences, and the University Research Council at La Sierra University, the Institut für Humangenetik at Würzburg University, and the Studienstiftung des Deutschen Volkes. I hope that my output will continue to be worthy of their input.

Suggested Reading

Kaiser, H., D. M. Green, and M. Schmid. 1994. Systematics and biogeography of Eastern Caribbean frogs of the genus *Eleutherodactylus* (Anura: Leptodactylidae) with the description of a new species from Dominica. *Can. Jour. Zool.* 72:2217–2237.

Kaiser, H. and R. W. Henderson. 1994. The conservation status of Lesser Antillean frogs. *Herpetol. Nat. Hist.* 2:41–56.

Kaiser, H., L. A. Coloma, and H. M. Gray. 1994. A new species of *Colostethus* (Anura: Dendrobatidae) from Martinique, French Antilles. *Herpetologica* 50:23–32.

Malhotra, A. and R. S. Thorpe. 1999. *Reptiles and Amphibians of the Eastern Caribbean*. Macmillan Education Ltd., London, United Kingdom.

Kaiser, H. 2003. Of copper eyes and ample nymphs, p. 249–254. *In* R. W. Henderson and R. Powell (eds.), Islands and the Sea: Essays on Herpetological Exploration in the West Indies. Society for the Study of Amphibians and Reptiles, Ithaca (New York). Contributions to Herpetology, volume 20.

In the Footsteps of French Naturalists, a "Battle" of Iguanas, and "Improvements" in Biodiversity

MICHEL BREUIL[1]

To understand the herpetological communities of the French West Indies, one must first understand precisely the geological history of these islands. We have to know when the volcanoes from which these islands originated emerged from the sea. We have to know the chronology and the topographical relations between the different parts of complex islands such as Basse-Terre, Grande-Terre, and Martinique. We have to know when the main islands separated from their satellites as polar ice caps melted about 12,000 years ago. All of this information is of great importance if we want to estimate the time of divergence between species, subspecies, or even distinctive populations on different islands. We then must endeavor to know precisely the distribution and the ecological niches of the many species of amphibians and reptiles. In addition, we must explore and search for fossils and archeological sites, and attempt to analyze the Amerindian way of life. Finally, to evaluate the European influence in the region, we must review the old records. Early fathers of the Catholic Church described a luxuriant natural paradise, and 18th and 19th century European naturalists, many working for the Muséum national d'Histoire naturelle (Paris), collected specimens and recorded their observations in field notes, catalogues, letters, and drawings.

My goal in this essay was to present a very personal account regarding the current state of understanding the composition of the herpetological communities of the French West Indies, particularly in the light of the poorly known and frequently misunderstood contributions of the first French chroniclers and naturalists to venture into the region. These historical perspectives provide a somewhat new vision of the French West Indian herpetofauna, and entice one to follow in the footprints of our predecessors.

How I Caught the Herpetofaunal Virus

As a 6 year-old, I became enamoured with and began to catch my first tadpoles and newts. My interest never waned, and I stayed in the field of herpetology throughout my academic training. My Ph.D. work focused on a European newt (*Triturus alpestris*) and involved both genetic and ecological methods. I now see considerable irony in the fact that this newt was described in 1768 by Josephus N. Laurenti in the same book in which he described *Iguana delicatissima*, a West Indian reptile which has subsequently become very important to me.

My first contact with the French West Indies was during my wedding trip in February 1987. What I found most striking when I first arrived in Guadeloupe was the ubiquity of anoles (*Anolis marmoratus*) and the piercing calls of whistling frogs (*Eleutherodactylus* spp.). This was so different from the study of newts, during which I had to search long and hard to find quiet ponds or lakes without fish. During that first trip to Guadeloupe, I took many pictures of these phenomenally common creatures, but, when I tried to identify them, the problems began. I don't know why, but at that time, I thought that the herpetological fauna of the French West Indies was depauperate because of the mongoose, but was otherwise well known.

At about that same time, I also had become interested in another island, Iceland. No amphibians or reptiles inhabit Iceland (except for the exceptional

[1] Laboratoire des Reptiles et Amphibiens, Muséum national d'Histoire naturelle, 25 rue Cuvier, 75005 Paris, France.

arrival of a Leatherback Sea Turtle which is now held in Reykjavik Natural History Museum). So, as a broad-minded zoologist, I had turned my attention to birds. This was not a great betrayal of my first love, because, as Huxley had said: "A bird is nothing more than an embellished reptile." Regardless, I came to the study of island biogeography with an historical perspective. I was mainly interested in the modifications of the avian fauna with respect to warming trends in the North Atlantic. Needless to say, these feathered reptiles have much less of a problem when colonizing new islands than do the amphibians and reptiles of the West Indies. In 1989, I wrote a book (*Les Oiseaux d'Islande*) which synthesized all that was known about Icelandic birds.

Iceland is a very fascinating country, but the climatic conditions, even in summer, are very harsh, especially when camping. We used to joke that: "In July, it is still winter, and in August, it is already winter again" (although this isn't true of every year). This is in stark contrast to the climate of the French West Indies, where only on the volcanic summits does the wind nearly always blow and where the constant threat of precipitation (nearly 10 meters a year, much more than on Iceland) creates conditions that cause one to long for warmth. When I studied whistling frogs on those cool peaks, I was far from the sunny beaches where iguanas nest and tourists roast in the sun. Maybe it was my Icelandic experiences that drove me down the slopes to the sandy shores that, fringed with coconut trees, represent the vision of paradise that lures from more temperate regions many of those who seek to work in the West Indies.

The Siren Call of the West Indies

While watching the film "Columbus, 1492," I was amused by the beautiful landscapes of the New World. What was wrong with this vision — nothing more than the emblematic symbol of nearly all tropical beaches: Coconut Palms were not present in the Antilles at that time. An old manuscript (*L'Anomyme de Carpentras*) described the life of some French adventurers who lived in friendly harmony with the Caribs of Martinique and Dominique in 1620, before the first major European settlements were established. The Caribs' way of life involved the extensive use of these islands' natural resources, including a number of palms. However, these trees had only little nuts which were used by the natives to make oil for body painting. Another now ubiquitous West Indian plant is also an exotic newcomer to these shores. Like the Coconut Palm, the Breadfruit Tree was imported intentionally; as a matter of fact, this was the principal goal of the famous voyage to Polynesia by H.M.S. *Bounty*, with the infamous Captain Bligh at the helm. These are only two of many conspicuous botanical examples that demonstrate clearly how the modern West Indian flora has become a polyglot of species originating from all over the Earth.

Is the situation similar for amphibians and reptiles? Can we determine with any precision what the natural herpetofauna of these islands was like before the Amerindian and European arrivals ? What might they have contributed in terms of either introductions or extirpations? What was the impact of trade between the Caribbean region and Africa? Generally, the scientists who work in this region exhibit two schools of thought. The first considers that the arrivals of plants and animals on these volcanic islands (Lesser Antilles), which were never connected with any continent, were due to dispersal by wind, water, or rafts. The second suggests that humans may have played an important role, bringing them to the islands intentionally or as stowaways. Whether or not the latter process was important historically, the process is ongoing. New invasive species are regularly arriving in the Lesser Antilles and some may pose a considerable threat to the original herpetological community. So, novelties are a constant in the West Indies — just one more excellent excuse to visit regularly.

In 1989, I participated in a bat survey in the National Park on Guadeloupe, created specifically to protect the tropical rain forest. For one month that summer, we used mist nets to catch bats — and, while waiting for the bats, I had time to look for frogs. These funny, sometimes very elusive animals were everywhere. But something was strange... most of these frogs were, according to my determinations, a species which was not known to occur on Guadeloupe. We know now that Johnstone's Whistling Frog (*Eleutherodactylus johnstonei*) arrived in Guadeloupe during the early 1970s and colonized the island very quickly. At that time, I mistakenly believed that herpetologists had nothing new to discover in the West Indies — obviously, I had to quickly change my mind.

As a consequence of the bat study, we proposed to the Parc National a study of seed dispersal by these flying mammals. Thus, in the summer of 1992 and again in April 1993, we returned to Guadeloupe, providing me with the perfect opportunity to study the herpetofauna. During that summer, we also went to Terre-de-Bas des Saintes, where I sought to observe Lesser Antillean Iguanas (*Iguana delicatissima*), a species I had previously encountered in 1989 on La Désirade. Once more, I was surprised to find that every iguana we saw was *I. iguana* (the Green or Common Iguana). What was going on? First, I thought I must be in error and had misidentified these animals — but back in Paris, with numerous slides of both species for comparisons, I could no longer doubt that *I. iguana* had taken the place of *I. delicatissima* on Terre-de-Bas.

Iguana delicatissima* versus *Iguana iguana

In association with the creation of a Man and Biosphere Reserve on Guadeloupe, I was asked by the National Park to study the herpetofauna of Les Saintes. In December 1992, I spent 15 days on the island with my colleague, Claude Sastre, a botanist at the Paris Museum and a specialist on the Antillean and South American floras. We worked every day, even on Christmas day — although we discovered that climbing the slopes was much harder this day, because at each house of the village we had to stop to drink the famous "shrub" (rum and orange peel). During that trip, I saw hundreds of *Iguana iguana*, two *I. delicatissima*, and some very strange individuals I was unable to identify. I had come to the conclusion that *I. iguana* was successfully competing with *I. delicatissima* and had nearly succeeded in eliminating it from the island. Later, looking at my slides, I had another idea. In the field, I had been unable to determine the identity of some iguanas because I came to different conclusions if I looked at the tail, at the head, or at color and pattern. Might these unconventional iguanas have been hybrids? Both of my tentative conclusions directly contradicted the prevailing wisdom, based on observations made during the 1960s.

At that time, I warned the Direction régionale de l'Environnement (both species of iguanas have been fully protected in Guadeloupe since 1989) of this potential problem, and I initiated a search for *I. delicatissima* on Basse-Terre and Grande-Terre, where this species was nearly unknown. In the summer of 1993, I spent four weeks, mainly on Basse-Terre, where I found a few populations of *I. delicatissima*, but some of them were coexisting with *I. iguana*, which was very abundant. In those areas, I again found individuals with characters of both species. Mark Day, a British colleague, analyzed tissue samples of iguanas from Les Saintes and Basse-Terre, and confirmed my hybridization hypothesis.

In the meantime, I had seen a picture of an unidentified iguana from the islands of Petite Terre (Figure 1) that appeared to be *I. delicatissima*. I subsequently visited these two small islets with Mark. Hurricane Hugo had hit the islands in September 1989 and many trees were uprooted. Consequently, we had a very tough time negotiating the tangle of spiny bushes. However, we found iguanas. With my wife Anne, who is a botanist, I studied their diet by analyzing fecal samples that had been collected by local ornithologists over the course of a full year. In the meantime, Mark was catching iguanas for his biometric study and tissue sampling, and I proceeded to investigate the swimming ability of *I. delicatissima* to see if active island-to-island dispersal was possible. The "winner" of my "contest" was a beautiful male that stayed under water for 28 minutes at a depth of three meters, suggesting that iguanas may well be able to swim from one island to another. I also tried to estimate the size of the population. In the woody part of the island, the density was incredible. On some big trees, I saw more than ten *I. delicatissima*. Iguanas were everywhere. For an island that was supposed to have none, this was unbelievable. My first estimate of population size was 4,000–6,000 adults.

As a consequence of these data, I proposed the creation of a natural reserve. These islands also are exceptional because Terre de Bas has about 100 Gaiacs (*Guajacum officinale*), a tree which was exploited intensely in historical times because

FIGURE 1. The two islands of Petite Terre which are inhabited by about 10,000 *Iguana delicatissima* and are the type locality of *Ameiva major*.

FIGURE 2. Dead *Iguana delicatissima* after the drought in 2001 on Petite Terre, Guadeloupe.

its wood was exceptionally hard. Furthermore, the species had long since disappeared from Guadeloupe. In addition, some salt marshes support legions of wading birds. The reserve was established in September 1998, a remarkably short time, primarily due to the support of local administrators in charge of nature conservation and the Ministère de l'Environement.

Unfortunately, due to technical, administrative, and financial problems, my scientific studies of *I. delicatissima* did not begin with the creation of the reserve. The only data we have at this time are rough estimates of population size based on methods proposed by local non-herpetologists, who did not consider some aspects of the iguanas' biology, mainly the migration of the females during summer months. I now believe that I've resolved this problem and that the next estimate will take this phenomenon into account. Nevertheless, our data suggest that the population dropped from about 10,000 adults to about 5,000 after the 1995 hurricanes, then recovered to about 10,000 in 1999. I didn't visit Petite Terre in 2000 because of hurricane warnings.

In August 2001, the weather conditions cooperated with me, but not with the iguanas. The drought in Guadeloupe was the most severe ever recorded. I had expected to see substantial mortality, and my prediction was sadly proven correct (Figures 2–4). As soon as I arrived at one of my study sites, I discovered that the situation was catastrophic. In the very hot air, the foul odor of decaying iguana carcasses was pervasive. Guided by my nose, I quickly found the first body. In about two and a half hours, I counted more than 320 dead adult iguanas. In the woody areas, I found as many as 25 dead iguanas per hectare. Extrapolated to the entire island area of 149 ha, between 2,500 and 4,000 iguanas had died. I quickly recognized the unique opportunity for scientific studies. I will be able to estimate the differential mortality in different habitats and the effects on different age classes. Unfortunately, I was unable to return with any scientific material because I had traveled to Petite Terre in a tourist boat with 40 delicate and sea-sick tourists, several of whom had already "fed the fish" on the way to the island. Even I had trouble imagining the impact of a dusty, stinking scientist arriving at the boat with about 100 kilograms of indescribably nauseous stuff. The next day, I proposed a plan for collecting these unique data. Nothing could have been done at the time because even the dead iguanas were protected and we needed authorization from the authorities. Sixteen months later, the collecting permits had still not been provided, even to the wardens of the area, mainly because of a series of local problems. This was a very bad day for science because, when the permits would finally be issued, if ever, most of the dead iguanas would have been recycled by hermit crabs — which are not known to be good herpetologists.

FIGURE 3. A starving and thirsty *Iguana delicatissima* after the drought in 2001 on Petite Terre, Guadeloupe.

Over the course of 66 weeks in the field since 1989, I became more and more concerned with this problem of competition and hybridization between the two *Iguana* species. When I discussed this issue with people who worked as naturalists in Guadeloupe or with Guadeloupean people who lived on Basse-Terre, all told me that no *I. iguana* were there except for a small population in Houëlmont in the south. Conversely, when I asked people living close to *Iguana* populations on Grande-Terre and Basse-Terre, all told me that they are new arrivals, probably having arrived from Les Saintes, where they have been protected since 1976. Apparently, depending on the region, the Common Iguana arrived sometime within the last 25 years, and I became convinced that the Common Iguana was not indigenous to Guadeloupe.

In Martinique, *I. iguana* was introduced in the 1960s from Les Saintes in Fort Saint-Louis, a military base in Fort-de-France which had a little zoo held by Father Robert Pinchon. Father Pinchon, a professor of biology, recorded a number of observations on the natural history of Martinique and Guadeloupe, and helped American naturalists visiting the island collect specimens and data. He published four books on natural history, and assembled a collection of natural objects ranging from stones to ferns, bats, insects, amphibians and reptiles, and human artifacts. A museum at Fort-de-France soon will be dedicated to this complete naturalist. Subsequently, when the zoo was closed, the iguanas were liberated. From this modest start, the species continues to colonize the island, where *I. delicatissima* survived in a few remote places.

FIGURE 4. A fortunate *Iguana delicatissima* that survived on a Caper Tree (*Capparis flexuosa*) that retained a few leaves during the drought in 2001 on Petite Terre, Guadeloupe.

In the Footsteps of French Naturalists

Stimulated by the problem of the introduced iguanas on Martinique and the competition and hybridization on Les Saintes and Basse-Terre, I began to study many of the old notes and manuscripts dealing with the French West Indies in an effort to find the data I would need to successfully address this problem. In the course of this library research, I ran across a number of previously unknown documents — which sent me on a journey back to the past. Little by little, I built a vision not only of iguanas, but of the entire herpetofaunas of Guadeloupe and Martinique with the help of the French fathers Bouton, Plumier, Du Tertre, and Labat, of 17th to 19th century naturalists working for Paris's museum (Jean-Baptiste Thibault de Chanvallon, Auguste Plée, Félix-Louis and Ferdinand-Joseph L'Herminier), and of Amerindians through the help of Father Raymond Breton and his 1,000-page dictionary. From the perspective of modern knowledge, I was able to recognize beyond any doubt the lizards (and other wildlife) described by my predecessors.

The Carib name for Martinique is *Iouanacaera*, which means "Iguana Island," not "Flower Island," as is suggested by many tourist leaflets and guides. Saint Lucia was named *Iouanalao*, which may have the same meaning. Thinking like a Carib is difficult, but it seems logical to assume that a species of *Iguana* was present on these two islands, and that these Amerindians coming from South America named the islands for their abundant wildlife, much as they named Marie-Galante *Tulucaera* ("Crab Island") or *Aulinagan* ("Cottonland," for a species of *Gossypium*) or Les Saintes *Caaroucaera* ("Parrot Island," although, sadly, parrots are no longer found on Les Saintes). Further, the Carib names for Saint Lucia and Martinique might be interpreted to suggest that they were the first islands, coming from the south, to have iguanas. *Iguana delicatissima* is not present on Saint Lucia, but *I. iguana* may have displaced them during historical times.

The oldest citation of an iguana in the Eastern Caribbean was made by Pierre Martyr d'Anghiera who, in 1511, chronicled the Columbus trips. During the second voyage, Columbus's boats arrived at Marie-Galante on 3 November 1493: "Ceux qui descendirent à terre pour visiter celle que l'on avait abordée racontèrent qu'ils n'y virent ni bêtes, ni gens hormis des lézards d'une grandeur extraordinaire" (those who went on land to visit the island where we stopped reported that they saw neither beasts nor people except lizards of an incredible size). From Anonyme de Carpentras and others, we now know that Caribs frequently carried roots, cotton, crabs, tortoises, and iguanas from island to island in order to establish a supply of readily available resources for exploitation during future trips. Thus, when the first Amerindians ventured north, they may very well have taken food with them. Any iguanas so relocated were surely *Iguana iguana*, and they may have been the source of the animals on Saint Lucia, even though the iguanas from Saint Lucia seem to be well differentiated from populations on the other islands. Later, they may have carried *I. delicatissima* from one island to the next. This hypothesis is supported by the fact that the genetic diversity in *I. iguana* from the Lesser Antilles is substantial, suggesting numerous introductions from different parts of South America or even Central America. Conversely, *I. delicatissima* exhibits little genetic variability, and this could be interpreted to suggest that all existing populations are recent descendants of a single founder colony. In addition, because of their great swimming ability, both species may very well have colonized some Lesser Antillean islands on their own.

My 1993 hypothesis that *Iguana iguana* is not indigenous to Guadeloupe and was, in fact, introduced by humans contradicted previously accepted ideas. However, all of the information gleaned from the descriptions and drawings of the old chroniclers (size, coloration, number of eggs, etc.) pertaining to both Martinique and Guadeloupe fit better for *I. delicatissima* than for *I. iguana*. Thus, the conclusion that *I. delicatissima* was the original inhabitant of Guadeloupe and that *I. iguana* arrived only recently is quite reasonable.

"Improvements" in Biodiversity

The Raccoon is the symbol of Parc national de la Guadeloupe. This mammal (*Procyon minor*) was thought to be an endemic species of Basse-Terre and Grande-Terre. A recent genetic study, however, showed that the Guadeloupean Raccoon is very closely related to Raccoons found along the eastern coast of North America. This population may have had its origin with a capsized boat during the 19th century. Ferdinand-Joseph L'Herminier noted in 1840 that "a été tué à Bois Goyave, un individu du genre *Felis*" (an individual which belongs to the genus *Felis* was killed at Bois Goyave). Although not a first hand observation, this was perhaps the first notation of a Raccoon from Guadeloupe (Raccoons at that time were considered to be *Felis montana americana*, or a small cat from South America). The Guadeloupean Raccoon has subsequently spread to Martinique, Saint-Martín, Marie-Galante, and La Désirade.

The Opossum (*Didelphis marsupialis*) is only known from Grenada, Martinique, and Dominica, but it has been found in archeological sites on Saint Lucia as well as Grenada. In the 17th century, no Opossums and no Raccoons were known to occur in the French West Indies, and no fossils of either species have been found. Du Tertre wrote that these two species occurred only on Grenada. For example, he said that when he was on Grenada in 1656 he saw and ate a "Tatoo" (an edentate, *Dasypus* sp.), which was very good, in contrast to the "Manicou" (*Didelphis marsupialis*), which had an awful taste, although it still is eaten locally in Grenada and sometimes in Martinique. Du Tertre also mentioned that several people tried to introduce the Tatoo (*Dasypus novemcinctus hoplites*) to Martinique, but that these animals died when they arrived on Saint-Vincent or Martinique. The oldest citations I was able to find recording either the Tatoo or the Opossum on Martinique were by Thibault de Chanvallon in a manuscript written between 1751 and 1756. In it, he noted that the Tatoo "venus de la Grenade" (came from Grenada) and that the Opossum "venu aussi de la Grenade" (came also from Grenada). The Manicou is very abundant in Martinique and was introduced to Dominica since the 1830s. The Manicou and the Raccoon both prey on amphibians and reptiles and may have depleted some of their populations (a single Raccoon is known to have decimated an iguana [*Cyclura*] population in the Bahamas in recent years). The Agouti (a cat-sized South American rodent, *Dasyprocta aguti*) that was present in the 18th century on Martinique was said to have come from other islands by Thibault de Chanvallon, but all were killed by Caribs. He noted also that monkeys (*Cercopithecus mona* from Grenada or *C. aethiops* from Barbados or Saint-Christopher) were present, but he did not give any information on their possible origins.

The West African Mud Turtle (*Pelusios castaneus*) is known only in Guadeloupe, whereas the Common House Gecko (*Hemidactylus mabouia*), also with an African origin, is found on nearly all of the islands, even on some of the smallest (0.3 ha). Scientists have not been able to determine for sure if these two species arrived with human assistance or if their presence in the West Indies is due to natural events. Father Pinchon mentioned the mud turtle in 1969 (under the name *P. subniger*, which was used in error by subsequent authors), but do we have any records from 17th through the 19th centuries of these two species? A letter by Félix-Louis L'Herminier noted that in 1821 he sent "une émyde de Guadeloupe" (an emydid turtle from Guadeloupe) to the Paris museum. Later his son, Ferdinand-Joseph, in 1837 sent from Guadeloupe to the Paris museum "une émyde à charnière" (an emydid with a hinge) along with "deux emydes de Porto Rico" (two emydid turtles from Puerto Rico). The "émyde à charnière" is probably *P. castaneus*. Unfortunately, the specimens were lost. Conversely, the two Puerto Rican turtles (*Trachemys stejnegeri*) remain in the collection. This suggests that L'Herminier knew the origins of his specimens, and it is not inconceivable that he released some Puerto Rican turtles in the ponds of Marie-Galante. Later, these Puerto Rican turtles may well have been carried to Les Saintes.

Trade between Africa and the West Indies began at the beginning of the 16th century, about 130 years before the first permanent French settlements. Du Tertre, in the first edition of his book (1654), described two "mabouyas" species (geckos), one is said to be very rare and the other one common. In the second edition (1667), he described only one species that seems to have characters of both *Thecadactylus rapicauda* (which is endemic to the Western Hemisphere) and *Hemidactylus mabouia* (which, as mentioned previously, has an African origin). The latter undoubtedly was present on Martinique at the beginning of 19th century (I find it interesting that this species was described from individuals collected outside their original African distribution). That this gecko may have traveled across the Atlantic as a stowaway is supported by the fact that I found an agamid lizard (a family known only from the Eastern Hemisphere) in the collection of Musée Père Pinchon in Fort-de-France. That lizard had been captured alive on a boat originating in the Mediterranean.

Kinixys erosa and *K. homeana* are terrestrial tortoises with hinged shells. These two species caused some problems for herpetologists in the 19th century. At the time of their description (1812, 1827), they were thought to come from South America and Guadeloupe. We now know that this genus is endemic to Africa. The confusion was due to the fact that L'Herminier had in Guadeloupe more than a dozen of these tortoises, which he sent alive to the Paris museum on four occasions between 1813 and 1821. In his accompanying letters, he always mentioned that these tortoises are from "la côte d'Afrique" (the coast of Africa). When Duméril and Bibron described these tortoises in 1835, they didn't look at the correspondence and, as the source of the specimens was known to have lived mainly on Guadeloupe, that was the origin attributed to these tortoises. L'Herminier may have bought animals from African slave traders, a possibility reinforced by the fact that they also had on Guadeloupe a guenon (a monkey), small antelope, and several birds from Africa, along with a number of animals from South America.

Is the French Island Herpetofauna Depauperate? To best understand the origins of a herpetological community, we must have an exact knowledge of the species present today as well as in the past. Because many herpetologists who have worked in the West Indies wanted to be sure to observe or catch their favorite species, they often returned to familiar places where those species were known to occur, not "wasting time" by exploring many remote areas. This approach is necessary when the goal of the trip is to collect animals for genetic or morphological studies. However, my involvement in West Indian herpetology is of a different nature. Herpetology is not my job (although I was asked to curate the museum for a time), but a sideline (I'm teaching biology and geology for biological, geological, and veterinary schools). Therefore, I can use my time to explore new areas or to go to the same places during dry and moist seasons and at night or day. I can afford to return to the sites visited by the early naturalists such as Auguste Plée, a famous but often misunderstood naturalist of the 1820s.

Plée worked at the Paris museum. He compiled extensive collections of plants, animals, and drawings from the West Indies, and also from South and North America. While in the field, one of his missions was to spy on the activities of the Spaniards in Puerto Rico. Because more than ten years passed between his death and the first descriptions of the species he collected, Duméril and Bibron, authors of *Erpétologie Générale*, overlooked his notes and catalogues despite a knowledge of their existence. At that time, most of the collections from Plée were considered to have been made in Martinique, where he had stayed for several months. Plée was an exceptional naturalist in the field, but his field notes also were outstanding, even by modern standards. All the animals, plants, and geological material he sent to the Paris museum are accompanied by vast information on morphology, behavior, and geographical origins. Unfortunately, some of his collections and manuscripts were lost. As I studied his writings, saw his drawings of the landscapes he visited, and studied his specimens, I realized what a pity it was that this great naturalist was not appreciated by the museum (mainly because he was unable to go to Puerto Rico for the Ministry of the Interior) and that Duméril and Bibron didn't use his precious information. Consequently, modern herpetologists have a totally wrong impression of his work.

Following Plée on some of his explorations, I, for example, on Martinique in July 1997, climbed the famous Pitons du Carbet to see if the whistling frogs which live there are the same as those collected by Plée (which Duméril and Bibron used to describe *Eleutherodactylus martinicensis*). In his field notes, Plée wrote that he visited these volcanoes and collected amphibians. In one note, he indicated that he had found "rainette sans palmures" (treefrogs without webbing), a very accurate description of the genus *Eleutherodactylus*. The day on which I followed in his footsteps up Pitons du Carbet was dreary and wet. The constant rain and wind made it very hard for people like me who wear glasses. To see frogs (or lizards), much less to catch them, was a nearly impossible task on the almost vertical slopes and precipitous paths. I caught only one male which, as expected, was consistent with Plée's sample. Thus, I was able to answer the question regarding the origin of these little frogs, which were thought in error to have come from Grande-Terre (where Plée never went).

In August 2000, I traveled to La Désirade, a mongoose-free island that was visited by Father Pinchon. He had mentioned the presence of two geckos (*Hemidactylus mabouia* and *Thecadactylus rapicauda*), but because of mistakes in some of his conclusions, I wanted to survey the herpetofauna. Since I was looking for species other than the *Iguana delicatissima* for which I usually came, I explored other places and other habitats — one of the most productive of which was a rubbish dump! The results were very interesting. At the time, the herpetofauna of La Désirade was considered to consist of only four species: *Eleutherodactylus martinicensis*, *Iguana delicatissima*, *Anolis* (*marmoratus*) *desiradei*, and *Sphaerodactylus fantasticus hippomane* (an aptly named dwarf gecko). After covering about 11 km in five hours and drinking five liters of water, I had added five new species to the list: *Eleutherodactylus johnstonei* (another whistling frog), *Gymnophthalmus underwoodi* (a microteiid lizard), and *Mabuya mabouya* (a skink), as well as confirming the presence of the two geckos, *H. mabouia* and *T. rapicauda*. The Turnip-tailed Geckos (*T. rapicauda*) were quite peculiar. They are usually gray-brown, but those from La Désirade are bluish-gray, orange, and white, a coloration pattern that was previously unknown. This called to mind other Turnip-tailed Geckos I have observed on Saint-Martín (Figure 5) and Saint-Barthélemy, which are uniform beige with small reddish patches. The skink was also of great interest, but unfortunately, at the very moment I was taking its picture, a car came by and this rare lizard dove into the vegetation and defied all of my best efforts to find it again. This was too bad, because this observation was only the second in a very long time and, to the best of my knowledge, the first for La Désirade (the other one was made in 1999 on Petite Terre, but the species was known to occur there from specimens observed in the 1960s).

FIGURE 5. *Thecadactylus rapicauda* from Sint Maarten.

If we look at the current herpetofauna of Martinique and compare it to those of nearby Dominica and Saint Lucia, that of Martinique has a considerably lower diversity. Saint Lucia and Dominica have a boa (*Boa constrictor*) and Dominica has a big frog (the Mountain Chicken, *Leptodactylus fallax*). Martinique, like Saint Lucia, has a venomous snake (*Bothrops lanceolatus* and *B. caribbaeus*, respectively), but neither has *Ameiva* (a ground-dwelling lizard), whereas Dominica has one species of *Ameiva* and the species known to have occurred on Guadeloupe was extirpated at the beginning of the 20th century. This is the conventional view.

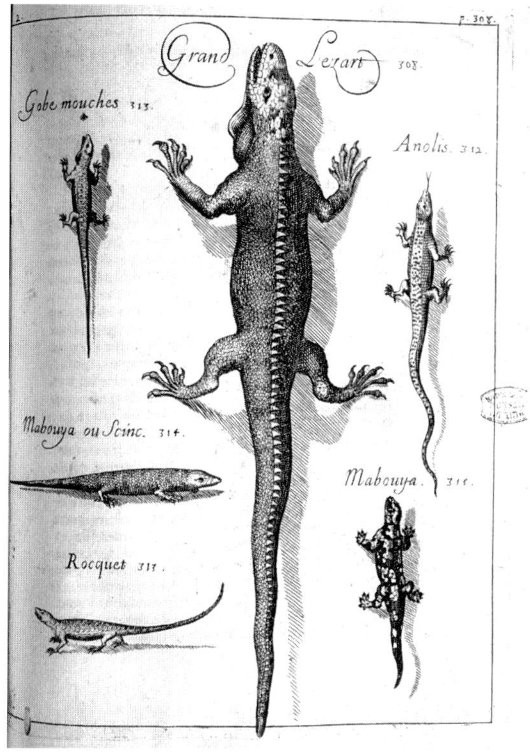

FIGURE 6. Plate from Du Tertre (1667): Anolis = *Ameiva*, Roquet = *Leiocephalus*, Gobemouche = *Anolis*, Mabuya = *Thecadactylus* or *Hemidactylus*, Scinc = *Mabuya* or *Diploglossus* (courtesy of the Bibliothèque Centrale, Museum National d'Histoire Naturelle).

When we examine the old texts, however, we run across some surprises. The Mountain Chicken was also present on Saint Lucia in the 17th century (Anonymous of Carpentras) as well as on Martinique where numerous chroniclers (Du Tertre, Labat, Leblond, Plée) mentioned it. Du Tertre provided a description of the foam nest of this species. Labat seems to have been the only witness to the presence of a boa on Martinique, and was fortunate enough to see one eating a Mountain Chicken. With the addition of these two large species, the herpetofauna of Saint Lucia, Martinique, and Dominica become substantially similar. Martinique and Guadeloupe were densely populated by Europeans and Africans, whereas the Caribs were sent to Dominica. The latter was not cultivated as intensely and endemic species were not exploited to the same degree as in the French West Indies. Consequently, Dominica retained a much larger proportion (maybe all?) of its pre-Columbian fauna.

From a careful reading of Rochefort and Du Tertre, I think at least four more species may have once occurred on Guadeloupe. The description of a snake species by Du Tertre leads me to the conclusion that a "Clibro" (*Clelia* sp.) may have been present. The possible presence of this ophiophagous snake in Guadeloupe is also supported by the fact that we have in our collection a juvenile *C. clelia* that differs considerably from those collected elsewhere. It may represent this lost population. Also, a *Clelia* or a closely related species was found in Recent fossil material on the Antigua-Barbuda Bank, suggesting a much broader distribution than what is known today. Rochefort described a lizard which could only be a "galliwasp" (anguid lizards of the genera *Celestus* or *Diploglossus*). As a rare and distinctive species lives on Montserrat, the presence of an anguid on Guadeloupe in the recent past is possible. Moreover, since this species looks like a skink, confusion between skinks and galliwasps was quite possible. The possible presence of a third species is testimony to the accuracy of Du Tertre's descriptions (Figure 6). His account of a lizard to which he referred as a "roquet" and its drawing led me to believe that it was a member of the genus *Leiocephalus*, which is mainly restricted to the Greater Antilles. In the Lesser Antilles, the genus is represented only by *L. herminieri*, which was said, without much certainty, to have come from Martinique (see below). Additional fossil remains from Anguilla and Antigua-Barbuda were assigned to the extinct form, *L. cuneus*. Since I originally came to the conclusion that a *Leiocephalus* may have occurred on Guadeloupe, fossils (*Leiocephalus* cf. *cuneus*) have been excavated from Guadeloupe. These discoveries strongly support the contention that the species described under the name "roquet" by Du Tertre was truly on Guadeloupe. Under the Caribbean name "anoli," the old father also described and drew a ground lizard that inhabited the small islets of the Cul-de-Sacs and some beaches. This "anole" is nothing other than an *Ameiva*. The last ameivas from Guadeloupe, described as *A. cineracea,* were seen on a small offshore island in 1914. They probably disappeared in 1928 after a major hurricane decimated the island.

So, Martinique and Saint Lucia are exceptions. They are the only two islands where an *Ameiva* was never discovered. The absence of this genus has been attributed to the mongoose, mainly because on Saint Lucia a snake (*Liophis ornatus*) and a relative of *Ameiva* (*Cnemidophorus vanzoi*, actually first described as *Ameiva vanzoi*) are known from small offshore islands where this introduced predator is not present. Consequently, both *Liophis* and *Cnemidophorus* may very well have had a greater distribution that included Saint Lucia. On Martinique, we have the same situation for a snake (*Liophis cursor*), which was well known in the 18th and 19th centuries but is now confined to Rocher du Diamant (Diamond Rock), a 1-ha island off the southern coast of Martinique. However, no ameivas are known to have occurred there and the question of whether one ever did remains unresolved. Sometimes, I wish I had been born before the arrival of the mongoose and the extensive habitat destruction that accompanied European colonization.

Backtracking with the Old Naturalists

Lizards in the genus *Ameiva* are very conspicuous, so we may assume that, if a naturalist or other people concerned with the fauna saw these ground lizards, they would have mentioned them. For example, we have the first records of ameivas by the first priests (fathers Breton and Du Tertre) who stayed on Guadeloupe. The descriptions and the drawing of these chroniclers demonstrate beyond a reasonable doubt that the lizards named "anoli," "anaoli," or "anolis" by the Caribs were, in fact, *Ameiva*. The small arboreal lizards (*Anolis* in the modern sense) were named "boulougoutou," according to Breton.

So, what about Martinique? Labat (1722) did not mention these ground lizards (he took most of his information from Du Tertre), but he saw a boa eating a Mountain Chicken. Thibault de Chanvallon listed Martinique's vertebrates with some comments and descriptions. He used the name "anoli" for all the small lizards found in the gardens (i.e., *Anolis* in the modern sense), noting that they were "the size of a finger," but elsewhere he used the same name for a ground lizard that probably was a curlytail (*Leiocephalus herminieri*).

Plée's only use of "anolis," which very well could have been applied to an ameiva, was: "Anolis de terre parce qu'il se tient toujours dans les trous qu'il y pratique. Cet animal s'enfle beaucoup quand on le poursuit" (a ground anolis because it is always in the hole it digs. This animal inflates itself when it is chased). Although I have seen many ameivas, I have never seen one that inflates itself. I think that this may have been another reference to *L. herminieri* and that this animal, caught by Auguste Plée, is still in the collections of Paris museum. Unfortunately, no other specimens have been found since the 1830s.

The name "anolis de terre" is used on Saint-Barthélemy for *Ameiva plei*, but it was not used by Plée for the specimens of this lizard that he caught on that island in 1821. Furthermore, during his stay on Saint-Barthélemy, Plée mentioned in his diary, under the number 14, a "espèce du genre rainette (*Hyla*) de Puerto Rico et de la Martinique, à doigts des pattes postérieures non palmées" (species of the treefrog genus [*Hyla*] from Puerto Rico and Martinique, with no webbed fingers on the hind feet). Also, under number 9, he mentioned a "Couleuvre: les serpents sont paraît-il très rares dans l'île car tout le monde courrait pour voir celui que j'avais" (grassnake: the snakes are said to be very rare because everybody rushed to see the one I had), and under number 10 and numbers 3/4, he listed "anolis." Unfortunately, the only specimens that arrived in the Paris museum were the three "anoles," all of which are *Ameiva plei*. Specimens of the other species he mentioned have unfortunately been lost (*Eleutherodactylus* sp., *Anolis gingivinus*, *Mabuya* sp., *Alsophis rijgersmaei*).

L'Herminier (probably the son) was also credited with sending two *L. herminieri* to Paris. One is now in the Leiden collection and is labeled "Trinidad," the other is still in Paris and is labeled "Trinité." Still another specimen is credited to Doctor Guyon, who lived on Martinique and wrote his doctoral dissertation on *Bothrops lanceolatus*. In a letter dated 1824 to Georges Guvier, the director of the Paris museum, he wrote: "Je vous envoie: le mabouia, l'anolis de terre et l'anolis proprement dit" (I sent to you: a mabouia, a ground anolis, and an anolis proper). The "mabouia" is the gecko *Hemidactylus mabouia*, the "ground anolis" is *L. herminieri*, and the "anolis proper" is *A. roquet*. Guyon's *Leiocephalus* is still in the collection. How were L'Herminier's localities mistakenly attributed to Trinidad? The answer is very simple, Trinidad at that time was a French island named Trinité and had a very active harbor. A lot of boats traveled from Paris to Trinité and Cayenne, and then stopped on their way back at Martinique and Guadeloupe to take on merchandise (and naturalists' collections). Moreover, in the manifests that were given to the forwarding agents, the harbor of departure was often listed as either Cayenne or Trinité. So, I believe that the conclusion that *L. herminieri* was endemic to Martinique is fully justified. Although I have explored the most probable habitats of this lizard (beaches), I have not found it. Perhaps it survives on one or more of several small offshore islets which I have yet to explore.

Since 1994, when I first visited Martinique in search of *I. delicatissima*, I have spent considerable time and effort looking for "lost" species. I have yet to find any, but I have found species not previously recorded (*Scinax* cf. *ruber*, *Gymnophthalmus underwoodi*). Still, the absence of an *Ameiva* species on Martinique, an island which appears to be suitable for this genus, seems strange from a biogeographic context — thus was born the hypothesis that an *Ameiva* once lived there but was extirpated by the mongoose.

An *Ameiva*, named *A. major*, was described by Duméril and Bibron from Trinité and Cayenne. Subsequently, Baskin and Williams (1966) thought that this lizard was caught in Trinité, a town in Martinique, instead of on the island of Trinidad. For reasons that are difficult to explain precisely, I was not convinced by the arguments developed by these authors. As one individual was credited to L'Herminier, I tried to find any mention of this animal in the National Archives and the Paris museum's library. In fact, a letter from F.-L. L'Herminier (1825) mentioned that he caught "deux lézards pilorii" in les îles de la Petite Terre (Guadeloupe) and will send them to the Paris museum. Initially, I believed that a "lézard pilorii" was *Leiocephalus herminieri*, but then I studied the Plée and Guyon documents and changed my mind. *Ameiva major* was considered at the beginning of the 20th century to belong to the same species as *A. erythrocephala* from Saint-Kitts. As I read more carefully the text in which Daudin (1803) first described the Saint-Kitts ameiva, I noted that Daudin took most of his information from Lacepède, who had previously described this species under the names "tête-rouge" and "lézard pilori" (both based on local names). When L'Herminier went to Saint-Christophe, he recognized the lizard from Petite Terre as the same species, so he merely used the existing name in his letter. The same explanation I noted earlier for the mistakes made in recording the type localities for *L. herminieri* also apply to *A. major*. Moreover, some confusion occurred at the Paris museum, resulting in incorrect transcriptions of locality data associated with specimens sent by F.-L. L'Herminier and Auguste Plée. For example, *Anolis marmoratus* and *Sphaerodactylus fantasticus* are two lizards which were said by Duméril and Bibron to live on Martinique and to have been collected by Plée — instead, these two common lizards are from Guadeloupe (Basse-Terre). In the National Archives, I found a letter (1821) in which L'Herminier wrote: "lézards Anolis de la Capesterre" and "Petit sauriens (salamandre?) des bois de Houëlmont" (anolis lizards from Capesterre and small saurians [salamander?] from Houëlmont's Woods). It is evident that the samples on which Duméril and Bibron based their two descriptions are those collected on Guadeloupe by L'Herminier. This error provides additional support for the idea that Plée collected the historical sample of *Eleutherodactylus martinicensis* on Guadeloupe. So, in this context, what are my conclusions pertaining to *Ameiva major*? Beyond a doubt, *A. major* is not the *Ameiva* from Martinique, instead it is the specialized form from Petite Terre. This conclusion fits well with the biogeographical evidence, because its closest relative is *A. cineracea*, which lived on an offshore islet belonging to the same bank.

Since I discovered the origin of *A. major*, each time I visit Petite Terre, I ask myself why the species is no longer there. I think the answer is rather simple. Petite Terre is a very flat island (maximum elevation, 8 m). I observed the destruction incurred by Hurricane Hugo in 1989, and I have found compelling evidence that, as a consequence of earthquakes and volcanic activity in the area, Petite Terre may have been completely submerged many times since 1825. Because they are beach dwellers, ameivas are exceedingly vulnerable to flooding (along with curlytail lizards); conversely, other lizard species may survive in trees to subsequently repopulate the island.

The only old citation of an ameiva I was able to find for Martinique is from Lacepède in his text on *Bothrops lanceolatus*. He wrote that: "Le Fer-de-lance se nourrit de lézards Améiva, & même de rats, de volaille, de gibier & de chats" (the Lancehead feeds on Ameiva lizards, and even rats, poultry, game, & cats). I tried to find the text from Bonodet de Foix, from which Lacepède drew all of his information on this snake. Unfortunately, it seems that the Bibliothèque Nationale in Paris no longer possesses a copy of this text. Lacepède knew well that the "anolis" from Du Tertre was an *Ameiva*. He undoubtedly recognized in the description of the diet of *B. lanceolatus* such a species. However, I think Lacepède was mistaken. For example, he erroneously named his anole, *Anolis roquet*. In fact, the name "roquet" was coined by the first Frenchmen to describe the *Leiocephalus* from Guadeloupe. Du Tertre, Rochefort, and Breton had used this name for *Leiocephalus*. Thus, Lacepède had a true *Anolis* in his hands, in the modern sense of the name, but he used in his monograph a number of details drawn from Rochefort and Du Tertre that these authors had applied to their own "roquet," i.e., *Leiocephalus* cf. *cuneus*. We know beyond a doubt that Thibault de Chanvallon collected the samples used by Lacepède to describe *Liophis cursor*, and he probably also collected the *Anolis* sent to Lacepède. As I noted earlier, Thibault de Chanvallon used the name "roquet" rather indiscriminately, and thus Lacepède was misled. If Bonodet de Foix used the name "anolis" to *describe* the prey of *Bothrops lanceolatus*, Lacepède must have considered that the "anolis" in question was actually an *Ameiva*. Moreover, if that "anolis" was mentioned as a prey item of *B. lanceolatus*, we could assume that this lizard was very abundant on Martinique. As a consequence, it should have been caught or at least mentioned by Labat, Thibault de Chanvallon, Guyon, or Plée. As Guyon used all of the amphibians and reptiles he could find for his research on *Bothrops* envenomation, if an *Ameiva* was present on Martinique, he surely would have caught one for his tests. To me, Bonodet's "anolis," which is Lacepède's "ameiva," is *L. herminieri*, of which five specimens were caught in 19th century. I think it is very improbable that Bonodet's "anolis" is an *Anolis* in the modern sense. So, once again, evidence for the existence of an *Ameiva* species on Martinique in historical times is lacking.

If my conclusion is correct, that is, no *Ameiva* existed historically on Martinique, the biogeographical data should provide some support. The widespread *A. ameiva* from South and Central America is present on nearly all of the islands from Trinidad and Tobago to Saint-Vincent. Saint Lucia has no *Ameiva*, but a close relative (*Cnemidophorus vanzoi*) occurs on two satellite islands, and the absence of this species on the main island could be due to predation by mongooses and rats.

Amphibians and reptiles appear to have invaded the Lesser Antilles from both the south and the north. For example, anoles in the *Anolis bimaculatus* group are related to Greater Antillean forms and inhabit the islands from Anguilla south to Dominica. In contrast, anoles from Martinique to the Grenadines belong to the *A. roquet* group, which originated in South America. These two groups are not known to coexist. Similarly, snakes of the genus *Liophis* are invaders from the south and those of the genus *Alsophis* are from the north. However, snakes in these two genera live together on Guadeloupe, Dominica, and perhaps on Les Saintes. For *Ameiva*, one may suggest that the northern *Ameiva* did not reach Martinique and that the southern forms stopped at Saint Lucia. The progression of *Ameiva* to the north may have been blocked by the existence of *Cnemidophorus* or by the fact that the necessary ecological niche was already occupied on Martinique by *L. herminieri*, another beach dweller. Such an explanation is quite reasonable because, according to Du Tertre, on Guadeloupe, *Leiocephalus* cf. *cuneus* inhabited the beaches of the Cul-de-Sac, whereas *Ameiva* sp. occupied mainly the small islets where it apparently differentiated into at least two species (*Ameiva major* and *A. cineracea*, plus the forms known only from fossil material collected on Les Saintes and La Désirade).

How Do We Protect What's Left?

We have already lost a huge portion of the historical biodiversity in the eastern Caribbean, and Martinique and Guadeloupe are at the top of the list of affected islands. With the apparently ongoing arrival of new invasive species, the situation has become critical. Only the herpetofaunal biodiversity of Dominica, which lies between these two French islands, is relatively pristine — and is consequently of great interest for purposes of comparison.

On the basis of my research in the archives, published results, and personal observations, the French West Indies have definitely lost at least 11 species of amphibians and reptiles in historical times: *Eleutherodactylus* sp. (Saint-Barthélemy), *Leptodactylus fallax* (Martinique), *Ameiva major*, *A. cineracea*, and *Ameiva* sp. (Guadeloupe and satellites), *Leiocephalus herminieri* (Martinique), *Leiocephalus* cf. *cuneus* (Guadeloupe), *Diploglossus* sp. (Guadeloupe), *Boa constrictor* (Martinique), *Alsophis antillensis* (Basse-Terre, Grande-Terre), and *Clelia* sp. (Guadeloupe). Other species have been extirpated from some islands: *Iguana delicatissima*, *Mabuya mabouya*, *Alsophis antillensis* ssp., *Liophis juliae*, and *L. cursor*.

Conversely, additions to the herpetofauna from inadvertent or intentional introductions include *Eleutherodactylus johnstonei* (all main islands), *E. planirostris* (?)(Basse-Terre), *Bufo marinus* (Basse-

Terre, Grande-Terre, Martinique), *Osteopilus septentrionalis* (Saint-Martín, Saint-Barthélemy), *Scinax* cf. *ruber* (Martinique), *Pelusios castaneus* (mainly Grande-Terre), *Trachemys scripta elegans* (Grande-Terre, Basse-Terre, Marie-Galante, Martinique), *Iguana iguana* (Martinique, Grande-Terre, Basse-Terre, Les Saintes), *Hemidactylus mabouia* (everywhere), *Gecko gekko* (Martinique), *Ramphotyphlops braminus* (Saint-Martín, Saint-Barthélemy), and *Elaphe guttata* (Saint-Barthélemy, although the documented waifs have failed to establish a self-perpetuating population). These introductions may result in very critical consequences. If nothing is done, additional endemic species may very well be lost through competition, predation, and sometimes hybridization with invasives.

Populations of tortoises, *Chelonoidis* (= *Geochelone*) *carbonaria* (and *C. denticulata*?) on Martinique, Grande-Terre, Basse-Terre, and Les Saintes, may have multiple origins. Some may have dispersed naturally, but others undoubtedly were introduced by Caribs, early Europeans, or current inhabitants of the islands. *Colostethus chalcopis* was described in 1994 as an endemic frog species from Martinique and, although unlikely, it might also be an introduced species. Many plants have been imported to the island from Venezuela, where the closest relatives of this species occur.

Les Iles de la Petite Terre are now a natural reserve created for *Iguana delicatissima*. With the expansion of *I. iguana* into Les Saintes, Basse-Terre, and Grande-Terre, a large pool of Common Iguanas has been generated. They could soon arrive naturally or with the help of humans on Petite Terre and La Désirade. Interestingly, a poster produced in 1999 in Guadeloupe to promote conservation of the Lesser Antillean Iguana didn't mention that competition and hybridization with *I. iguana* was the main threat, after habitat destruction, for the endemic species. I was told that this decision was made because the poster was directed to the general public, and that too much detail might be confusing. If this is the case, why did an article in the local newspaper, "*France-Antilles*," speak to the problem of competition and hybridization in Martinique each time an iguana is caught?

The colonization of new islands by *Iguana iguana* through over-water dispersal is now well documented. Hurricane Luis (1995) rafted *I. iguana* from Les Saintes to Antigua-Barbuda and Anguilla. Also, documented observations exist of active dispersal from Les Saintes to Basse-Terre. In early 2001, at least three *I. iguana* with a "*rhinolopha*" phenotype and probably from Saint Lucia were found in the south of Martinique. They were caught by park wardens and liberated in Fort Saint-Louis, where an introduced population was already established in the 1960s from Les Saintes.

On Martinique, to compensate for the expansion of Common Iguanas, an effort has begun to increase, for possible future reintroductions, the size of the population of *I. delicatissima* from Ilet Chancel, which is the species' last stronghold in the archipelago. Even this population may have resulted from an introduction by humans to establish a food resource for the slaves who worked in the pottery factory. This 80-ha island is inhabited by only about 300 adults, whereas Petite Terre, with an area of 149 ha, supports more than 10,000 iguanas. The feeding capacity of Chancel is capable of sustaining more iguanas than currently exist, even in dry conditions. The limiting factor seems to be the existence of only two small colonial breeding sites. Thus, due to spatial constraints, females laying later destroy the eggs already deposited. In July–August, i.e., in the main laying period, more than ten nests can be excavated almost every day. To address this issue, I have proposed to the Office national des Forêts and Direction régionale de l'Environnement a plan to improve one of the breeding sites. In April 2000, the area between two small patches of one of the laying sites was tilled to a depth of about 50 centimeters to create a smooth, bare, and slightly inclined substrate of sand and earth. Large roots and stones were removed, and the area was enclosed by a fence to prevent the 10,000 annual visitors from collapsing the nests.

Results during the subsequent summers of 2000, 2001, and 2002 were very positive. Very few eggs were destroyed by late-nesting females and numerous hatchlings were observed. This first effort suggests that we can increase the reproductive success and thus the size of the existing population — possibly to the extent of supplying a pool for use in future reintroductions of the species into suitable habitats.

On Chancel, individual *I. delicatissima* have been uniquely tagged since August 1993. Every time I go there to check on the population, I hope to find some of the "old fellows." This past summer (2002), we found number 348, which was tagged by Mark Day in 1993. He is now a beautiful male (the iguana,

FIGURE 7. Male (left) and female *Iguana delicatissima* fed by Mr. and Mrs. Jean-Claude Plassais, the owners of Villa Yuana, an iguana "hotel" on Saint-Barthélemy.

not Mark!), about 12–13 years in age. This is a good example of international cooperation to study and protect these fabulous lizards.

In August 1996, I traveled to Saint-Barthélemy for the first time. Among other things, I discovered in a garden center some tadpoles of the Cuban Treefrog (*Osteopilus septentrionalis*). Adults of this species arrived with imported vegetation from Florida (as did *Elaphe guttata* and, probably, *Ramphotyphlops braminus*). Plants were used to restore the gardens after the destruction caused by the 1995 hurricanes Luis and Marilyn. In September 1996, I warned the Direction régionale de l'Environnement about this invasive species. Two years later, the same Guadeloupean office to which I had sent my warning informed me of the presence of a huge frog on the island. In spite of my recommendations, nothing was done. Finally, in August 2000, I was asked to go and evaluate the situation. Cuban Treefrogs were everywhere, even on the highest tree on the highest point on the island. Five years after my first observations, no effort has been made to eradicate this treefrog, excepting two articles I wrote for the local newspaper and a leaflet that will be published in the near future. At the Rio Conference, invasive species were determined to be the second leading cause of decreases in biodiversity (after only habitat destruction).

Saint-Bart's iguanas are fortunate. They have their own "hotel" (Figure 7), the Villa Yuana was created by Mr. and Mrs. Jean-Claude Plassais. The "visitors'" accommodations are situated over a ravine which is in a natural state compared to the surrounding tropical gardens. Mr. and Mrs. Plassais love their iguanas. About 40 adults live here. When food is depleted after a hurricane or during a drought, Mr. and Mrs. Plassais buy food for their residents. This careful attention allows the iguanas to survive during hard times. These iguanas are very tame, entering the house to sunbathe and copulate on their deck or to eat from their hands. Moreover, this iguana hotel supports small numbers of nearly all of Saint-Bart's herpetofauna. Since August 2000, with the assistance of Mr. and Mrs. Plassais, I began to tag this population. This will allow me to compare the growth of the southern *Iguana delicatissima* population (on Chancel) with one of the northern populations. This is of particular interest, because the maximum sizes in the two regions differ by about 20 cm in total length.

Moreover, this last stronghold will perhaps enable us to repopulate the offshore islets that were swarming with iguanas in the 1960s. Although part of the Saint Barts Marine Nature Reserve, created in 1998, the islands today support fewer than ten iguanas (Figures 8–9), primarily because feral goats ate all of the vegetation. With the help of the staff of the reserve, we plan to remove these mammals and to fence some areas to allow plants to grow and iguanas to feed.

FIGURE 8. The author with an *Iguana delicatissima* at Anse des Cayes, Saint-Barthélemy.

FIGURE 9. An *Iguana delicatissima* at Anse des Cayes, Saint-Barthélemy.

Epilogue

So I come to the conclusion of my backward tale. A tale in which I attempted to examine the processes responsible for the current state of the French West Indian herpetofauna. Backward because I had to explore the old archives for the information necessary to understand current conditions. Sadly, many species that were present at the time of European colonization have been lost and are known only from specimens collected (and sometimes incorrectly labeled) by the early naturalists. Unfortunately, many of the recent changes have been negative, habitat destruction is rampant, and a number of potentially harmful invaders are established.

The next step in this venture is to go even farther into the past in an effort to imagine how the original herpetofauna of Guadeloupe and Martinique originated after the islands' emergence from the sea many millions of years ago, and to see what happened before and after the sea level rises resulting from melting polar ice caps some 12,000 years ago. These islands provide an excellent model to study speciation, adaptation, and colonization, largely because the geology of the French West Indies is better known than that of other Caribbean Islands.

When I am in the field, especially in the volcanic areas, I try to imagine the landscapes of the different periods of the past and the very nature of this world as it existed before the European onslaught. The Lesser Antillean herpetofauna is at a crossroads of geology, zoology, botany, archaeology, and history — and much remains to be discovered. I can only hope to be a part of that adventure.

Acknowledgments

I thank Robert Henderson and Robert Powell for their invitation to contribute to this collection and for their immense help in improving the text. Many thanks also to all the people who have assisted me in the field, given me precious information, and provided accommodations, and to the administrators who fund my work and believe in my projects.

Suggested Reading

Anonyme de Carpentras. 1994. *Un flibustier français dans la mer des Antilles en 1618–1620, présenté par J.-P. Moreau.* Payot et Rivages, Paris.

Baskin, J. N. and E. E. Williams. 1966. The Lesser Antillean *Ameiva*. Stud. Fauna Curaçao Carib. Isl. 89:143–176.

Breton, R. 1647. *Relations de l'Ile de la Guadeloupe contenant l'histoire des choses naturelles les plus rares de cette île, des façons de faire, et des mœurs des anciens habitants, appelés communément sauvages, et de ce qui s'est passé de plus remarquable en cette mission, depuis que l'île est habitée des François*. Société d'Histoire de la Guadeloupe (1978), Basse-Terre.

Breton, R. 1665–1667. *Dictionnaire caraïbe françois. Dictionnaire françois caraïbe meslé de quantité de remarques historiques pour l'éclaircissement de la langue. Grammaire caraïbe.* 3 vols., Gilles Bouquet, Auxerre.

Breton, R. 1999. *Dictionnaire caraïbe-français* (with CDrom). IRD-Karthala. Edition of 1665 edited and annotated by CELIA and GEREC, Paris.

Breuil, M. 2002. *Histoire naturelle des Amphibiens et des Reptiles terrestres de l'Archipel Guadeloupéen.* Guadeloupe, Saint-Martín, Saint-Barthélemy. Patrinoines Naturels, vol. 54. Publications Scientifiques du Muséum, Paris.

Du Tertre, J.-B. 1654. *Histoire générale des isles de Saint-Christophe, de la Guadeloupe, de la Martinique et autres de l'Amérique où l'on verra l'établissement des colonies françaises dans ces isles, leurs guerres civiles et étrangères et tout ce qui se passe dans le voyage et retour des Indes*. J. et E. Langlois, Paris.

Du Tertre, J.-B. 1667. *Histoire générale des Antilles habitées par les François*. 2 vols., Thomas Jolly, Paris.

Labat J.-B. 1722. *Voyage aux isles. Chronique aventureuse des Caraïbes 1693–1705*. Phébus libretto, Paris (1993, from the edition of 6 volumes of 1722).

Labat J.-B. 1724. *Voyage du Père Labat, aux isles de l'Amérique contenant: une exacte description de toutes ces isles; des arbres, plantes, fleurs et fruits qu'elles produisent; des animaux, oiseaux, reptiles et poissons qu'on y trouve; des habitants, de leurs mœurs et coutûmes, des manufactures et du commerce qu'on y fait etc.* 2 vols. P. Lusson et al., La Haye (from the edition of 1722 in 6 volumes).

Leblond, J.-B. 1813. *Voyages aux Antilles. D'île en île, de la Martinique à Trinidad (1767–1773)*. Karthala, Paris (edited and annotated by Monique Pouliquen, 2000).

Lescure, J. 1992. L'épopée des voyageurs naturalistes aux Antilles et en Guyane, p. 59–70. *In Voyages aux îles d'Amérique. Muséum national d'Histoire naturelle*. Société des Amis des Archives, Archives Nationales, Paris.

Lescure, J. 2001. Les voyageurs et les naturalistes français aux Antilles (XVIIe–XIXe siècle), p. 108–133. *In* Histoire naturelle (ed.), *C.T.H.S., A.-G. 123e Congrès national des sociétés historiques et scientifiques*, 1998, Paris.

Pons, J.-M., V. Volobouev, J. F. Ducroz, A. Tillier, and D. Reudet. 1999. Is the Guadeloupean raccoon (*Procyon minor*) really an endemic species? New insights from molecular and chromosomal analyses. *Jour. Zool. Syst. Evol. Res*. 37:101–108.

Rochefort, C. D. 1658. *Histoire naturelle et morale des Antilles de l'Amérique. Histoire générale des Antilles habitées par les Français*. 2 vols., Arnould Liers, Rotterdam.

Schwartz, A. and R. W. Henderson. 1991. *Amphibians and Reptiles of the West Indies: Descriptions, Distributions, and Natural History*. Univ. Florida Press, Gainesville.

Thésée, F. 1989. *Auguste Plée: ses travaux et ses tribulations aux Antilles, au Canada, en Colombie*. Éd. Caribéennes, Paris.

Thibault de Chanvallon, J.-B. 1763. *Voyage à la Martinique, contenant diverses observations sur la Physique, l'Histoire naturelle, l'Agriculture, les Mœurs et les Usages de cette isle, faites en 1751 et les années suivantes*. Bauché Libraire, Paris.

Breuil, M. 2003. In the footsteps of French naturalists, a "battle" of iguanas, and "improvements" in biodiversity, p. 255–270. *In* R. W. Henderson and R. Powell (eds.), Islands and the Sea: Essays on Herpetological Exploration in the West Indies. Society for the Study of Amphibians and Reptiles, Ithaca (New York). Contributions to Herpetology, volume 20.

Night Shift

Robert W. Henderson[1]

Over the past two decades, I have had the good fortune of visiting more than 40 islands in the West Indies. I have studied the ecology of anoles (*Anolis*) and tree snakes (*Uromacer*) on Hispaniola and dwarf geckos (*Sphaerodactylus*) on Anguilla; searched for racers (*Alsophis*) on Cuba, St. Barthélemy, Saba, St. Eustatius, Montserrat, and other islands; and hunted for snakes long believed extirpated or extinct on Antigua, Guadeloupe, and St. Vincent. The opportunity to do this research has been wonderful, but to do it on sun-drenched, palm-fringed islands surrounded by the turquoise waters of the Caribbean has provided the ultimate "icing on the cake."

Long, Sharp Teeth

For more than a decade, beginning in 1988, I have focused my research on West Indian treeboas, and especially the species that occurs on the Grenada Bank — *Corallus grenadensis*. A treeboa has a large, chunky head and a mouth armed with lots of long, sharp teeth. That head is attached to an amazingly slender neck, followed by a laterally compressed body, and a prehensile tail. On Grenada, treeboas display an intriguing array of colors and patterns. The predominant color is associated with the environment in which the treeboas occur (Figure 1). Those snakes that occur in open, sun-drenched areas, such as Acacia forest near sea level, are often yellowish. At elevations above about 400 m, in forests or orchards that are frequently blanketed in clouds and misting rain, the treeboas are usually very dark brown. In more heterogeneous habitats (often a combination of mixed orchards and native forest) at middle elevations, the snakes show more variation in color and pattern, but taupe predominates (Figure 2). Treeboas are strongly arboreal, rarely encountered at ground level, and have irascible tempers. They are quick to bite, but their lunging strikes are usually avoidable, and a bite is nothing to worry about (merely wipe the blood on your trousers and look for more snakes). Treeboas spend most daylight hours resting coiled in trees and begin foraging for lizards, small rodents, and, occasionally, birds and bats, at dusk.

Figure 1. Treeboas (*Corallus grenadensis*) from different areas on Grenada vary considerably in color and pattern. The yellow boa is from Pearls (30 m above sea level), whereas the dark individual was found at Spring Gardens Estate (460 m above sea level)(top photograph by Richard A. Sajdak, bottom photograph by author).

[1] Section of Vertebrate Zoology, Milwaukee Public Museum, Milwaukee, Wisconsin 53233, USA.

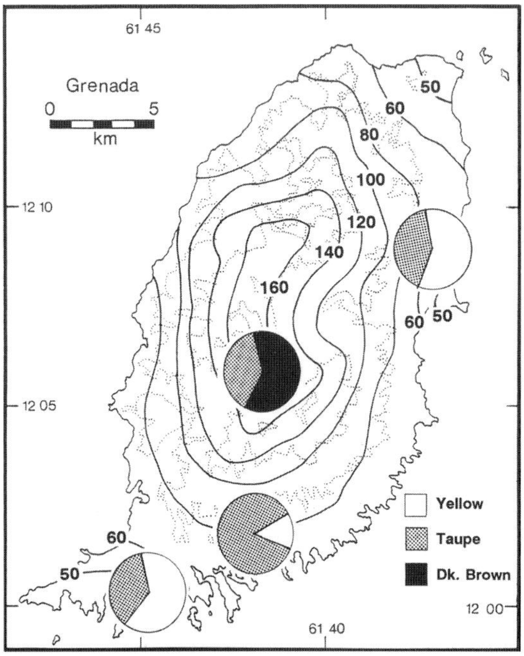

FIGURE 2. Map of Grenada showing contour (dotted lines at 400, 1200, and 2000 ft.), rainfall (solid lines, in inches), and the relative frequency of occurrence of the primary treeboa color phases. English units of measurement were used for the original base maps created by and for the government of Grenada.

Nocturnal Activities

Because treeboa activity isn't initiated until nightfall, my daily activity is largely nocturnal as well. By day, I can sit on the balcony of a small hotel in Mourne Rouge, just south of St. George's, and work on my field notes, transcribing data collected the night before. Before me is the translucent blue Caribbean, edged by graceful Coconut Palms that so effectively symbolize the prototypical tropical paradise. Turning inland, I look at the mountains I will cross later to get to a study site on the windward (Atlantic) coast. Looking at the forest-covered mountains, I imagine treeboas coiled neatly on branches, sleeping the day away, possibly assimilating a meal captured the night before. Gazing towards those distant mountains always triggers an intense anticipation for the work to be done that night: Will my companions and I find treeboas? How many? Will we observe feeding activity? Tonight we're heading for one of my favorite study sites on Grenada. I initially visited Pearls in 1988 during my first day in Grenada, having stumbled upon it quite by accident. It was a hotbed of treeboa activity, and subsequent visits over the next 12 years have consistently reinforced that impression.

In the mid-to-late afternoon, before leaving our living quarters for the night's work, equipment and supplies are checked. We make sure our headlamps have adequate battery power and spare bulbs, and we gather collecting bags, clipboard and data sheets for recording information, and binoculars for identifying and watching treeboas on high branches. We also carry mosquito repellent, scales and a tape measure for weighing and measuring snakes, and paint pens for marking them. A jar of peanut butter, a loaf of bread, some cookies, soda pop, and bottles of water will provide the evening's nourishment.

Although only about 19 km "as the crow flies," we will need over an hour of steady driving to get to Pearls, and this afternoon we'll probably make a couple of stops along the way in order to check other sites. Our route takes us through St. George's: along Lagoon Road, around the picturesque harbor (considered by many to be the most beautiful in the Antilles), through the Sendall Tunnel, and into a quagmire of mini-buses and fish-mongers and a cacophony of bleating horns and shouting voices. Leaving that confusion behind, we swing east onto River Road. Initially, the narrow road parallels the St. Johns River where a few large trees stand as stark reminders of forests long gone. Treeboas once foraged along this stretch of the river, and a few may still. We pass through Gretna Green and Sans Souci and begin our slow ascent into the mountains. The road, one of the primary thoroughfares on the island, is narrow and heavily pock-marked, with some potholes of magnificent proportions. Dodging the potholes and oncoming traffic is made more interesting by the blind curves, numerous switchbacks, and alarming drop-offs that characterize the ascent and descent of the central range of mountains.

As we climb ever higher, we pass Snug Corner, and then to Vendôme at about 350 m. Here, in 1988, the headlights of my jeep elicited the eyeshine of a treeboa foraging in a roadside nutmeg tree for the first time. Soon we enter the Grand Étang Forest Reserve, where night-time vision is often limited to a few meters in front of one's vehicle as it becomes cloaked in dense fog and driving rain. The road becomes very steep and slippery and the switchbacks in places approach 180 degrees. However, the narrow road is bordered on either side by high walls of spectacular vegetation and, in some places, a ceiling as well. The impression is of a green tunnel, and it may elicit feelings of claustrophobia in some. The walls

FIGURE 3. Sunset on the abandoned runway at Pearls, Grenada (photograph by Joel Friesch).

FIGURE 4. Wreckage of a Russian plane at Pearls, Grenada (photograph by Joel Friesch).

are comprised of palms, bamboo, tree ferns, and bromeliads; in short, a botanical wonderland. Rain frogs (*Eleutherodactylus*) serenade enthusiastically, trying to distract me from the job at hand, but just concentrating on driving is critical.

Eventually we arrive at Grand Étang, a lake situated at 530 m in a dormant volcano. This lake and surrounding forest, networked with hiking trails, is a favorite destination for both tourists and Grenadians. A visitor center provides information about the lake, the forest, and its wildlife, as well as the area's history and culture. I have searched for treeboas at this site dozens of times over a period of 13 years, and have encountered only four caught over three nights. Thus, dozens of hunts resulted in no snakes. Nevertheless, the site is intriguing, anticipation is always high, and despite almost predictable failure, it's worth the effort. Finding a Grand Étang treeboa is a momentous occasion, and I recall each encounter as though it happened yesterday. Unlike the dingy yellow or taupe-colored treeboas at lower elevations, those from Grand Étang and nearby Spring Gardens Estate are very dark brown with an attractive network of lacy narrow white markings.

Leaving Grand Étang, we begin the long descent to the windward coast. Again, the narrow road is a seemingly endless series of astonishing switchbacks and potholes of stunning magnitude. We pass long stands of BananaTrees, interspersed with Nutmeg, Breadfruit, Citrus, and Mango. Towns and villages appear frequently, with Birch Grove one of the largest (but you're still in and out in a matter of a minute or two). The road eventually levels off and straight stretches materialize as we approach Balthazar Estate, 75 m above sea level. This has always been one of my favorite treeboa sites. I can't tell you how many times the tree under which I parked my jeep has yielded a snake. The gentle hills sustain a variety of orchard crops, and one particular hillside harbors a majestic stand of Royal Palms. The Balthazar River parallels the study area, with dense stands of treeboa-inhabited bamboo on its banks. This scenic and productive site has remained relatively unchanged over the 12-year period I've been visiting, and I hope it remains so for many more years to come.

It's getting late and we still have not reached the site of the night's work. We press on from Balthazar, and eventually arrive in the coastal town of Grenville, Grenada's "second city." Other than to drive through, I've spent no time there, and today is no exception. Past Grenville, the countryside becomes even more interesting. Orchards are commonplace as we pass through Telescope, cross the Great River and arrive at — and exit from — Paradise. A short time later we near our final destination. We drive onto the abandoned airstrip at Pearls Airport, at one time Grenada's principal site for air traffic (Figure 3). Today it is in ruin, pock-marked, and serving only as a thoroughfare for pedestrians, bicycles, automobiles, and livestock. The wreckage of two airplanes, one Cuban and one Russian (Figure 4), offer silent testimony to the conflict in 1983. The airstrip ends only a few meters from the beach and

FIGURE 5. Treeboa habitat at Pearls, Grenada (photograph by Richard A. Sajdak).

surf of the windswept coast. We check in with military personnel stationed at the airport to let them know we will be active in the area that night.

Leaving the airstrip, our jeep weaves through a maze of tethered livestock and young boys playing cricket (certainly one of the most confusing games invented by humans). We reach a narrow, deeply rutted dirt road and enter a large area of mixed agriculture: bananas, mangos, citrus, nutmeg, coconuts, and breadfruit — simply stated, this is superb treeboa habitat (Figure 5).

"Sarpents?"

Establishing contact with people living in the area (Figure 6) is important. Letting them know what we're up to allays suspicions once we start moving around at night with our headlamps. We talk to two or three people and they will probably spread the word. After a greeting, a handshake, and introducing myself and my companions, the conversation usually goes something like this:

"Have you seen any sarpents recently?"

"Sarpents?"

"Yeah, sarpents. Snakes."

"Oh sure, sure, but they not so plentiful as they used to be."

"Really. Why do you think that is?"

"Oh, well, I don't know for sure. Maybe mongoose, maybe insecticides."

"That's too bad. Do you think anyone will mind if we walk around tonight looking for sarpents? We won't bother the fruit trees."

"No man, that's okay. How're you going to find them at night?"

"We use lights and when the light hits their eyes it shines back at us."

"Yeah, that's right. How you gonna capture them?"

"Well, we just use our hands or a long branch. Sarpents aren't venomous. They can't hurt you. Even if they bite, it's not dangerous.

"Is that true?"

"Yeah, they're harmless to people."

"Oh. Well, that's good to know."

"Thanks for your help. Good night."

Developing a Search Image

Although the kind, polite people with whom I speak may, in reality, think I'm an idiot, they never appear suspicious of my queries, seemingly take them seriously, and often offer advice on where to look

FIGURE 6. The author, here learning some fine points of treeboa hunting from a local in his Pearls study site, parks his jeep at the home of Mr. Charles (photograph by R. A. Sajdak).

for treeboas. Most likely, my eccentricity is attributed to my being white and from the United States. So, aside from letting someone know what I'm up to, speaking with the locals can lead to worthwhile information regarding the animals in which I am interested. Having lived all or most of their lives at or near the area in which I'll be working, these people have probably had numerous treeboa encounters over a span of many years. They can provide information on where snakes have been recently seen, their perception as to whether the snakes are decreasing in abundance, the kinds of situations in which they are most often encountered, who is the most snake-savvy person in the neighborhood, etc. This has been brought home to me on more than one occasion as I've searched for treeboas during the day on Grenada, approached a passerby, and asked if he had seen "sarpents" recently.

"Oh sure. I saw one just now."

"You mean today?"

"Yes, man. Right now."

"Can you show me where?"

We walk a few meters from where we'd been conversing and from where I'd been looking for treeboas only minutes before, and the person points a finger into the roadside vegetation.

"Right there."

"Where. I don't see it."

"Right there. See where I'm pointing?"

Straining my eyes, not wanting to appear any more inept than necessary, I try to pick a snake out of a three-dimensional, multi-colored matrix of vegetation. Suddenly, I separate the boa from its complex environment and see it coiled neatly on a branch at a height slightly above my head, sprigs of vegetation casting a mosaic of shadows over its yellow-brown body. I had walked past the same tree several times, seeing only leaves, vines, and branches. Having lived with treeboas all of their lives, Grenadians have, without really trying, developed an enviable "search image" for the snakes. In contrast, I have a very poor treeboa search image by day, rarely able to spot one resting on a tree branch. I'm better than I was ten years ago, but have nowhere near the skills of most Grenadians who live and work where treeboas occur.

Palms Silhouetted Against a Purple Sky

The afternoon is waning. We make a quick reconnaissance over the rutted dirt road that we will travel tonight, making sure to identify any potential hazards that we might overlook with our headlamps while our attention is focused on the trees. Back at the jeep, peanut butter is slathered on bread, cans of pop are opened, and we relax before the adventure starts. As dusk settles in, bats swoop past as we munch on cookies. Then begins the intermittent, high-pitched call of male *Eleutherodactylus johnstonei*, an introduced species of Rain Frog that has successfully invaded many West Indian islands. Despite the fact that it's not supposed to be on Grenada, I like hearing the melodious attempts of the males to woo any female in the neighborhood.

Having finished our most excellent dinner, we douse ourselves in mosquito repellent, attach battery packs to belts, and slip on our headlamps. I have a clipboard and data sheets for recording observations, and carry a canvas pack containing scales, measuring tape, cloth and plastic bags, flagging material, pens and pencils, and spare bulbs for our headlamps.

Darkness is upon us and the Coconut Palms, evocativelyly romantic by day, are even more so now as their fronds perform a graceful ballet against the purple hue of the tropical night sky. Silhouetted against a full moon they're even more wonderful. I generally plan fieldwork around phases of the moon, trying to avoid a full moon since experience indicates that treeboas seem less active on bright, moonlit nights. Whether because of diminished activity of rodent prey, or increased likelihood of being predated themselves, I don't know, but treeboas do seem to lay low during a full moon.

Don't Drop the Snake

Our headlamps are flicked on, moths, attracted to the lights, flutter annoyingly in our faces, and we begin scanning the vegetation with our lights. I can think of few things more exciting than reducing one's visual world to a narrow beam of light as you walk through a tropical forest on a moonless night. We know from experience that treeboas usually forage slowly and methodically on the distal ends of branches, tongue-flicking leaf and branch surfaces, and that's where we direct the beams of light. When searching for treeboas at night, we don't look for snake shapes in the vegetation. Rather, we look for the bright red-orange eyeshine that is obvious when a beam of light reflects back from a treeboa's eye. It appears almost as a red-hot ember glowing in the above-ground vegetation, and an eyeshine can be visible from distances of 50 m or more. Other animals produce eyeshines, and moths, opossums, and birds occasionally fool us. If a sprig of vegetation is covering an eye, or if the snake's head is oriented into the interior of a tree crown, our headlamp beam may well miss the eye and, therefore, no reflection will be elicited — but, if we miss the treeboa this time, we might get it on the way back.

Concentration is important for finding treeboas at night, and I don't invite lots of conversation. I've been accused of being obsessive about finding snakes, but it's what I do. My son, Ky, recounts how, as a high school student and on his first trip to Grenada with me, I demonstrated a certain degree of insensitivity when I asked him to find a suitable stick for bringing a boa down out of a tree.

"My father had disappeared in the white mist and dense greenery, so I called out to let him know I had found a suitable stick... I had to crane my neck

FIGURE 7. Ky Henderson being careful not to "drop the snake."

FIGURE 8. Rich Sajdak and a local Pearls youth up-close and personal with a treeboa (photograph by Joel Friesch).

straight up as I coaxed the snake onto my branch and so I'm not sure why I happened to glance down at my hands and arms. Perhaps it was a long-dormant survival instinct that years of living in the Milwaukee suburbs had made duller than opera.

"'Ants!'

"I was crawling with refugees from the branch. The insects were so small I couldn't even feel them, yet I was certain I heard their razor-sharp mandibles clicking with gleeful bloodlust. The troop relentlessly continued its march north towards my face, now frozen in a look of horror. My dad peered at me, and in a voice filled with concern said, 'Whatever you do, don't drop the snake'" (Figure 7).

Insincerity and Abandonment

At Pearls, someone usually spots the first treeboa within the first minute of searching. My search strategy is to move through an area quickly, moving my light rapidly over the vegetation. By so doing, I can cover lots of ground and, I feel, encounter more treeboas than if I move very slowly, looking at virtually every branch of every tree. Al Winstel, one of my field companions, and my wife, Rose, employ a branch-by-branch strategy, and they do occasionally find treeboas that I missed. Rich Sajdak (Figure 8) and Ky don't work as fast as I do, but they're not as slow as Al and Rose. Regardless of the strategy, all are excellent spotters of treeboas. Still, over a period of time, I encounter more snakes. Working with one or more companions, more treeboas are encountered and more data accumulated. We always try to work within earshot of one another, as I want to see each snake in order to document time, activity of the snake, kind of tree or bush, and certain perch characteristics (height, diameter, etc.). We may or may not capture the snake, depending on our goals for a particular trip. If we want to capture the snake, we "fish" it out of the tree with a long branch (Figure 9), or climb the tree and grab it by hand. Joel Friesch is especially good at the latter. He will occasionally ascend to great heights to get a treeboa, despite my insincere attempts to try and talk him out of it. Then, it's great fun to abandon Joel after we have the snake in-hand and he's still several stories up in a tree in near-total darkness.

Incredible Numbers

Tonight we're doing a census of sorts. I want to document an encounter rate for treeboas (that is, how many snakes do we find per unit of time). This rate can be compared to those determined at other sites, and to rates observed at Pearls at other times (different seasons or years). When working with Al, Rich, Ky, or Joel, they get to spend more time searching while I'm busy writing as snakes are encountered. On a good night, between two or three of us, we may find a treeboa on the average of every

FIGURE 9. The author "fishing" a boa from a tree at Westerhall Estate, St. David Parish, Grenada (photograph by R. A. Winstel).

three minutes. For example, on 4 April 2000 at Pearls, Ky and I started searching at 7:05 PM, and here are the times at which treeboas were encountered during the next hour: 7:06, two at 7:11, 7:18, 7:22, 7:27, 7:30, 7:32, 7:34, 7:38, 7:44, 7:47, 7:52, 8:00, 8:03, and 8:05. That's pretty darned amazing, and it's one of the main attractions of the West Indies for me. Although we encountered only one species of snake, we saw it in numbers almost unheard of on the Neotropical mainland of Central and South America. There you may see one or two snakes per night, each a different species. And the next night may yield another snake or two, different from those of the night before. Similarly, Grenada harbors only a handful of lizard species, but on a given day I may see hundreds of anoles of one or two species. On the mainland, I might encounter a dozen different species of lizards in the course of a good day, but, at best, only one or a few individuals of each. The West Indies have yielded the highest population densities of any terrestrial vertebrate, which just happens to be a lizard (the diminutive gecko *Sphaerodactylus macrolepis* in the British Virgin Islands: 6.76 per m^2).

After an hour at Pearls we have encountered more than 20 treeboas. We leave that incredible site and start the trek back to the leeward coast, planning to make stops at Balthazar Estate and Grand Étang in order to conduct additional censuses. We encounter fewer treeboas at Balthazar than we did at Pearls and, as usual, we come up empty at Grand Étang. Nevertheless, it's been a good night. Treeboa activity starts to decline after midnight, so at about one in the morning, having started the descent from Grand Étang to St. George's and our living quarters, I reflect on the night's work. How many treeboas were encountered? Was the number more or less than I had anticipated? If fewer, why? If numbers are low, could rainfall, season, or moon phase have contributed to the decline, or might this be a signal of a greater problem such as habitat alteration? Were any snakes doing anything unexpected or unusual based on hundreds of previous observations?

More Questions Than Answers

The next morning I review my notes from the night before. I add them to those taken on previous nights and try to recognize patterns in what we observe on our nightly hunts. When do the snakes become active and when does activity start to decline? What kinds of habitat are the snakes exploiting, and how are they using the habitat? Do small (= young) treeboas exhibit different habitat preferences than adults? Do young snakes use different hunting strategies than adults? Why haven't we observed more feeding episodes (only one in 700–800 observations)? What is the impact of human activity on treeboa ecology, and how bright is the future for treeboas on Grenada? As usual, fieldwork generates more questions than answers.

Total Immersion

Early in the afternoon, I go to Grand Anse Beach. There are few tourists or Grenadians with whom I must share the wide expanse of glistening white sand. I sit in the sun for a few minutes, thinking about where I am and why I am here. Being on a beach, I try to envision the ancestral *Corallus grenadensis* that must have washed up on an ancient Grenadian shore, perhaps over a million years ago. After an overwater voyage on a raft of vegetation that had begun its journey in coastal northeastern South America, perhaps at the mouth of what is now Guyana's Essequibo River, it arrived hungry and thirsty, beaten by wind, waves, and sun. Little did that ancestor realize that the prevailing ocean currents had fortuitously carried it to a land of plenty. Here and elsewhere in the West Indies, *Anolis* lizards occur in incredible numbers, and, over time, the descendants of those first treeboa colonists adapted behaviorally and physically to efficiently exploit this trophic bounty.

Eventually, I stop daydreaming and wade cautiously into the wonderfully clear water before diving in and splashing around for a few minutes. When I return to the towel I've spread on the sand, I lay down for a few minutes to dry off and warm-up in the sun, basking, albeit briefly, like a lizard, before leaving the beach. While lying there, I again ponder the joy of doing fieldwork in the West Indies. Being in Grenada, divorced from all other aspects of my life, I immerse myself in treeboa biology as completely as I immerse myself in the Caribbean Sea itself — no television or radio and, if I so choose, no telephone or newspaper. My only concern is treeboas: where are they and how many can I find each night; the more encounters I have, the more I learn. And every day and night I'm in the field, I learn something new about treeboa biology.

Although the past decade I've spent working on treeboas has been the most enjoyable and productive of my career, if I was not studying treeboas

I'd be working on some other West Indian reptile. The same elements that make the West Indies so alluring to tourists from the United States and Europe, the clear water of the Caribbean Sea, white sand beaches, Coconut Palms, and bright sunshine, in concert with remarkable lizards and snakes occurring in amazing population densities, made the West Indies appealing to me more than 20 years ago, and the same is true today.

Dried by the sun, I head back to the jeep and then to our living quarters. It's time to start preparing for tonight's work.

Epilogue

Although we still experience fantastic nights and encounter phenomenal numbers of treeboas, things in Grenada have changed. When Rose and I made our initial trips in order to assess the potential for doing ecological work with *Corallus*, after a week we were convinced that we would find treeboas wherever we stopped our jeep, and we were rarely disappointed. I'd choose what I thought was the most unlikely *Corallus* habitat (e.g., Acacia scrub, the grounds of a hotel or the botanic garden, under streetlights at the edge of town), and there'd be a treeboa; perhaps not many, but they'd be there. We could occasionally detect the eyeshine of *C. grenadensis* in the headlights of our vehicle, and driving slowly along the back roads of Grenada I'd be at the wheel and Rose would shine a headlamp out of the window and she'd find treeboa after treeboa along the roadside vegetation — but no more.

Besides changes in the distribution of *C. grenadensis*, Grenada has entered the 21st century: stoplights have replaced white-gloved policemen at busy intersections, St. George's has ATMs, and an enclosed shopping mall entices both residents and visitors. The nice folks of Grenada no doubt perceive these as enhancements to their quality of life, but I know better. My jeep has been broken into the last three years, and that had not happened during the previous decade — but that's progress.

Although I no longer find treeboas wherever I stop my jeep, I can't imagine a more productive or intriguing place to study *Corallus* than in Grenada. I have just started injecting microchips into snakes (something I should have started long ago — six of ten snakes collected in June 2002 had been so marked four months before) and nocturnal videotaping has provided new insights into treeboa foraging behavior. In a few years we should be able to address aspects of treeboa natural history not covered in the recently published progress report.

Acknowledgments

This essay is dedicated to the memory of Robert W. Bourgeois (1923–2002), a friend and volunteer-without-equal in herpetology at the Milwaukee Public Museum. Recent field work on Grenada has been funded by the Windway Foundation and the Milwaukee Public Museum. Grenada's Forestry and National Parks Department has been supportive of my work and it's much appreciated. While in the field in Grenada, I have been fortunate to be accompanied by hard-working, good-natured family and friends: Joel Friesch, Ky Henderson, Rose Henderson, Rich Sajdak, and Al Winstel.

Suggested Reading

Henderson, K. 2001. The snake hunters. *Milwaukee Mag.* 26(7): 48–55.

Henderson, R. W. 2002. *Neotropical Treeboas: Natural History of the* Corallus hortulanus *Complex*. Krieger Publ. Co., Malabar, Florida.

Rodda, G. H., G. Perry, R. J. Rondeau, and J. Lazell. 2001. The densest terrestrial vertebrate. *Jour. Trop. Ecol.* 17:331–338.

Henderson, R. W. 2003. Night Shift, p. 271–278. *In* R. W. Henderson and R. Powell (eds.), Islands and the Sea: Essays on Herpetological Exploration in the West Indies. Society for the Study of Amphibians and Reptiles, Ithaca (New York). Contributions to Herpetology, volume 20.

The Mule Lizards of Trinidad: Jungle Dreams and Backyard Explorations

George C. Gorman[1]

Every academic probably suffers from inflating the importance of his or her research subject, but I simply can't understand why the adaptive radiation of West Indian *Anolis* lizards has not entered popular consciousness. If you want to see textbook examples of island evolution, drop in on any Caribbean island and look at the nearest tree. Anoles will probably be in plain sight — and they have a story to tell. Ecotourists make pilgrimages to the Galápagos Islands where they try to distinguish differences among Darwin's Finches. It's darn difficult — and Darwin didn't even spend that much time thinking about finches. John Gould pointed out that the collection contained multiple species. Darwin did point out differences among tortoises ("Galápagos" means tortoise), but the famed tortoise populations have generally been reduced to mere remnants and are observed by most tourists in the zoo-like confines of the Charles Darwin Research Station.

The Hawaiian islands are no better. *Drosophila* fruit flies have undergone a fabulous adaptive radiation, but that is so far beyond the possibility of entering the consciousness of the intelligent layperson's mind that it's barely worth mentioning. Another extraordinary Hawaiian radiation involved a group of fascinating birds known as the Honeycreepers, but these have been systematically decimated by human invaders, starting with the Polynesians. Consequently, they cannot serve as a textbook example of anything other than subjects for threatened species conservation. So, get thee to the West Indies and, after your rum punch and your reef snorkeling, be entertained and enthralled by island evolution starring *Anolis* — and when you do, this is what you'll find:

[1] 952 The Alameda, Berkeley, California 94707, USA.

Island Endemism

The anole fauna of each island bank is comprised almost completely of endemic species. The Bahamas are an exception, but they are geologically young islands. Each Lesser Antillean island harbors one or two species; the Greater Antillean islands have 7 species (Jamaica), 10 or 11 species (Puerto Rico), about 40 species (Hispaniola), and about 50 species (Cuba). Exact numbers are always in flux because of taxonomic revisions and discoveries of new forms, but the pattern is a fine example of the species-area relationship, a cornerstone of the theory of island biogeography.

Extraordinary Abundance

Most species tend to occur in dense populations wherever habitat is suitable. In one study, we estimated the density of *Anolis pulchellus* in a grassy roadside site on Puerto Rico to be 10,000 per hectare (one per square meter).

Conspicuousness

Anoles on islands tend to be bolder than mainland species. I will speculate (but not substantiate) that this is because fewer kinds of mammalian and avian predators are available to make their lives miserable. They are easily observed in the field.

Textbook "Niche-splitting"

Take a trip to Puerto Rico and walk around the campus of the University of Puerto Rico in Rio Piedras. You will see a chunky brown "trunk-ground" anole perched low on trees or on the ground (*Anolis cristatellus*); a more slender "trunk crown" gray, speckled anole higher up in large shady trees (*A. stratulus*), and a small, pale thin "grass bush" anole with a red throat fan in almost any patch of tall grass (*A. pulchellus*). Then, venture up into the eastern mountains towards the popular tourist site of El Yunque or the El Verde research station, and you will find these same species in open areas along the roadside. However, if you enter the forest just beyond the road, you will find that each is replaced by a somewhat similar species that has a different (shadier, cooler) preferred climatic niche. The chunky *A. cristatellus* is replaced by blue-eyed *A. gundlachi*; the gray speckled *A. stratulus* gives way to the green *A. evermanni*, and the red-dewlapped, slender *A. pulchellus* is replaced by a similar, but orange-dewlapped *A. krugi*. If you travel to the very dry southwest, another set of species replaces these "ecomorphs." The white-dewlapped slender grass-bush anole of the driest areas around Ponce (*A. poncensis)* replaces red-throated *A. pulchellus*, and, on grayish limestone rocky outcrops near Guanica, the widespread lowland trunk-ground *A. cristatellus* gives way to *A. cooki*.

A Striking Dichotomy

Though Trinidad is culturally and politically allied with the West Indies, zoogeographically it is a classic continental island with a moderately rich South American flora and fauna and few endemic species. Trinidad's only truly native anole is widespread in South America. The presence of two West Indian anoles on Trinidad is a seeming anomaly — and it is the subject of this essay.

Ten Years' Work in Two Paragraphs

Two populations of *Anolis* lizards, very similar in morphology, behavior, and ecology, but strikingly different in color and pattern co-exist on the island of Trinidad. Were these color morphs or two distinct species? If two species, did any behavioral or ecological isolating mechanisms keep them apart? Most populations seemed to be comprised of either a green form or a gray form. But, in those few places where they came into contact, I learned that they interbred freely. Later, I demonstrated that the hybrids were sterile, completely so or almost completely so. How did such a situation arise? Was it stable?

I gained insights through observations while traveling to the adjacent South American mainland and to the Lesser Antilles. I concluded that the two species had evolved in isolation from one another on separate Lesser Antillean islands. They lacked behavioral isolating mechanisms presumably because, where they originated, any anole of the right size and shape was a conspecific. Contact between the species on Trinidad was presumably recent, and likely caused by inadvertent human introductions. Where they met, they readily mated. In general, the more abundant species wiped out the less abundant species in any particular patch of habitat because of a simple "advantage of numbers." The rare species couldn't locate mates. This explained the typical pattern of pure enclaves of one or the other species.

The Story

In the process of working this out, I first encountered the West Indies, West Indian biogeography, and a variety of laboratory techniques far removed from the field study I had envisioned when the research began. This then is a personal story of how one biologist was introduced to a research problem and how unraveling the threads of this specific problem broadened my education in evolutionary biology. Unfortunately, this is not a swashbuckling tale of adventure. My dreams of Trinidad were of jungles with Ocelots, swamps with Scarlet Ibis and anacondas (the world's largest snake), of bushmasters (the largest venomous snake in the Western Hemisphere), caimans, and Vampire Bats. The reality was spending 90% of my field time studying lizards whose largest, most robust and macho males reach a body length of 3 inches (75 mm) and a weight of a quarter of an ounce (7 grams). Females are considerably smaller and even less intimidating! And, what I found even more frustrating, these lizards seem to be found only in towns around houses and gardens and abandoned lots. My experiences were more in fence post rather than tropical rainforest ecology — but it was all pretty wonderful.

The Autobiographical Setting

I can recall precisely two seemingly trivial but linked events that were to influence profoundly my professional career as a zoologist. The first occurred when I was a suburban New York high-school student, the second about eight years later as I was contemplating graduate school. As a ninth grader, at the Albert Leonard Junior High School, in New Rochelle, New York, we had to do "occupation units"; you know, those "what I'd like to do when I grow up" exercises to which all of us were subjected. We had to choose two careers about which we would write. I chose a generic "chemist," and a more specific "herpetologist" — in fact, quite specifically, the "Curator of Reptiles at the Bronx Zoo." Now, I have not paid and will not pay to sit for hours on someone's couch to understand *why* I wanted to be a herpetologist, so I won't spend too much time analyzing the career of my dreams. I can only share the following information: (1) absolutely no one in my immediate family had any interest in any organisms or anything to do with nature; (2) as a pre-schooler in Brooklyn, my grandfather took me on nearly daily walks to the Prospect Park Zoo (because it was convenient); and (3) I was hooked. Non-human creatures have fascinated me ever since, and reptiles took precedence among them.

In the seventh grade, I discovered the books of Raymond Ditmars. My fate was sealed. In fact, the first emotional loss that I can recall (other than the Brooklyn Dodgers annually losing the baseball World Series to the New York Yankees) occurred when I was about to write Ditmars a fan letter — and I discovered that he had died some 15 years earlier.

In conjunction with the aforementioned "occupation unit," I reached a lineal successor of Ditmars. Dr. James Oliver held the position of Curator of Reptiles at the Bronx Zoo. He patiently agreed to be interviewed for my school assignment. The most significant result for me was to learn about and to be invited to attend the "Reptile Club" (it probably had a more formal name, but if it did, memory fails me). The club met more or less monthly at the zoo. Oliver, and subsequently his successor, Herndon Dowling, put up with young enthusiasts. I have no way of knowing whether alumni of the club would have independently become professional biologists had the club not aided and abetted our teen-age passions. However, a significant number of the participants in those club meetings have made important contributions to herpetology or other fields of biology. Alumni from my era include Drs. Harvey Pough of Cornell University and, later, Arizona State, Jay Cole of the American Museum of Natural History, Doug Futuyma, now at the University of Michigan, and the late Tim Brown of Cal Poly, Pomona. I'm sure many others shared a similar history, and I apologize for any omissions. Nevertheless, at one of these Reptile Club meetings I saw the film which pointed me to the West Indies for much of my herpetological research career. The film told the story of William Beebe and his tropical research station "Simla," located in Trinidad's Northern Range above the town of Arima. Something about that film infected me. Alternately, I may have already caught the Trinidad germ from reading Ditmars's accounts of hunting bushmasters. Regardless, I resolved that one day I would go to Trinidad to study creatures in a tropical forest paradise.

Several years later, as a zoology major at Cornell University, I wandered dangerously close to a cell physiology track rather than one leading to a career as an evolutionary biologist/ecologist. While interviewing for graduate schools in the autumn of 1961, I flew from Ithaca, New York, to Western Reserve University in Cleveland. During the interview, I met a new faculty member who had just arrived from the University of the West Indies (UWI), Mona, Jamaica. His name was Michael Locke. He was an electron microscopist. Professor Locke is somebody to whom I've never spoken again, nor have I thought about him in any other context, but our 10 minute interview gave birth to my research career. I confided that I wasn't particularly enthusiastic about the cell biology and neurobiology programs his university was offering. What I really wanted was to study reptiles in Trinidad. His eyes lit up. He told me that I must contact Garth Underwood. Professor Underwood, from the United Kingdom, had spent much of his career at the Jamaican campus of the UWI, and had relocated to the Trinidad campus a few years earlier. My formal exposure to academic herpetology was still quite minimal at that stage in my training, and I had no idea that Underwood was an internationally known herpetologist who had published major papers on the origin and classification of snakes, nor did I know of his publications on Caribbean herpetology.

The immediate result of that meeting was my decision to enter a graduate zoology program that emphasized ecology and field studies. In the autumn of 1962, I enrolled as a Ph.D. candidate in the Zoology Department of the University of California, Berkeley. I was drawn by the reputation of the department and the Museum of Vertebrate Zoology. But, I will admit, a bit of non-academic romanticism was involved in my institutional choice. I had never been west of anywhere in the United States, and the mystique of California more than a particular mentor drove my choice of graduate institution. Soon after the academic year began, I became the graduate student of Peter Marler, whose field is animal behavior.

During that academic year at Berkeley, I entered into correspondence with Professor

Underwood, who kindly provided me with an idea for a field research project on Trinidad. The University of California awarded me a traveling fellowship and, in September 1963, I landed at Piarco Airport, Trinidad, to begin a year of field investigations. I landed in a colossal rainstorm, which was mild compared to the hurricane that was battering Trinidad's sister island Tobago that very night. Not dampened by the weather, my research career soon began.

Some Background on the Problem

Restricted to the Western Hemisphere, the lizard genus *Anolis* is comprised of hundreds of species of small to moderately sized "tree lizards," characterized by lamellae (toe pads with microscopic hairs) and a throat fan (or dewlap) that often is brightly colored and strikingly different from the green, brown, or gray body color. Many species are highly sexually dimorphic. Typically, males are much larger than females and have more highly developed and larger dewlaps. The dewlaps are extended during male-male territorial encounters and in male-female courtship displays. Many species are capable of rapid color change. Because of this, North Americans commonly call these lizards "chameleons" (although they are not related to the true chameleons of the "Old World"). One species, *A. carolinensis*, is abundant in the southeastern United States and in the pet trade. Anyone familiar with this lizard will have a good idea of the general appearance and behavior of my Trinidad lizards.

Differences between species often involve color or pattern. These tend to disappear in preservative. Other differences may be primarily behavioral or ecological, and are not amenable to study strictly from preserved specimens. Thus, definitions of species and patterns of relationship among species in the genus were difficult for museum taxonomists to unravel. In the 1950s, Richard Etheridge, then a graduate student at the University of Michigan, undertook the onerous task of sorting all the anoles into species groups and working out probable relationships among those groups. Etheridge relied upon osteological characters, and used X-rays extensively to examine dozens of species. His monumental Ph.D. thesis proved extremely influential in guiding subsequent research efforts on the adaptive radiation of anoles. At about the same time, Ernest Williams, Curator of Reptiles at Harvard University's Museum of Comparative Zoology, switched the primary focus of his research from turtles to *Anolis* lizards. He became the hub of research efforts on West Indian anoles, and would continue in that capacity for nearly four decades.

The published paper that was instrumental in launching my research on Trinidad anoles appeared in the *Bulletin of the Museum of Comparative Zoology* in 1959 under the title: "The Anoles of the Eastern Caribbean. Sauria, Iguanidae: Parts I–III." OK, it wasn't a blockbuster bestseller, but it had a large impact on my life.

In Part I (Preface), Ernest Williams set the stage: "It will be recalled that the segment of *Anolis* that is here reported on was believed to be, and may well be, the simplest areal segment of the genus in which more than one species occurs.... Yet we have found the complications formidable and not amenable to study on the basis of preserved specimens alone.... It is even more evident in this genus than in others that we cannot be prisoners of our conventional museum techniques, that a very wide approach will barely encompass the things we need to know. Knowledge of live animals is needed — of captives if necessary — but preferably field knowledge. *Anolis* must be known intimately — anatomically, ecologically, ethologically — in all its relationships within and between species, if the many puzzles the genus poses are to be solved."

In part II (Anoles of Trinidad), Quesnel and Kenny first proposed that two, not one, species of *Anolis* inhabited the gardens of Trinidad's cities and towns. In 1935, Parker had listed *A. aeneus* and *A. chrysolepis* as the two anoles on Trinidad. These species are not at all similar. *Anolis chrysolepis* is small, brownish, and has a red dewlap (with little ability for striking color change). It lives almost exclusively in deep shade and usually is seen on or close to the ground. *Anolis chrysolepis* would almost never be seen in towns and gardens. *Anolis aeneus* is larger, gray in color, with a pale yellowish to white dewlap. It is capable of darkening to browns but not turning solid green. It was known almost exclusively from open, disturbed areas. In 1955, the authors sent a peculiar anole to the British Museum for identification. It was identified as *A. aeneus*, but in life this specimen had been green, not a gray patterned lizard. Quesnel and Kenny noted that populations of gray patterned anoles and green anoles tended to occur in pure enclaves; in any given patch of suitable habitat, they would tend to find either a gray population or a green population (the green form could darken to deep brown and had a bright yellow dewlap). The two were similar in squamation (scales) and limb measurements, and, faded in preservative, could look quite alike. Although not quantified by measurements, Quesnel and Kenny noted that in life the gray form appeared to be more heavily built, and that slight differences in head shape existed between the two forms. They also noted one minor difference in habitat choice. The gray form rarely entered houses, whereas the green form commonly inhabited houses. In Port-of-Spain and St. Augustine, "where they occur, they are seldom found together." However, the authors did find two mixed populations and "in these one glance is usually enough to classify an individual as gray or green. Only one large male in one of these populations seemed to be intermediate in color. Finally, they suggested "perhaps it is the near identity of ecological niche that accounts for the usual absence of mixture of populations of the two forms in Trinidad. Or, perhaps there are differences in preferred habitats that we have not detected. These observations show the necessity for a thorough study of the anatomy and behavior of the two forms in order to substantiate or reject our view that they are distinct species."

In Part III of this paper, Underwood unraveled the species identity of the Trinidad anoles. He established that the green species had been described from Trinidad and named *A. trinitatis* in a Danish publication by Reinhardt and Lütken in 1863. This fact was apparently lost and forgotten for nearly a century. The name *A. aeneus* belonged to the gray form. J. E. Gray described it in 1840, but without locality data. *Anolis aeneus* was known from South America. Beebe (1944) discussed a population at Kartabo, British Guiana (now Guyana), and remarked that *A. aeneus* was fairly common at Kartabo but absent from Caripito, Venezuela (implicitly where he might have expected them). Beebe also referred to a "blue tinged individual" that was captured at a pitch lake at Guanoco a few miles down the road from Caripito. I do not know the whereabouts of that specimen and do not know what species of *Anolis* was found at that lake. Beebe listed the distribution of *A. aeneus* as British Guiana, Venezuela, Trinidad, and Tobago. Underwood posited that what he termed the *A. trinitatis* series of the Lesser Antilles almost certainly occurred on the mainland of South America, a reasonable assumption, given the biogeography of Trinidad.

A Nomenclatural Digression

Generic names.—Subsequent to much of the work described here, Craig Guyer and Jay Savage proposed splitting the genus *Anolis* into several genera. This generated heated controversy among the few of us who care about such things. Many workers would agree that the Guyer and Savage genera were based on natural species groups; but they also would agree that these groups would be nearly impossible to diagnose on the basis of external characters. The generic name *Norops* (representing Etheridge's "Beta" section of the genus *Anolis*) seems to be widely used in the literature, except by direct lineal descendents of Ernest Williams's group. For example, Trinidad's *A. chrysolepis* would be called *N. chrysolepis* in many publications, and *A. trinitatis* and *A. aeneus* would be assigned to the genus *Dactyloa*. In contrast, in a recent book on the reptiles of Trinidad, John Murphy conservatively treated all the species as *Anolis*. I shall stick with *Anolis* as the generic name of choice, as that name is used in all of my published papers on the subject under discussion.

Specific and subspecific names.—Underwood referred to the Trinidad green anole as *A. trinitatis trinitatis* and the small green anole of the Lesser Antillean island of St. Vincent as *A. trinitatis vincenti*. On St. Lucia, he recognized two subspecies of anoles that he also placed in *A. trinitatis* (*A. trinitatis luciae* and *A. trinitatis procuratoris*). He treated *A. aeneus* as *A. roquet aeneus*, and listed its distribution as British Guiana, Trinidad, and the Grenadine Islands (the chain of islands between Grenada and St. Vincent). The small anole of Grenada and Barbados was called *A. roquet cinereus*, and *A. r. roquet* was from Martinique.

With respect to the green anole, subsequent studies revealed that no valid separation existed between *A. t. trinitatis* and *A. t. vincenti*. The small green anole of St. Vincent is not distinguishable from the green anole of Trinidad. The endemic anole of St. Lucia is in the same species group, as are the other southern Lesser Antillean anoles, but it is not especially close to *A. trinitatis*; instead, it is generally treated as *A. luciae* and will not re-enter our story.

With respect to the gray anole of Trinidad, the nomenclature gets a bit trickier. That a close relationship exists between the anoles of Martinique, Barbados, and the small species on the Grenadines and Grenada seems very clear. If one accepts a definition of species as "populations that actually

FIGURE 1. Arima Valley, Trinidad, where neither *Anolis trinitatis* nor *A. aeneus* are found in the forest.

or potentially exchange genes under natural conditions," a conundrum exists. We do not know whether populations isolated on separate island banks would or could exchange genes should the populations somehow be reunited. For convenience, then, I have used binomials for island banks. *Anolis aeneus* is the small anole of Grenada and the Grenadines. The non-Lesser Antillean presence of *A. aeneus* on Trinidad and Guyana and *A. trinitatis* on Trinidad is the subject of this tale.

Biogeographic Background

Biogeography quite simply may be defined as the scientific study of the distribution of organisms. Approaches to biogeography may focus on the historical, what organisms are where and how they got there, or on the ecological, how the interaction of organisms at a certain place in time influences the structure of communities. Islands have played a major role in the development of biogeographic thinking from the mid-19th century papers of Alfred Russel Wallace through and beyond the mid-20th century theoretical work of Robert MacArthur and E. O. Wilson.

Wallace was one of the first to point out the classic distinction between "continental" and "oceanic" islands. Trinidad is a classic "continental island." It lies close to the South American mainland (about 15 km), from which it is separated by shallow water no more than 15 meters deep. Within the past 10,000 years, Trinidad was part of South America as we know it. Its flora and fauna are typically South American, and little endemism exists. Almost all species of mammals, birds, reptiles, amphibians, freshwater fishes, or flowering plants on Trinidad also are found in adjacent South America.

The Lesser Antilles, in contrast, are classic oceanic islands. Most are volcanic and presumably arose from the sea with no mainland connection. Animals and plants got there through dispersal. They tend to have a flora and fauna much lower in species richness than Trinidad. They also tend to have a high degree of endemism. The lizard or snake or frog that did get there evolved in isolation, and may well be distinct from the mainland form.

Another characteristic of oceanic island faunas is that, given the total absence of certain potentially competing (or predating) species, certain taxa undergo adaptive radiations so that the number of species per genus exceeds that on a continental island, but far fewer genera occur. Other taxa do not radiate and are impoverished. For example, comparing Trinidad to Puerto Rico (oceanic and about twice as large as Trinidad), both have roughly the same number of lizard species — but Puerto Rico has ten species of *Anolis* and about 10 species of *Sphaerodactylus* lizards, compared to Trinidad's one native anole and one *Sphaerodactylus*. Virtually every genus of lizard found on Puerto Rico is found on Trinidad, but Puerto Rico is missing Trinidadian lizard genera such as *Plica*, *Polychrus, Tupinambis*, *Cnemidophorus*, *Kentropyx*, *Gymnophthalmus*, *Bachia*, *Gonatodes*, *Thecadactylus*, etc. Similarly, the native Puerto Rican frog fauna is almost entirely a radiation of the genus *Eleutherodactylus* (about a dozen species) plus a couple of toads and a *Leptodactylus*. The Trinidad frog fauna is much richer in families and genera, but is not much richer in species. Only one *Eleutherodactylus* occurs on the continental island, but ranids, dendrobatids, microhylids, and hylids are present — which are frequently absent from the oceanic Antillean Islands.

This comparison of continental and oceanic islands is important to our story for one reason: no

FIGURE 2. Anoles frequently exploit such extensively altered habitats on Trinidad.

FIGURE 3. An *Anolis trinitatis* painted pure white and dangled in the territory of a resident elicited a typical territorial display.

one had pointed it out, but the presence of both *A. trinitatis* and *A. aeneus* on Trinidad was quite peculiar. Although *A. aeneus* was known from Guyana and *A. trinitatis* was asserted to be on the mainland, these were typical Lesser Antillean species, with nothing known about them from *undisturbed* mainland localities. This is in stark contrast to much of the remainder of Trinidad's herpetofauna (or mammal or bird fauna), which is typically South American and not West Indian. Thus, a high priority for me was to try to become familiar with *A. trinitatis* and *A. aeneus* on the mainland and in the Lesser Antilles as well as on Trinidad.

Getting Started

In the first few months after my arrival on Trinidad, I began looking for *Anolis aeneus* and *A. trinitatis* in the field, focusing on behavioral and ecological differences between them. I was disappointed to find that I could never locate either of these species when I searched in relatively undisturbed areas (Figure 1). I was thrilled to see Simla, Beebe's field site that had stimulated my desire to work on the island, but venturing into the forest was not the way to find my lizards. I soon had to accept the reality that the best place to look for them was around people's houses in towns. Fence posts, hibiscus bushes, ornamental trees, and the like seemed to be the only places where I could find these lizards (Figure 2). Soon after my arrival, I met the brothers Hans and Julius Boos, keen naturalists with a special interest in reptiles. They were extremely enthusiastic partners in many of my ventures in the field. Julius lived in the south in San Fernando. He pointed out that both *A. trinitatis* and *A. aeneus* were abundant there. In his home town, we confirmed the previously observed pattern that in a given localized area one would find either one species or the other, but rarely both.

My first behavioral studies involved comparison of the stereotyped display action patterns between the two species. At first I began looking at male-male territorial responses. I had little trouble evoking a territorial response in the field; I merely captured an "intruder" male and tethered him near a territory holder who likely was perched on a tree trunk (Figure 3). The "resident" responded with a series of reasonably stereotyped movements that included distension of the dewlap, head-bobbing, and tail-waving. I also studied these responses in the laboratory; again, this was a very simple matter. I set up series of terraria that were divided in two with an opaque partition. Within a couple of days, the males were ready to defend their territories. Removal of the partition almost invariably elicited stereotyped displays. In the pre-video tape/digital days, these displays were filmed in 16 mm and later analyzed frame by frame (Figure 4).

Charles Carpenter had popularized this method of comparing populations of iguanid lizards by analyzing so-called display action patterns. He commonly found species-specificity. I identified clear-cut differences in the displays of my two anoles. The basic patterns of dewlap extension, head-bobbing, and tail-waving were shared by both species, but they differed in the frequency and amplitude of the head bobs and tail waves. The fact that the displays were different did not mean that they were functionally different in any biologically meaningful way. That is, I had no evidence that a difference in pattern in head-bobbing would serve as an isolating mechanism. One could easily elicit a stereotyped territorial display from a resident inde-

FIGURE 4. The author studying display behavior without an assistant by dangling a tethered male in the territory of a resident and filming the elicited display.

FIGURE 5. Soufriere, St. Vincent; anoles are found all over the island and to the very top of the mountain.

pendent of whether the tethered or caged intruder was a conspecific or a member of the other species.

On to the Lesser Antilles

A few months after settling in on Trinidad, I made my first trips to the Lesser Antilles. My initial focus was on Grenada and St. Vincent. Life on these islands was gloriously relaxed compared to the bustle of Trinidad. In the early 1960s, no American medical students were in need of protection, no communist threats demanded squelching, and no drug dealers required interception. These Lesser Antillean islands were known more for their nutmeg (Grenada) and arrowroot (St. Vincent) than for tourism or anything else.

Landing first on Grenada at the airport near Grenville in the northeast, I rented a car and headed to the capital city of St. George's, in the Southwest. The drive was spectacular, through verdant forest, but much slower going than I had anticipated. Along the way, I learned that while I speak English and Grenadians speak English, we didn't always fully comprehend one another. I saw a roadside sign warning of a "Sleeping Policeman" and started smiling at what I thought was West Indian humor about the pace of life. However, when I banged my head hard against the roof of my car, I realized that a Sleeping Policeman was a speed bump.

Toward dusk, I reached St. George's, the picture-postcard perfect harbor and capital city. Someone had given me the name of an inexpensive guest-house/boarding house. I checked in and soon learned that it should have been termed a Rest Home. I seemed to be at least 50 years younger than the next youngest guest. A bell rang for dinner. I was seated at a table for one. When my waiter brought out the first course, I attempted to engage him in some polite, if meaningless conversation, and said, "In my country we call this a lively crowd." To which he replied, as he placed a bowl in front of me, "Oh? Here we call it soup."

Reaching St. Vincent's capital, Kingstown, from its only commercial airport was an experience quite the opposite of my experience in Grenada. Instead of a picturesque drive across the island, St. Vincent located the airport just out of town, with a major road bisecting the runway. I couldn't believe my eyes. It looked like a railroad crossing. Small wooden barriers were lowered, traffic stopped, and the plane landed and taxied to the offloading area. Barriers were raised, and the traffic resumed. After spending a few days looking at anoles in Kingstown's lovely botanic gardens, I hiked up the Soufriere (Figure 5), where the Lesser Antilles' volcanic oceanic origin was made palpable (Figure 6). From the summit, I gazed into a gorgeous crater lake.

On each island I was introduced to a contact who would prove invaluable in helping me with logistics. On Grenada, Dr. J. R. Groome represented a remnant of British colonialism. On St. Vincent, Dr. Earl Kirby, the only veterinarian on the island, was a man of boundless enthusiasm and knowledge about all aspects of his island's archeology, natural history, history, and politics.

Both islands are small (a little over 300 km^2), reasonably steep, and of volcanic origin. Both have a pair of *Anolis* species living on them. As I traveled the islands, I immediately noticed three striking differences between the *Anolis* situation of the Lesser Antilles and Trinidad:

FIGURE 6. Cold and miserable on top of Soufriere in July 1968. Julius Boos (standing) provided assistance and companionship while studying mule lizards. Amy Wiltchik Gorman, the author's wife, is seated to his right. The late Dr. James F. Lynch is seated to Boos's immediate left. Lynch was a Smithsonian ecologist on his first trip to the tropics; he would later work extensively in tropical biology. Mike De Freitas (far right) is a Vincentian who provided considerable assistance on our monthly sampling of reproductive cycles.

(1) Variable populations: Both *A. aeneus* and *A. trinitatis* showed a great deal of microgeographic variation in color and pattern throughout the islands of Grenada and St. Vincent. James Lazell had described numerous within-island subspecies for the endemic anoles of Dominica and Guadeloupe. Although those examples are perhaps more marked, and no one has formally named geographic races of the Vincentian and Grenadian anoles, a trained eye can easily discern differences in local populations. In Grenada, this was particularly marked; anoles in the wet montane uplands were decidedly greenish in color and populations from the xeric areas were much grayer. In contrast, on Trinidad, I had observed no local differentiation among populations of *A. trinitatis* or *A. aeneus* (Figure 7).

(2) Broad distributions: *A. aeneus* and *A. trinitatis* were found throughout Grenada and St. Vincent, respectively. Although easiest to observe and collect around homes and gardens, I had absolutely no problem finding them in deep forest, whether natural or disturbed (e.g., cocoa plantations). Rather than being patchily distributed, I would call them virtually ubiquitous.

(3) Niche differences in sympatry: Geographic overlap with a congener, rather than being rare, was widespread. Both *A. aeneus* and *A. trinitatis* co-occurred extensively with a sympatric species that was much larger and that appeared to be ecologically differentiated. On Grenada, Tom Schoener and I documented that the larger species (*A. richardii*) utilizes significantly higher perch sites, has significantly lower body temperatures, and eats significantly larger prey. The same pattern appears to be repeated on St. Vincent, where the large *A. griseus* overlaps extensively with *A. trinitatis*.

FIGURE 7. *Anolis trinitatis* and *A. aeneus* are distinctly different on Trinidad. *Anolis trinitatis* (left) is typically green (it can turn brown) with blue around the eyes and yellow on the head. *Anolis aeneus* (center) is speckled gray. However, considerable microgeographic variation is evident in the Lesser Antillean islands; in verdant areas, *A. aeneus* (e.g., from Grenada; right) has a deep green ground color.

Searching the South American Mainland

Using William Beebe's report as my guide and Underwood's certainty that the *A. trinitatis* series would be found on the South American mainland, I set off to find the Trinidad anoles on the mainland. A trip to the area around Carapito in eastern Venezuela proved completely fruitless. I saw many familiar lizards that I had met on Trinidad, but no anoles. I could not find a trace of Beebe's bluish-tinged presumptive *A. aeneus* — and, to my knowledge, no subsequent accounts in the last 40 years or so have reported anything resembling *A. aeneus* or *A. trinitatis* in Venezuela. In British Guiana, I did find *A. aeneus* as had been reported. They were abundant in Georgetown and down the coast to New Amsterdam. Leaving the coast, I traveled upriver to Bartica (near Beebe's site at Kartabo) where I found *A. aeneus* abundant at my guest house — but I failed to find *A. aeneus* in the forest surrounding Bartica. Once again, it was a garden anole.

Finding "Mules"

During the first half of 1964, with much help from Julius Boos, I located several areas both in St. Augustine in the north of Trinidad and San Fernando in the south, where *Anolis aeneus* and *A. trinitatis* not only overlapped but also seemed to be interbreeding. We began finding a number of "funny-looking" individuals that appeared to be intermediate between the parental forms in color and pattern — and these intermediates would be found only in these areas of overlap. We suspected they were hybrids. I believe Julius Boos first began calling them "mules," although, at the time, we knew nothing of their reproductive capacity.

We often attracted considerable crowds while we were keenly mapping and tracking an overlap area in a San Fernando backyard. Gawking at urban reptile collectors is understandable. This can be amusing at times. It can also be exasperating. The weirder the activity, the more likely a crowd will gather. What would look stranger than "fishing" high in a tree in the middle of a busy town? Yet one easy way to collect anoles is to noose them with a loop of dental floss or fishing leader attached to a long fishing pole (Figure 8). On one particular afternoon, a group was watching us collect anoles and put them in bags. "Hey mon, why you catchin'

FIGURE 8. A "typical" lizard market; when kids hear that you are interested in lizards, they make grass nooses and bring in specimens.

lizard?" came a query from the crowd. Boos quickly replied "We makin' curry." Any answer would do. And "'makin' curry," however strange, was a lot easier to comprehend than "we will be taking tissue samples for starch gel electrophoresis because we are interested in the evolutionary relationships of island populations." The answer sufficed — for a while. Then we collected an adult *Polychrus marmoratus*. *Polychrus* is an arboreal lizard that, while not rare, was not very frequently seen. We examined the lizard, then, having no need for it, let it go. Of course, this triggered the obvious question — we had just passed up what appeared to be a juicy morsel for our curry. "Hey mon, why you let that one go?" Without missing a beat or looking up, Julius answered, "The skin too hard."

Into the Lab

Suspecting that we had hybrids was different from demonstrating that we had hybrids. I sent a letter to Wade Fox and Herb Dessauer at the Louisiana State University Medical School in New Orleans. Wade was a herpetologist teaching anatomy; Herb was a professor of biochemistry. They had co-authored several papers that employed biochemical techniques to sort out taxonomic and evolutionary problems. In my letter, I asked whether they would be willing to compare *Anolis aeneus*, *A. trinitatis*, and the putative hybrids. The response was an unexpected invitation to Herb's lab to learn the techniques myself. Herb probably still regrets that invitation of nearly 40 years ago. If the term "all thumbs" defines clumsiness, then my lab nickname would probably be "All Thumbs Cubed." Nevertheless, I took Dessauer and Fox up on their offer and spent a marvelous summer in New Orleans (although I couldn't conceive of human beings inhabiting that city in summer months before the advent of air conditioning — and this coming from someone who had just spent nearly a year close to the equator!).

Long before the days of DNA analysis became routine, we used electrophoresis of blood proteins to look at differences in gene products between populations and species. Essentially, a drop from a protein mixture (blood plasma or solid tissue extract) is placed in a starch gel. A half dozen or more samples, each from different individuals, may be compared at once. A direct current is applied to the gel, and the proteins migrate varying distances based on their net charge, molecular weight, and three dimensional shape. The gel is then stained, and one compares the pattern of banding. We were able to demonstrate quite conclusively that several genetic markers separated *A. trinitatis* from *A. aeneus*, and all the funny looking individuals appeared to have the bands of both species. The fact that hybridization was occurring was incontrovertible (Figure 9).

In terms of my own professional development, accepting the invitation to work in a lab with techniques about which I knew virtually nothing was one of the best decisions I ever made. Despite my lack of the gene necessary for working effectively in a laboratory, I overcame that handicap and went on to use various biochemical techniques to study population genetics for many years to come. I also gained the confidence to try new avenues that at first might have seemed daunting.

To Harvard

I completed my one-year traveling fellowship in the summer of 1964. My original plan was to return to

FIGURE 9. The author (standing) and Herbert Dessauer in the laboratory at the University of the West Indies, St. Augustine, Trinidad, summer 1965. Early electrophoretic studies in Dessauer's lab clearly demonstrated hybridization between *Anolis trinitatis* and *A. aeneus*.

Berkeley that autumn. During the course of the academic year, Professor Marler informed me that he would be on sabbatical in Africa for the 1964–1965 academic year. Rather than returning to California with a year of field and lab data but no mentor, I decided to try to spend the year at Harvard's Museum of Comparative Zoology. The herpetology department of the MCZ had a strong tradition in West Indian herpetology, extending back the better part of a century under the curatorships of Samuel Garman and Thomas Barbour. Ernest Williams followed suit. I enrolled as a "special student" at the MCZ while maintaining my official enrollment at U.C.-Berkeley. For a young evolutionary biologist, this was an exciting time to be around Harvard, the MCZ, and Williams. P. J. Darlington's classic *Zoogeography* had appeared at the end of the 1950s. Ernst Mayr's *Animal Species and Evolution* had just been published. MacArthur and Wilson's *Island Biogeography* was about to transform biogeography from a purely historical science to a more dynamic, even experimental science. All of these people (except MacArthur) were part of the MCZ. In the herpetology department, Stan Rand had recently completed a thesis on the ecological distribution of the Puerto Rican anoles and a post-doctoral year-long field study of the behavior and ecology of a Jamaican anole. Skip Lazell, one of Williams' graduate students, was producing papers on the systematics of Lesser Antillean anoles. Tom Schoener would soon begin using anoles as models in his ecological studies of animal communities.

Early in my first semester at Harvard, I came across a 1949 French-language review paper by R. Matthey on the chromosomes of vertebrates. It included information on only one species of anole, *Anolis carolinensis*. The literature on reptilian karyotypes was small and scattered, with very little having been added since Matthey's publication. Wouldn't it be interesting, I thought, to see whether karyotypes could tell us something about relationships among anoles. I decided to try to get chromosome data from some *Anolis*.

In the early 1960s, techniques to culture human leukocytes in order to obtain karyotypic information were used commonly in clinical laboratories. I decided to try this with lizards. My earliest efforts were very lucky on two counts. First, I simply had beginner's luck and was able to get nice chromosomal information immediately (I later discovered that such a high success rate was very unusual). Had I failed initially, I might have given up, assuming that human/mammalian techniques probably wouldn't work for reptiles. The second stroke of luck was that I didn't initially use the species of interest to me (*A. trinitatis* and *A. aeneus*) for my early chromosome efforts. The *A. trinitatis* karyotype is very similar to the well known karyotype of *A. carolinensis*. Both have six pairs of large biarmed metacentric macrochromosomes and twelve pairs of tiny microchromosomes. Because *A. trinitatis* and *A. carolinensis* were not thought to be particularly closely related, had I used *A. trinitatis* I probably would have concluded that karyotypes were conservative, and all anoles would be the same (despite variation within true chamaeleons, the general impression one got from Matthey and the other chromosome literature up to the early 1960s was that karyotypes were relatively conservative within lizard genera, and one might not expect to find much variation within a genus). Had I used *A. aeneus*, whose karyotype we now know differs from that of *A. trinitatis* (and *A. carolinensis*) only in having 11 instead of 12 pairs of microchromosomes, I am sure that I would have concluded that the karyotype was identical to *A. carolinensis*, and that I was simply having trouble obtaining good preparations and good counts of microchromosomes.

In fact, in my very first published paper, I refused to believe precise chromosome counts and overlooked an interesting case, not previously known for any lizard, of male sex chromosome heteromorphism. I merely pointed out that there were striking differences in chromosome complements among the various species sampled.

My expendable samples included North American *A. carolinensis*, a member of the northern Lesser Antillean *A. bimaculatus* group, and several members of Etheridge's beta series (*Norops* to some) from Curaçao and Jamaica. The three had strikingly different karyotypes, and started me on extensive studies of lizard chromosome evolution in general, with an emphasis on *Anolis*. Dr. Leonard Atkins collaborated in much of this work at the Massachusetts General Hospital. In addition to being a forensic pathologist who liked to regale his lab staff about tracing the pathway of bullets in last night's homicide, Atkins ran a clinical cytogenetics laboratory at the hospital. He and his staff were quite adept at tissue culture, which they used to screen parents and newborns for cytogenetic anomalies. The team enjoyed the diversion of looking at non-human karyotypes. Although the Atkins lab had little interest in evolution or herpetology, we developed a productive relationship — and we determined that karyotypes proved very powerful for defining certain species groups of *Anolis* and for looking at relationships within species groups.

FIGURE 10. *Anolis roquet* from Martinique.

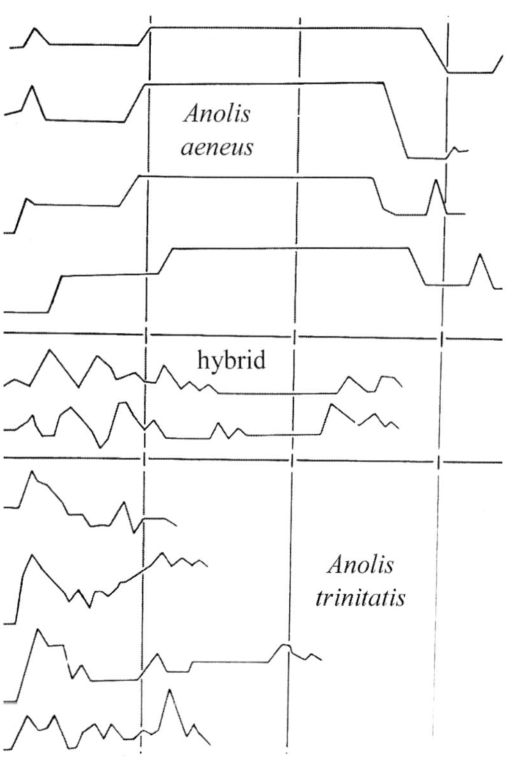

FIGURE 11. Display action patterns of *Anolis trinitatis*, *A. aeneus*, and a hybrid (modified from Gorman 1969).

Once we developed confidence in the analyses of our chromosome preparations, we were able to demonstrate that all the forms that Underwood had considered subspecies of *A. roquet* (from Martinique, Barbados, and the entire Grenada Bank; Figure 10) shared an identical karyotype of 11 microchromsome pairs; every other species in the species group had 12 microchromosome pairs. Thus, chromosomes and biochemical analyses were concordant in supporting the species-level distinction between Trinidad anoles.

Those first few months at Harvard were very stimulating because of the interaction among people interested in my study animals and those interested in the West Indies. Toward the end of the semester, Allen Greer, a rare E. E. Williams graduate student who managed to avoid working on *Anolis*, asked me a simple question: "Why don't you stay and finish your degree here?"

I hadn't thought of it, but it made perfect sense. I had spent the past four academic years as a nomad, having been based at four different academic institutions (Cornell, Berkeley, UWI Trinidad, Harvard). My research was using behavioral data, but my thesis was not going to be ethological. At the time, the MCZ was the West Indian *Anolis* capital of the world. With Williams' sponsorship, Harvard formally admitted me as a graduate student. Over the next few years, as student and postdoctoral fellow, I was able to wrap-up, for my own satisfaction at least, the story of the hybridizing anoles of Trinidad. I completed all of the work in the 1960s and 1970s. If this essay stimulates just one reader to follow-up on a dynamic interaction between two competing species, it will have served its purpose — and, be sure to let me know what you find, if I'm still here!

Summary

The Fact of Hybridization.—The evidence for hybridization in areas of overlap between *Anolis trinitatis* and *A. aeneus* on Trinidad is incontrovertible. Every individual that appeared to be intermediate or "funny-looking," based on color or pattern, had a hybrid karyotype and had electrophoretic protein markers representing both parental species. I also found that the male territorial display of a hybrid was intermediate to the parental types (Figure 11).

FIGURE 12. Taking lizard body temperatures with a quick-reading "Schultheis" thermometer. On Trinidad, *Anolis trinitatis* and *A. aeneus* had very similar thermal niches, but on St. Vincent and Grenada, respectively, they each maintained significantly higher temperatures than much larger congeners.

The Ecology of Overlap.—In a study of head-size of Lesser Antillean anoles, Tom Schoener demonstrated a reasonably consistent picture. Independent of species group or taxonomic relationship, where two species of anoles co-existed on an island (e.g., St. Vincent, Grenada, St. Kitts, Antigua), they differed strikingly in size. No Lesser Antillean island had two sympatric species that overlapped significantly in adult male head size or body size. On islands with only one species (e.g., Dominica, Martinique, St. Lucia, Barbados), the solitary species tends to be intermediate in size between the large and small forms of the two-species islands (the correlation is not perfect, but this is not the place to enter into a lengthy discussion). The Trinidad situation was atypical. The two species are very similar in body size and general habits. They may be unable to co-exist. Thus, the pure single-species enclaves may reflect the only stable state (Figure 12).

Deduction that Both Species Were Introduced.—With hindsight, everything points to the fact that both species were introduced. They are commensal with people on Trinidad. They show less inter-population variation on Trinidad than on their respective West Indian islands. The populations may have come from single point-source introductions. The *A. aeneus* populations in Trinidad are quite uniform and resemble most closely a Grenada Bank population from a xeric area, such as Bequia in the Grenadines. The Trinidad *A. trinitatis* would look at home in the St. Vincent Botanic Gardens in Kingstown. *Anolis trinitatis* is completely unknown from the South American mainland, again very unusual for Trinidad reptile species. In South America, *A. aeneus* is definitively recorded only from Guyana which, significantly, was *British* Guiana, and which historically was treated as part of the British West Indies. As is the case on Trinidad, *A. aeneus* is known only in disturbed areas around towns. As these lizards readily inhabit homes and lay eggs in flowerpots, inadvertent human transport seems reasonably likely.

Hybrids Have Greatly Reduced Reproductive Fitness, but May Occasionally Succeed.—A study of male meiosis in hybrids demonstrated that the reduction-division was impaired, with most cells getting "stuck" at meiotic metaphase. A collection of large numbers of both sexes of both parental species and the hybrids revealed that, except for color, hybrid males appeared normal in body size. Female hybrids tended to be much larger than females of either parental species. A comparison of reproductive condition between hybrid females and the parental species indicated that a vastly lower percentage of hybrids than parentals produced mature eggs. The female "mules" tended to have large fat bodies (Figure 13). These fat bodies are absent in reproductively active females. Why the hybrid females are giants remains unknown. Paul Licht and I performed an experiment (on readily available *A. carolinensis*) to try to create "giant"

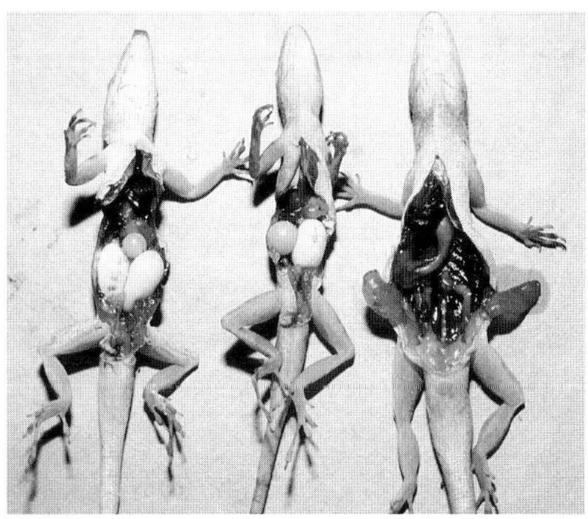

FIGURE 13. Dissection of *Anolis trinitatis*, *A. aeneus*, and hybrid *A. trinitatis* x *aeneus* females (left to right) from San Fernando, Trinidad. Hybrid females are typically much larger than either of the parental species and rarely produce eggs. Note the lack of developing eggs and huge fat pads (abundance of fat is inversely correlated with reproductive activity). The two pure forms are usually gravid throughout all but the driest months.

females by suppressing reproductive activity, but found no correlation between adult body size and extent of egg production. Further, an electrophoretic study of allele frequencies revealed the probable existence of a very low level of backcrossing. The "mules" are presumably not completely sterile.

Finally, an Ecological Difference that Might Confer a Significant Competitive Advantage on One of the Species . . . and the Winner is . . . Anolis aeneus *(Maybe)*.—An early study of reproductive cycles of several species of West Indian *Anolis* indicated that, where a significant dry season occurred, females ceased egg production in the driest months. This presumably conserves energy at times not favorable for reproduction. In contrast, in moist localities with a less marked dry season, females would be reproductively active throughout the year.

Anolis trinitatis from the St. Vincent Botanic Gardens was a population that showed the least significant seasonality in breeding. On Trinidad, we found that *A. trinitatis* was non-seasonal, but that *A. aeneus* from identical habitats was highly seasonal in egg production. This would be consistent with a genetically determined reproductive cycle. The Trinidad *A. aeneus* resemble *A. aeneus* from the dry (seasonal) portions of the Grenada Bank. This is the only significant difference between these two species of which I am aware. Because the rainfall pattern in Trinidad is seasonal, I presume that *A. trinitatis* females waste energy by producing eggs at a non-propitious time. This might confer a slight competitive advantage on *A. aeneus*. This conjecture is worthy of further study.

A Look to the Future?

As I write this essay, about 30 years have passed since anyone has carefully examined the "mule" lizards of Trinidad. Are the two species still established or did one (*A. aeneus*?) drive out the other? If both are present, how extensively do they overlap? Are they still producing mules?

Suggested Reading

(Annotations are specifically related to the hybridization story)

Beebe, W. 1944. Field notes on the lizards of Kartabo, British Guiana and Caripito, Venezuela. Part 2. Iguanidae. *Zoologica* 29:195–216.

Carpenter, C. C. 1962. A comparison of the patterns of display of *Urosaurus*, *Uta* and *Streptosaurus*. *Herpetologica* 18:145–152.

Etheridge, R. 1960. The relationships of the anoles (Reptilia: Sauria: Iguanidae): An interpretation based on skeletal morphology. Ph.D. Thesis, University of Michigan, Ann Arbor (University Microfilms, Inc., Ann Arbor, Michigan).

Gorman, G. C. 1965. Interspecific karyotypic variation as a systematic character in the genus *Anolis* (Sauria: Iguanidae). *Nature* 208:95–97.

Gorman, G. C. 1968. The relationships of *Anolis* of the *roquet* species group III. Comparative study of display behavior. *Breviora* 284:1–31.

Gorman, G. C. 1969. Intermediate territorial display of a hybrid *Anolis* lizard (Sauria: Iguanidae). *Zeitschr. Tierpsychol.* 26:390–393 — demonstrates the intermediacy of hybrid males' stereotyped territorial displays.

Gorman, G. C. and L. Atkins. 1966. Chromosomal heteromorphism in some male lizards of the genus *Anolis*. *Amer. Nat.* 100:579–583.

Gorman, G. C. and L. Atkins. 1968. Natural hybridization between two sibling species of *Anolis* lizards: Chromosome cytology. *Science* 159:1358–1360 — shows intermediate chromosome complement of hybrids; more importantly, the fact that meiosis in male hybrids is abnormal — cells seem to get "stuck" in the reduction division.

Gorman, G. C. and J. O. Boos. 1972. Extinction of a local population of *Anolis* lizards through competition with a congener. *Syst. Zool.* 21:441–442 — traces the relative population density of parental species and hybrids over several years.

Gorman, G. C. and H. C. Dessauer. 1966. The relationships of *Anolis* of the *roquet* species group (Sauria: Iguanidae). *Comp. Biochem. Physiol.* 19:845–853 — blood protein markers determined by electrophoresis demonstrate hybridization; suggests that both parental species reached Trinidad through human introduction.

Gorman, G. C. and P. Licht. 1973. Sexual dimorphism in body size and ovarian activity in the lizard *Anolis carolinensis*. *Amer. Midl. Nat.* 90:235 — attempts, but fails, to explain hybrid female gigantism on the assumption that energy otherwise expended for egg production is converted to growth.

Gorman, G. C. and P. Licht. 1975. Differences between the reproductive cycles of sympatric *Anolis* lizards on Trinidad. *Copeia* 1975:332–337 — demonstrates a difference in seasonality of egg production in the two parental species and posits that *A. trinitatis* is genetically "programmed" to be reproductively active throughout the year, which is appropriate for its native habitat but inappropriate on Trinidad; this unnecessary energy expenditure may provide a competitive advantage to *A. aeneus*.

Gorman, G. C., P. Licht, H. C. Dessauer, and J. Boos. 1971. Reproductive failure among hybridizing *Anolis* lizards of Trinidad. *Syst. Zool.* 20:1–18 — quantifies reproductive failure in female hybrids as well as male hybrids; demonstrates "gigantism" of female hybrids; suggests the reason that we generally find only one species or the other at any particular site is simply an "advantage of numbers"; given both a seeming lack of isolating mechanism and hybrid sterility, the rare species becomes exponentially rarer in each subsequent generation because of its inability to find conspecific mates.

Guyer, C. and J. Savage. 1986. Cladistic relationships among anoles (Sauria: Iguanidae). *Syst. Zool.* 35:509–531.

Lazell, J. D., Jr. 1962. The anoles of the eastern Caribbean: Part V. Geographic differentiation in *Anolis oculatus* on Dominica. *Bull. Mus. Comp. Zool., Harvard Univ.* 125:466–475.

Lazell, J. D., Jr. 1964. The anoles (Sauria: Iguanidae) of the Guadeloupeen Archipelago. *Bull. Mus. Comp. Zool., Harvard Univ.* 131:359–401.

Matthey, R. 1949. *Les Chromosomes des Vertébrés*. Librairie de l'Universite, F. Rouge, Lausanne.

Quesnel, V. C. and J. S. Kenny. 1959. Anoles of the Eastern Caribbean: two sibling species of anoles in Trinidad. *Bull. Mus. Comp. Zool., Harvard Univ.* 121(2):189–191.

Schoener, T. 1969. Size patterns in West Indian *Anolis* lizards: I. Size and species diversity. *Syst. Zool.* 18:386–401.

Schoener, T. and G. C. Gorman. 1968. Some niche differences in three Lesser Antillean lizards of the genus *Anolis*. *Ecology* 49:819–830.

Underwood, G. 1959. Anoles of the Eastern Caribbean: revisionary notes. *Bull. Mus. Comp. Zool., Harvard Univ.* 121(3):191–226.

Williams, E. E. 1959. Anoles of the Eastern Caribbean: preface. *Bull. Mus. Comp. Zool., Harvard Univ.* 121(1):187–189.

Williams, E. E. 1972. The origin of faunas: evolution of lizard congeners in a complex fauna: a trial analysis. *Evol. Biol.* 6:47–89.

Williams, E. E. 1989. A critique of Guyer and Savage (1986): cladistic relationships among anoles (Sauria: Iguanidae): are the data available to reclassify the anoles? p. 433–478. *In* C. A. Woods (ed.), *Biogeography of the West Indies: Past, Present, and Future*. Sandhill Crane Press, Inc. Gainesville, Florida.

Gorman, G. C. 2003. The mule lizards of Trinidad: Jungle dreams and backyard explorations, p. 279–293. *In* R. W. Henderson and R. Powell (eds.), Islands and the Sea: Essays on Herpetological Exploration in the West Indies. Society for the Study of Amphibians and Reptiles, Ithaca (New York). Contributions to Herpetology, volume 20.

Indices

•

References to Greater Antillean islands are indexed only for pages outside of the section concerned with that particular island (e.g., pages on which reference is made to the island of Jamaica are not indexed for essays in that section of the book dealing specifically with Jamaica). Common names for taxa are indexed only if they were accompanied by a scientific name.

Illustrations are denoted by **boldfaced type**.

Index 1: Places

Index 2: People

Index 3: Animals

PUBLICATIONS OF THE
SOCIETY FOR THE STUDY OF AMPHIBIANS AND REPTILES

SOCIETY PUBLICATIONS may be purchased from:
Breck Bartholomew, SSAR Publications Secretary
P.O. Box 58517
Salt Lake City, Utah 84158–0517, USA

Telephone and fax: area code (801) 453–0489
E-mail: ssar@herplit.com
Web: http://www.ssarherps.org/

Prices are effective through December 2003. Make checks payable to "SSAR." Overseas customers must make payment in USA funds using a draft drawn on American banks or by International Money Order. All persons may charge to MasterCard, American Express, Discover Card, or VISA (please provide account number and expiration date).

Shipping and Handling Costs

• *Shipments inside the USA*: Shipping costs are in addition to the price of publications. Add an amount for shipping of the first item ($4.00 for a volume costing $15.00 or more or $3.00 if it costs less than $15.00) plus an amount for any additional items ($3.00 each for a volume costing over $15.00 and $2.00 for each item costing less than $15.00).

• *Shipments outside the USA*: Determine the cost for shipments inside USA (above) and then add 6% of the total cost of the order.

• *Large prints* (marked *): Shipments only inside the USA, add $5.00 for any quantity; regrettably, prints cannot be shipped outside the USA due to their large size.

CONTRIBUTIONS TO HERPETOLOGY

Book-length monographs, comprising taxonomic revisions, results of symposia, and other major works. Pre-publication discount to Society members. Missing volumes are out-of-print and no longer available from the SSAR.

Vol. 2. *The Turtles of Venezuela,* by Peter C. H. Pritchard and Pedro Trebbau. 1984. Covers half of the turtle species of South America. 414 p., 48 color plates (25 watercolors by Giorgio Voltolina and 165 photographs of turtles and habitats) measuring 8½ × 11 inches, keys, 16 maps. Regular edition, clothbound $50.00; patron's edition, two leatherbound volumes in cloth-covered box, signed and numbered by authors and artist $300.00.

Vol. 4. *Studies on Chinese Salamanders*, by Ermi Zhao, Qixiong Hu, Yaoming Jiang, and Yuhua Yang. 1988. Evolutionary review of all Chinese species with keys, diagnostic figures, and distribution maps. 80 p., 7 plates (including 10 color photographs of salamanders and habitats). Clothbound $17.00.

Vol. 6. *Snakes of the* Agkistrodon *Complex: A Monographic Review*, by Howard K. Gloyd and Roger Conant. 1990. Comprehensive treatment of four genera: *Agkistrodon*, *Calloselasma*, *Deinagkistrodon*, and *Hypnale*. Includes nine supplementary chapters by leading specialists. 620 p., 33 color plates (247 photographs of snakes and habitats), 20 uncolored plates, 60 text figures, keys, 6 charts, 28 maps. Clothbound $75.00. (*Also*: separate set of the 33 plates, in protective wrapper. $30.00; limited-edition print of the book's frontispiece illustrating snakes of all four genera, from watercolor by David M. Dennis, and signed by Dr. Conant and the artist $25.00.)

Vol. 7. *The Snakes of Iran*, by Mahmoud Latifi. 1991. Review of the 60 species of Iranian snakes, covering general biology, venoms, and snake bite. Appendix and supplemental bibliography by Alan E. Leviton and George R. Zug. 167 p., 22 color plates of snakes (66 figures), 2 color relief maps, 44 species range maps. Clothbound $25.00.

Vol. 9. *Herpetology: Current Research on the Biology of Amphibians and Reptiles*. 1992. Proceedings of the First World Congress of Herpetology (Canterbury, 1989), edited by Kraig Adler and with a foreword by H.R.H. Prince Philip, Duke of Edinburgh. Includes the plenary lectures, a summary of the congress, and a list of delegates with their current addresses. 225 p., 28 photographs. Clothbound $28.00.

Vol. 10. *Herpetology of China*, by Ermi Zhao and Kraig Adler. 1993. Comprehensive review of Chinese amphibians and reptiles. 522 p., 48 color plates (371 photographs illustrating all 164 genera and half of the 661 species), portraits, text figures, maps, indices. Clothbound $60.00.

Vol. 11. *Captive Management and Conservation of Amphibians and Reptiles*, by James B. Murphy, Kraig Adler, and Joseph T. Collins (eds.). 1994. Includes chapters by 70 leading specialists. Foreword by Gerald Durrell. 408 p., 35 photographs, 1 color plate. Clothbound $58.00.

Vol. 12. *Contributions to West Indian Herpetology*, by Robert Powell and Robert W. Henderson (eds.). 1996. Results of a Society-sponsored symposium, including research chapters by 59 authors and a checklist of species with complete citations. Foreword by Thomas W. Schoener. 457 p., 98 photographs (70 in color), index. Clothbound $60.00.

Vol. 13. *Gecko Fauna of the USSR and Contiguous Regions*, by Nikolai N. Szczerbak and Michael L. Golubev. 1996. Covers the systematics, natural history, and conservation of the gecko fauna of the former Soviet Union and related species in surrounding regions from Mongolia through Pakistan, the Middle East, and northern Africa. 245 p., 24 colored and 60 black-and-white photographs, spot distribution maps, bibliography, index. Clothbound $48.00.

Vol. 14. *Biology of the Reptilia, vol. 19 (Morphology G)*, by Carl Gans and Abbot S. Gaunt (eds.). 1998. Chapters by 11 authors cover the major organs situated in the coelom: lungs, heart, liver, and spleen. 660 p., 145 figures, indices. Clothbound $58.00. (A complete list of the earlier volumes in this series [vols. 1–18], with names of publishers, is given in the book.)

Vol. 15. *The Lizards of Iran*, by Steven C. Anderson. 1999. Comprehensive summary, including systematics, natural history, and distribution. 450 p., 103 maps, 190 color photographs of lizards and habitats. Clothbound $65.00.

Vol. 16. *Slithy Toves: Illustrated Classic Herpetological Books at the University of Kansas in Pictures and Conversations*, by Sally Haines. 2000. A treasure trove of some of the finest illustrations of amphibians and reptiles ever produced, dating from the 16th to early 20th centuries. 190 p., 84 color photographs. Stiff paper cover $60.00.

Vol. 17. *The Herpetofauna of New Caledonia*, by Aaron M. Bauer and Ross A. Sadlier. French translations by Ivan Ineich. 2000. 322 p., 47 maps, 63 figures, 189 color photographs of animals and habitats. Clothbound $60.00.

Vol. 18. *The Hylid Frogs of Middle America*, expanded edition, by William E. Duellman. 2001. Review of the 165 hylid species from México through Panamá, with paintings by David M. Dennis. Foreword by David B. Wake. 1180 p., 443 figures and maps, 94 plates (46 in color). Clothbound in 2 volumes $125.00. (*Also*: separate set of the 46 color plates, in protective wrapper $45.00.)

Vol. 19. *The Amphibians of Honduras*, by James R. McCranie and Larry David Wilson. 2002. Comprehensive summary of 116 species, including systematics, natural history, and distribution. Foreword by Jay M. Savage. 635 p., 126 figures, 33 tables, 154 color photographs of animals and habitats. Clothbound $60.00.

Vol. 20. *Islands and the Sea. Essays on Herpetological Exploration in the West Indies*, by Robert W. Henderson and Robert Powell (eds.). 2003. Personal recollections of field experiences by 31 scientists who are or were active in the West Indies. 312 p., 316 photographs, 14 maps, indices. Clothbound $48.00.

FACSIMILE REPRINTS IN HERPETOLOGY

Exact reprints of classic and important books and papers. Most titles have extensive new introductions by leading authorities. Prepublication discount to Society members. Missing volumes are out-of-print and no longer available.

ANDERSON, J. 1896. *Contribution to the Herpetology of Arabia*. Introduction and new checklist of Arabian amphibians and reptiles by Alan E. Leviton and Michele L. Aldrich. 160 p., illus. (one plate in color), map. Clothbound $25.00.

BARBOUR, T. and C. T. RAMSDEN. 1919. *The Herpetology of Cuba*. The first comprehensive review of Cuban amphibians and reptiles with a complete descriptions of all species. Foreword by Rodolfo Ruibal. 200 p., 15 plates, measuring 8½ × 11 inches. Clothbound $55.00.

BOULENGER, G. A. 1877–1920. *Contributions to American Herpetology*. A collection of papers (from various journals) covering North, Central, and South American species, with an introduction by James C. Battersby. Complete in 18 parts totalling 880 p., numerous illustrations, index. Complete set: 18 parts plus index and two tables of contents (for binding in two volumes), in parts as issued $55.00.

COPE, E. D. 1864. *Papers on the Higher Classification of Frogs*. Reprinted from Proceedings of the Academy of Natural Sciences of Philadelphia and Natural History Review. 32 p. $3.00.

—. 1871. *Catalogue of Batrachia and Reptilia Obtained by McNiel in Nicaragua; Catalogue of Reptilia and Batrachia Obtained by Maynard in Florida*. 8 p. $1.00.

—. 1892. *The Osteology of the Lacertilia*. 44 p., 6 plates. $4.00.

COWLES, R. B., and C. M. BOGERT. 1944. *A Preliminary Study of the Thermal Requirements of Desert Reptiles*. With extensive review of recent studies by F. Harvey Pough. Reprinted from Bulletin of American Museum of Natural History. 52 p., 11 plates. Paper cover $5.00.

ESCHSCHOLTZ, F. 1829–1833. *Zoologischer Atlas* (herpetological sections). Descriptions of new reptiles and amphibians from California and the Pacific. Introduction by Kraig Adler. 32 p., 4 plates (measuring 8½ × 11 inches). $3.00.

ESPADA, M. JIMÉNEZ DE LA. 1875. *Vertebrados del Viaje al Pacifico: Batracios*. A major taxonomic work on South American frogs. Introduction by Jay M. Savage. 208 p., 6 plates, maps. Clothbound $20.00.

FAUVEL, A.-A. 1879. *Alligators in China*. Original description of *Alligator sinensis*, including classical and natural history. 42 p., 3 plates. Paper cover $5.00.

FERGUSON, W. 1877. *Reptile Fauna of Ceylon*. First comprehensive summary of the herpetofauna of Sri Lanka. Introduction by Kraig Adler. 48 p. $8.00.

FITZINGER, L. 1826 & 1835. *Neue Classification der Reptilien* and *Systematische Anordnung der Schildkröten*. Important nomenclatural landmarks for herpetology, including Amphibia as well as reptiles; world-wide in scope. Introduction by Robert Mertens. 110 p., folding chart. Clothbound $30.00.

FRANCIS, E. T. B. 1934. *The Anatomy of the Salamander*. Standard text of salamander anatomy, including osteology, myology, nervous system, and internal organs. Introductions by F. J. Cole and James Hanken. 465 p., 25 highly detailed plates, color frontispiece. Clothbound $60.00.

GRAY, J. E. 1825. *A Synopsis of the Genera of Reptiles and Amphibia*. Reprinted from Annals of Philosophy. 32 p. $3.00.

—. 1831–1844. *Zoological Miscellany*. Introduction by Arnold G. Kluge. 86 p., 4 plates. Paper cover $6.00.

GRAY, J. E., and A. GÜNTHER. 1845–1875. *Lizards of Australia and New Zealand*. The reptile section from "Voyage of H.M.S. Erebus and Terror," together with Gray's 1867 related book on Australian lizards. Introduction by Glenn M. Shea. 82 p., 20 plates (measuring 8½ × 11 inches). Clothbound $20.00. (*Also*: set of the 20 plates, in protective wrapper $12.00.)

GÜNTHER, A. 1885–1902. *Biologia Centrali-Americana. Reptilia and Batrachia*. The standard work on Middle American herpetology with 76 full-page plates measuring 8½ × 11 inches (12 in color). Introductions by Hobart M. Smith, A. E. Gunther, and Kraig Adler. 575 p., photographs, maps. Clothbound $50.00. (*Also*: separate set of the 12 color plates, in protective wrapper $18.00.)

HOLBROOK, J. E. 1842. *North American Herpetology*. Five volumes bound in one. The classic work by the father of North American herpetology. Exact facsimile of the definitive second edition, including all 147 plates, measuring 8½ × 11 inches (20 reproduced in full color). Introduction and checklists by Richard and Patricia Worthington and by Kraig Adler. 1032 p. Clothbound $60.00.

KIRTLAND, J. P. 1838. *Zoology of Ohio* (herpetological portion). 8 p. $1.00.

LECONTE, J. E. 1824–1828. *Three Papers on Amphibians*, from the Annals of the Lyceum of Natural History, New York. 16 p. $2.00.

MCLAIN, R. B. 1899. *Contributions to North American Herpetology* (three parts). 28 p., index. Paper cover $2.00.

ORBIGNY, A. D' [and G. BIBRON]. 1847. *Voyage dans l'Amérique Méridionale*. This extract comprises the complete section on reptiles and amphibians from this voyage to South America. 14 p., 9 plates measuring 8½ × 11 inches. Paper cover $3.00.

PETERS, W. 1838–1883. *The Herpetological Contributions of Wilhelm C. H. Peters (1815–1883)*. A collection of 174 titles, world-wide in scope, and including the herpetological volume in Peters' series, "Reise nach Mossambique." Biography, annotated bibliography, and synopsis of species by Aaron M. Bauer, Rainer Günther, and Meghan Klipfel. 714 pages, 114 plates, 9 photographs, maps, index. Clothbound $75.00.

RAFINESQUE, C. S. 1820. *Annals of Nature* (herpetological and ichthyological sections), 4 p. $1.00.

—. 1822. *On Two New Salamanders of Kentucky*. 2 p. $1.00.

—. 1832–1833. *Five Herpetological Papers from the Atlantic Journal*. 4 p. $1.00.

SCHMIDT, K. P., and G. K. NOBLE. 1919–1923. *Contributions to the Herpetology of the Belgian Congo*. Essential reference for the Congo rain forest and Sudanese savanna. Introductions by Donald G. Broadley and John C. Poynton. 780 pages, 141 photographs, maps, indices. Clothbound $65.00.

SHAW, G. 1802. *General Zoology, vol. 3: Amphibia*. Herpetological section from the first world summary of amphibians and reptiles in English. Introduction by Hobart M. Smith and Patrick David. 1014 p., 140 plates. Clothbound $75.00.

SOWERBY, J. DEC., E. LEAR, and J. E. GRAY. 1872. *Tortoises, Terrapins, and Turtles Drawn From Life*. The finest atlas of turtle illustrations ever produced. Introduction by Ernest E. Williams. 26 p., 61 full-page plates (measuring 8½ × 11 inches). Clothbound $40.00.

STEJNEGER, L. 1907. *Herpetology of Japan and Adjacent Territory*. Introduction by Masafumi Matsui. Also covers Taiwan, Korea, and adjacent China and Siberia. 684 pages, 35 plates, 409 text figures, keys, index. Clothbound $58.00.

TROSCHEL, F. H. 1850 [1852]. Cophosaurus texanus, *neue Eidechsengattung aus Texas*. 8 p. $1.00.

TSCHUDI, J. J. VON. 1838. *Classification der Batrachier*. A major work in systematic herpetology, with introduction by Robert Mertens. 118 p., 6 plates. Paper cover $18.00.

—. 1845. *Reptilium Conspectus*. New reptiles and amphibians from Peru. 24 p. $2.00.

VANDENBURGH, J. 1895–1896. *Herpetology of Lower California*. Herpetology of Baja California, Mexico (collected papers). 101 p., 11 plates, index. Paper cover $11.00.

—. 1914. *The Gigantic Land Tortoises of the Galapagos Archipelago*. The most extensive review of Galapagos tortoises. Foreword by Peter C. H. Pritchard. 290 pages, 205 photographs, maps, index. Clothbound $55.00.

WAITE, E. R. 1929. *The Reptiles and Amphibians of South Australia*. Introduction by Michael J. Tyler and Mark Hutchinson. 282 p., color plate, portrait, 192 text figures including numerous photographs. Clothbound $35.00.

WRIGHT, A. H., and A. A. WRIGHT. 1962. *Handbook of Snakes of the United States and Canada, Volume 3, Bibliography*. Cross-indexed bibliography to Volumes 1 and 2. 187 p. Clothbound $18.00.

JOURNAL OF HERPETOLOGY

The Society's official scientific journal, international in scope. Issued quarterly as part of Society membership. All numbers are paperbound as issued, measuring 7 × 10 inches.

Volume 1 (1968), numbers 1–4 combined.

Volumes 2–5 (1968–1971), numbers 1–2 and 3–4 combined, $8.00 per double number.

Volume 6 (1972), numbers 1, 2, and double number 3–4.

Volumes 7–36 (1973–2002), four numbers in each volume, $8.00 per single number.

The following volumes and numbers are out-of-print and are *no longer available*: Volumes 1, 2, 3, 4, 5, 6, 7(1), 8(1, 2), 9(1, 2, 4), 10(1, 4), 11(1, 4), 12(1, 2, 4), 18(1), and 24(1).

Cumulative Index for Volumes 1–10 (1968–1976), 72 pages, $8.00.

HERPETOLOGICAL REVIEW AND H.I.S.S. PUBLICATIONS

The Society's official newsletter, international in coverage. In addition to news notes and feature articles, regular departments include regional societies, techniques, husbandry, life history, geographic distribution, and book reviews. Issued quarterly as part of Society membership or separately by subscription. All numbers are paperbound as issued and measure 8½ × 11 inches. In 1973, publications of the Herpetological Information Search Systems (*News-Journal* and *Titles and Reviews*) were substituted for *Herpetological Review*; content and format are the same.

Volume 1 (1967–1969), numbers 1–9, $5.00 per number.

Volumes 2–33 (1970–2002), four numbers in each volume (except volumes 3–4, with 6 numbers each), $5.00 per number.

The following volumes and numbers are out-of-print and *no longer available*: Volumes 1(number 7), 3(2), 4(1, 4, 6), 5(1, 2), 6, 7(1, 3, 4), 10(2), 19(3), 21(1), 24(1), and 25(2).

Cumulative Index for Volumes 1–7 (1967–1976), 60 pages, $5.00.

Cumulative Index for Volumes 1–17 (1967–1986), 90 pages, $8.00.

H.I.S.S. Publications: News-Journal, volume 1, numbers 1–6, and Titles and Reviews, volume 1, numbers 1–2 (all of 1973–1974), complete set, $10.00.

Index to Geographic Distribution Records for Volumes 1–17 (1967–1986), including H.I.S.S. publications, 44 pages, $6.00.

CATALOGUE OF AMERICAN AMPHIBIANS AND REPTILES

Loose-leaf accounts of taxa (measuring 8½ × 11 inches) prepared by specialists, including synonymy, definition, description, distribution map, and comprehensive list of literature for each taxon. Covers amphibians and reptiles of the entire Western Hemisphere. Issued by subscription. Individual accounts are not sold separately.

CATALOGUE ACCOUNTS:

Complete set:	Numbers 1–760, $460.00.
Partial sets:	Numbers 1–190, $75.00.
	Numbers 191–410, $85.00.
	Numbers 411–760, $320.00.

INDEX TO ACCOUNTS 1–400: Cross-referenced, 64 pages, $6.00; accounts 401–600: Cross-referenced, 32 pages, $6.00.

IMPRINTED POST BINDER: $35.00. (*Note*: one binder holds about 200 accounts.)

SYSTEMATIC TABS: Ten printed tabs for binder, such as "Class Amphibia," "Order Caudata," etc., $6.00 per set.

HERPETOLOGICAL CONSERVATION

A new series of book-length monographs, including symposia, devoted to all aspects of the conservation of amphibian and reptiles. Prepublication discount to Society members.

Vol. 1. *Amphibians in Decline: Canadian Studies of a Global Problem*, by David M. Green (ed.). 1997. Chapters by 52 authors dealing with population dispersal and fluctuations, genetic diversity, monitoring of natural populations, as well as effects of temperature, acidity, pesticides, UV light, forestry practices, and disease. 351 p., numerous photographs, figures, and tables, index. Paperbound $40.00.

HERPETOLOGICAL CIRCULARS

Miscellaneous publications of general interest to the herpetological community. All numbers are paperbound, as issued. Prepublication discount to Society members. Missing numbers are out-of-print and no longer available from SSAR.

No. 1. *A Guide to Preservation Techniques for Amphibians and Reptiles* by George R. Pisani. 1973. 22 p., illus. $3.00.

No. 6. *Longevity of Reptiles and Amphibians in North American Collections* by J. Kevin Bowler. 1977. 32 p. $3.00. (See also number 21.)

No. 7. *Standard Common and Current Scientific Names for North American Amphibians and Reptiles* (1st ed.) by Joseph T. Collins, James E. Huheey, James L. Knight, and Hobart M. Smith. 1978. 36 p. $3.00. (See also numbers 25 and 29.)

No. 8. *A Brief History of Herpetology in North America Before 1900* by Kraig Adler. 1979. 40 p., 24 photographs, 1 map. $3.00.

No. 10. *Vernacular Names of South American Turtles* by Russell A. Mittermeier, Federico Medem, and Anders G. J. Rhodin. 1980. 44 p. $3.00.

No. 11. *Recent Instances of Albinism in North American Amphibians and Reptiles* by Stanley Dyrkacz. 1981. 36 p. $3.00.

No. 13. *Silver Anniversary Membership Directory*, including addresses of all SSAR members, addresses and publications of the herpetological societies of the world, and a brief history of the Society. 1983. 56 p., 4 photographs. $3.00.

No. 14. *Checklist of the Turtles of the World with English Common Names* by John Iverson. 1985. 14 p. $3.00.

No. 15. *Cannibalism in Reptiles: A World-Wide Review* by Joseph C. Mitchell. 1986. 37 p. $4.00.

No. 17. *An Annotated List and Guide to the Amphibians and Reptiles of Monteverde, Costa Rica* by Marc P. Hayes, J. Alan Pounds, and Walter W. Timmerman. 1989. 70 p., 32 figures. $5.00.

No. 18. *Type Catalogues of Herpetological Collections: An Annotated List of Lists* by Charles R. Crumly. 1990. 50 p. $5.00.

No. 21. *Longevity of Reptiles and Amphibians in North American Collections* (2nd ed.) by Andrew T. Snider and J. Kevin Bowler. 1992. 44 p. $5.00. (See also number 6.)

No. 22. *Biology, Status, and Management of the Timber Rattlesnake* (Crotalus horridus): *A Guide for Conservation* by William S. Brown. 1993. 84 p., 16 color photographs. $12.00.

No. 23. *Scientific and Common Names for the Amphibians and Reptiles of Mexico in English and Spanish / Nombres Científicos y Comunes en Ingles y Español de los Anfibios y los Reptiles de México* by Ernest A. Liner. Spanish translation by José L. Camarillo R. 1994. 118 p. $12.00.

No. 24. *Citations for the Original Descriptions of North American Amphibians and Reptiles*, by Ellin Beltz. 1995. 48 p. $7.00.

No. 25. *Standard Common and Scientific Names for North American Amphibians and Reptiles* (4th ed.) compiled by Joseph T. Collins (coordinator for SSAR Common and Scientific Names List). 1997. 44 p. $9.00. (See also numbers 7 and 29.)

No. 26. *Venomous Snakes: a Safety Guide for Reptile Keepers*, by William Altimari. 1998. 28 p. $6.00.

No. 27. *Lineages and Histories of Zoo Herpetologists in the United States*, by Winston Card and James B. Murphy. 2000. 49 p., 53 photographs. $8.00.

No. 28. *State and Provincial Amphibian and Reptile Publications for the United States and Canada*, by John J. Moriarty and Aaron M. Bauer. 2000. 56 p. $9.00.

No. 29. *Scientific and Standard English Names of Amphibians and Reptiles of North America North of Mexico, with Comments Regarding Confidence in Our Understanding*, by the Committee on Standard English and Scientific Names (Brian I. Crother, chair). 2000 [2001]. 86 p. $11.00. (See also numbers 7 and 25.)

No. 30. *Amphibian Monitoring in Latin America: a Protocol Manual / Monitoreo de Anfibios en America Latina: Manual de Protocolos*, by Karen Lips, Jamie K. Reaser, Bruce E. Young, and Roberto Ibañez. 2001. 121 p. $13.00.

No. 31. *Herpetological Collecting and Collections Management.* (Revised edition), by John E. Simmons. 2002. 159 p. $16.00.

No. 32. *Conservation Guide to the Eastern Diamondback Rattlesnake*, Crotalus adamanteus, by Walter W. Timmerman and William H. Martin. 2003. 74 p., color photographs. $13.00.

PUBLICATIONS OF THE OHIO HERPETOLOGICAL SOCIETY

OHS was the predecessor to the Society for the Study of Amphibians and Reptiles. All publications international in scope. Paperbound as issued.

Volume 1 numbers 1–4, plus Special Publications 1–2 (all 1958), facsimile reprint, all out-of-print.

Volume 2 (1959–1960), four numbers, $2.00 per number; numbers 3 and 4 out-of-print.

Volume 3 (1961–1962), four numbers, $1.00 per number; numbers 1 and 3 out-of-print.

Volume 4 (1963–1964), four numbers; double number 1–2, $4.00, numbers 3 and 4 $2.00 each.

Volume 5 (1965–1966), four numbers, $2.00 per number.

Special Publications 3–4 (1961–1962), $2.00 per number; number 3 out-of-print.

OTHER MATERIALS AVAILABLE FROM THE SOCIETY

The following color prints and brochures may be purchased from the Society. (*Extra postage required; see "Shipping and Handling Costs" at the beginning of this list.)

*SILVER ANNIVERSARY COMMEMORATIVE PRINT. Full-color print (11½ × 15¼ inches) of a Gila Monster (*Heloderma suspectum*) on natural background, from a watercolor by David M. Dennis. Issued as part of Society's 25th Anniversary in 1982. Edition limited to 1000. $6.00 each or $5.00 in quantities of 10 or more.

*WORLD CONGRESS COMMEMORATIVE PRINT. Full-color print (11½ × 15 inches) of an Eastern Box Turtle (*Terrapene carolina*) in a natural setting, from a watercolor by David M. Dennis. Issued as part of SSAR's salute to the First World Congress of Herpetology, held at Canterbury, United Kingdom, in 1989. Edition limited to 1500. $6.00 each or $5.00 in quantities of 10 or more.

GUIDELINES FOR THE USE OF LIVE AMPHIBIANS AND REPTILES IN FIELD RESEARCH, by George R. Pisani, Stephen D. Busack, Herbert C. Dessauer, and Victor H. Hutchison, representing a joint committee of ASIH, HL, and SSAR. 1987. Brochure covers animal care, regulations, collecting, restraint and handling, marking, housing and maintenance in field, and final disposition of specimens. 16 p. $2.00 ($1.00 each in quantities of five or more copies).

GRANTS AND AWARDS FOR HERPETOLOGISTS, by Joan C. Milam. 1997. A detailed listing, with descriptions and addresses, for about 100 research award programs. 106 p. $8.00.

•

PRODUCTION SPECIFICATIONS

ENTRY AND FORMATTING OF NEW TEXT: Avila University, Kansas City, Missouri, USA (Robert Powell). Type was set in Times New face using Adobe PageMaker on an Apple Macintosh computer. Entry of preliminaries and reformatting of text: Cornell University, Ithaca, New York, USA (Timothy D. Perry).

PRINTING OF TEXT AND BINDING: Thomson-Shore, Inc., Dexter, Michigan, USA (Lana Paton, Dawn Rice). The text is printed on 70-pound Fortune Matte paper, a recycled stock. The volumes are covered in Roxite C cloth (vellum finish) with Multicolor Antique endpapers.

PAPER: All paper in this book is acid- and groundwood-free. It meets the guidelines for permanence and durability of the Committee on Publication Guidelines for Book Longevity of the Council on Library Resources.

DATE OF PUBLICATION: 20 August 2003.

PLACE OF PUBLICATION: Ithaca, New York, USA.

NUMBER OF COPIES: 700.